Trace Metals in the Westerschelde Estuary: A Case-Study of a Polluted, Partially Anoxic Estuary

Developments in Hydrobiology 128

Series editor
H.J. Dumont

Trace Metals in the Westerschelde Estuary: A Case-Study of a Polluted, Partially Anoxic Estuary

Edited by

Willy F.J. Baeyens

Reprinted from Hydrobiologia, volume 366 (1997/8)

Springer-Science+Business Media, B.V.

Library of Congress Cataloging-in-Publication Data

A C.I.P. Catalogue record for this book is available from the Library of Congress.

ISBN 978-90-481-5062-5 ISBN 978-94-017-3573-5 (eBook)
DOI 10.1007/978-94-017-3573-5

Printed on acid-free paper

Only after the last tree has been cut down
Only after the last river has been poisoned
Only after the last fish has been caught
Only then will man find
That money cannot be eaten.

(MOHAWK)

To my family Dany, Wendy and Audrey who had to put up with my mental absence during the writing of this book

Hydrobiologia **366:** vii, 1998.
W.F.J. Baeyens (ed.), Trace Metals in the Westerschelde Estuary.

Contents

Hydrobiologia **366:** ix, 1998.
W.F.J. Baeyens (ed.), Trace Metals in the Westerschelde Estuary.

To Professor Ivan Elskens

Professor Ivan Elskens graduated from and taught at the U.L.B. (the French-speaking Brussels University), but when the V.U.B. (the Dutch-speaking Brussels University) separated from the U.L.B. and both universities became independent, he moved over. At that time he had a quite impressive career behind him. After he had graduated, he had become a much-demanded expert and collaborator in numerous research projects which were not only scientifically sound, but also socially relevant.

As a result of a detailed study of Lake Kivu in the fifties, he was the first to identify large reserves of methane in that lake. It was at this time that he met volcanologist Haroun Tazieff, who was to invite him frequently to analyse the gases escaping from the craters of active volcanoes. Tazieff, who was looking for a method to predict volcanic eruptions, was only too happy to be able to rely on Elskens' scientific know-how.

His arrival at the V.U.B. coincided with the start of the North Sea R&D Project, a research programme with international recognition. The fact that J. Nihoul, the co-ordinator of the project, was to receive the Franqui price, Belgium's greatest scientific distinction a few years later, was in part due to Elskens' contribution. At the same time, he continued as an ambitious researcher to improve and refine in a professional way the analytical methods used in his research projects. All the students who had the privilege of following his courses learned the precise meaning of analytical chemistry.

Outside the scope of his scientific activities, he also had a wide interest in culture. His passion for the painter Hieronymus Bosch, his friendship with the Brel family, and his knowledge of and love for French wines, were not only known to insiders. Ivan Elskens was a much appreciated colleague who leaves not only at the V.U.B. but also far beyond an everlasting impression, as a person and as a free-thinker. He was my thesis supervisor, my former department head and my friend.

W. BAEYENS

Preface

Oceans and lakes are characterised by great residence times and can thus be considered to be in a state of equilibrium or quasi-equilibrium. Estuaries, by contrast, which constitute the interface between freshwater and marine ecosystems, are biogeochemically and physically very dynamic. Strong gradients in physical and chemical variables such as temperature, pH, salinity, dissolved oxygen, nutrients and the amount and composition of particulates result from the mixing of freshwater and saline constituents. These gradients provoke increased biogeochemical processes, generally resulting in a geochemical filter (the area of high turbidity, showing enhanced adsorption–desorption processes) and a biological filter (the area of high plankton activity showing a high accumulation of trace metals as well as a high uptake of nutrients). Both filters disturb the conservative mixing pattern of the trace metals in an estuary.

Four additional aspects make the Scheldt estuary very particular and different from other estuaries:

(1) the Scheldt is a tide-governed estuary, which means greater residence times;
(2) the upper estuary receives large inputs of biodegradable organic matter which leads to oxygen depletion in winter and to anoxic conditions in summer;
(3) inputs of toxic pollutants from the industrial zone of Antwerp occur in the upper estuary;
(4) the anoxic zone, the zone of pollutant input and the zone of the turbidity maximum coincide geographically.

The Scheldt estuary is therefore an ideal field laboratory for the study of biogeochemical processes which do not occur elsewhere, or occur only partially, to a very slight extent or in a specific compartment such as the sediments. More specifically, redox processes involving trace metals, the formation of non-labile metal complexes or the methylation of mercury occur as a result of anoxic conditions and the high load of organic matter. Several models describing in detail the transport, sedimentation and speciation of the trace metals in this well-mixed estuary (all major estuaries in EC countries adjacent to the North Sea or the Atlantic Ocean are well-mixed) are presented in this volume. A comparison of the metal concentrations over a 15-year period shows a net reduction in the pollution level.

Finally, I also wish to acknowledge and thank all the scientific contributors to this book, the personnel of the Analytical Chemistry Department at the V.U.B. for their constructive comments on all the articles and for their support and assistance in the editing of the volume, in particular R. Van Ryssen, and F. Winter for the language correction of the forewords. The Management Unit of the Mathematical Models of the North Sea and Scheldt Estuary and their director Dr G. Pichot, have helped us during this time by providing ship space on board the research vessels *Mechelen* and *Belgica*. Essential research grants were obtained from the Department of Science Policy (North Sea R&D Project, Joint Actions in Oceanology, the EUROTRAC Prograrnme, the Global Change Programme).

W. BAEYENS

Figure 1. The Western Scheldt is an important shipping route to the Antwerp harbour. The navigation channel is constantly dredged in order to keep the shipping route open for large sea vessels.

Figure 2. The area around the harbour is heavily industrialised. On this figure the nuclear plant 'Doel', located on the left bank of the river near the Antwerp harbour, is shown.

Figure 3. The city of Antwerp, located on the right bank of the Scheldt at approximately 78 km from the mouth, is the commercial heart of the river and has attracted traders and tourists for many years.

Figure 4. The intertidal flat 'Groot Buitenschoor' is located in the area of maximum turbidity of the estuary and receives large amounts of mud deposits. As the Scheldt water becomes anoxic a large part of the year, trace metals precipitate and are deposited on the mud flats.

Figure 5. Sampling of sediment cores for the analysis of trace metals in sediments, pore waters and benthic organisms provides information on the bioavailability of the pollutants.

Figure 6. The research vessel R V *Belgica* plays an important role in the assessment of pollution levels of the Scheldt estuary and the North Sea. A clean lab container is installed on the ship for the handling of water samples for trace metal analyses.

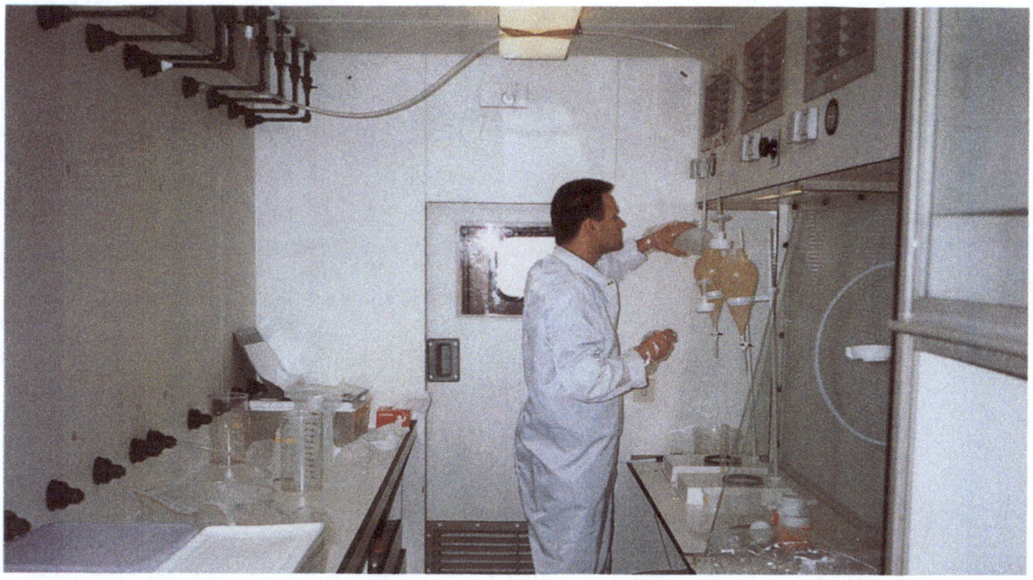

Figure 7. An inside view of the clean lab container showing the filtration of water samples using contamination free protocols for trace metal analyses.

Figure 8. Sampling of surface water from aboard an inflatable boat for trace metal analyses in order to avoid contamination of samples from the ship itself (paint, propellers, Zn anodes, etc.). Arm length gloves are worn during sampling.

Figure 9. The intertidal flat Groot Buitenschoor, one of the largest intertidal flats of the 7 Scheldt estuary, located in the vicinity of the industrial area of Antwerp.

Figure 10. Sampling for water quality variables (salinity, temperature, dissolved oxygen, turbidity) and for nutrient analysis.

Hydrobiologia **366:** 1–14, 1998.
W. F. J. Baeyens (ed.), Trace Metals in the Westerschelde Estuary.

General description of the Scheldt estuary

Willy Baeyens[1], Bert van Eck[2], Claude Lambert[3], Roland Wollast[4] & Leo Goeyens[1]
[1] *Department of Analytical Chemistry, Free University of Brussels, Pleinlaan 2, 1050 Brussels, Belgium*
[2] *Dienst Getijdewateren, Rijksinstituut voor Kust en Zee, Rijkswaterstaat, 4330 EA Middelburg, the Netherlands*
[3] *Centre National de la Recherche Scientifique (CNRS), 91198 Gif Sur Yvette CEDEX, France*
[4] *Laboratory of Oceanology, Université Libre de Bruxelles, 1050 Brussels, Belgium*

Key words: Scheldt estuary, description, hydrology, sedimentation, oxygen profiles, productivity

Abstract

A general description of the Scheldt estuary, including the hydrology, the sediment transport, the productivity and the biodegradation with respect to their influence on the trace metal behaviour in the Scheldt estuary, is given.

The river basin can be divided in several sections according to their morphological, hydrodynamical and sedimentary properties. The zone from km 78 to 55, which corresponds roughly with the salinity zone from 2 to 10 psu, is the zone of high turbidity, high sedimentation and of oxygen depletion, especially in the summer period. That area is called the geochemical filter because the solid/dissolved distribution of the trace metals is controlled by redox, adsorption/desorption, complexation and precipitation/coprecipitation processes. The sedimentation rate in that area is estimated at 280 Mkg y^{-1}.

In the downstream estuary the phytoplankton activity increases due to the restoration of oxygen and to the much lower turbidity values. That area is called the biological filter because incorporation of trace metals by the plankton communities lowers the trace metal concentrations during the productivity period, while transformation of metal species, especially observed with mercury, occurs during that period too.

Introduction

Like other branches of the Zeeland Delta, the Scheldt estuary fulfils numerous and diverse ecological functions. Total productivity is elevated (Saeijs, 1977) and it is not restricted to the lowest trophic level (higher plants and algae). Secondary production (e.g. zooplankton and shrimps) as well as production of higher trophic levels (e.g. benthic organisms and fish) are significant due to the availability of large amounts of detritus. Moreover, it is quite possible that secondary production is more important than primary production, and that the Scheldt ecosystem showed this feature already in the past.

Benthic organisms contribute largely to the total biomass. Especially the populations on tidal flats and sand plates are a food supply for wintering and migratory stilt-birds but also for ducks. Additionally, the Scheldt estuary serves as a nursery room for demersal fish species such as sole. The pronounced productivity makes the Scheldt an important 'natural resource'.

The species diversity per surface unit is relatively low, but the diversity in the whole estuary is fairly high. This results from the presence of many different environmental types (from fresh water over brackish to marine water types), each of them contributing to the species diversity. Generally, the number of individuals of each species is high, as a result of the high food abundance and the fact that in a given salinity range some species have a favourable, competitive position.

The estuary includes many wetlands and marshes (called schorren and polders in Dutch), Saeftinge and the Ballastplaat being the most important ones. These wetlands are very different, not only geomorphologically, but also with respect to their bottom compositions and plant communities. The qualitative and quantitative vegetation composition differ from one marsh to the other according to the salinity gradient (Saeijs,

1977). All marshes together determine the diversity in the vegetation along the estuary.

Pollution by toxic metals is one of the major threats to the estuarine ecosystem. Concentrations of dissolved Pb, Hg, Zn, Cu and Cd in the coastal estuarine water mass are about 2 times higher than in the marine water mass of the Belgian coastal zone, which is a part of the Southern Bight of the North Sea (Baeyens et al., 1987), and an order of magnitude higher than ocean values. In the Scheldt estuary itself, the heavy metal concentrations are still significantly higher than in the coastal estuarine water mass (Baeyens et al., 1998: see related paper on 'Sea inputs'). Therefore, a better understanding of the respective biogeochemical cycles of heavy metals in the Scheldt estuary with emphasis on the basic governing processes is essential. Amongst the major estuarine processes are the geophysical (water and sediment circulation) and the biological ones (production and biodegradation). Human activities such as discharging of liquid waste and dredging of sludge, are superimposed on the natural processes and, moreover interfere with them. Four aspects make the Scheldt estuary very peculiar and distinct from other estuaries:

(1) the Scheldt is a tide-governed estuary due to the low river flow leading to large residence times;
(2) the upper estuary receives large inputs of biodegradable organic matter inducing anoxic conditions in the water column during summer;
(3) considerable and direct supply of toxic pollutants occurs in the upper estuary as a result of the diverse activities by the industrial park around Antwerp;
(4) the anoxic zone, the area of pollutant inputs and the zone of maximum turbidity coincide geographically, making it very difficult to distinguish between their individual effects on the metal distribution and behaviour.

In this paper hydrology, sediment transport, production and biodegradation are discussed with respect to their possible influence on the trace metal behaviour in the Scheldt.

General description of the Scheldt river basin

The river Scheldt (Figure 1) is a lowland-river, which takes its rise in the northern part of France (St. Quentin), and flows into the North Sea near Vlissingen (the Netherlands). The total catchment area is 22 10^3 km^2. The total length of the river is 355 km, the fall over the total river length is at most 100 meters and the mean depth of the Scheldt estuary is about 10 m.

The river Scheldt as well as all its branches are rain-fed. The discharge of these rivers varies considerably with minimal discharges occurring in summer and autumn, and maximum ones in winter and spring. During winter and spring, the lower riverine part from Gent to Antwerp is a tidal fresh water river. During summer and autumn, on the contrary, when the discharge of the river is reduced, the water becomes brackish between Antwerp and the Rupel mouth.

In the total river basin of the Scheldt live approximately 7 million people. Urban areas with population densities of over 1000 inhabitants per km^2 are found near Lille (France), Gent, Brussels and Antwerp. The largest industrial areas are concentrated near Lille, Antwerp, along the canal from Gent to Terneuzen, and near Vlissingen. The river Scheldt and its branches are used as a major drain for industrial and domestic wastes. A substantial part of these is not treated in a waste water treatment plant. This gives rise to very poor water quality in the larger part of the river and the eastern part of the estuary.

Valuable natural areas, some of them are protected natural reserves, are situated in the estuarine part of the basin and along the tributaries. In the estuarine part of the basin the intertidal areas are very valuable, e.g. the fresh, brackish and salt water marshes.

Hydrological description

Four sections can be distinguished in the Scheldt river basin:

- the non-tidal fresh water river corresponding to the upper-Scheldt and a considerable part of the branches;
- the mostly fresh water tidal river extending from Gent to Rupelmonde with the lower parts of some branches;
- the upper estuary (brackish) or the Sea Scheldt between Rupelmonde and the Belgian-Dutch border; and
- the lower estuary (brackish and salt) called the Western Scheldt.

The predominant factors determining the hydrological characteristics of the Scheldt river are the fresh water flow rate and the tidal influence.

Fresh water flow

The mean discharge rate of the Scheldt, determined at Schelle (90 km from the mouth), amounted to 104 m^3

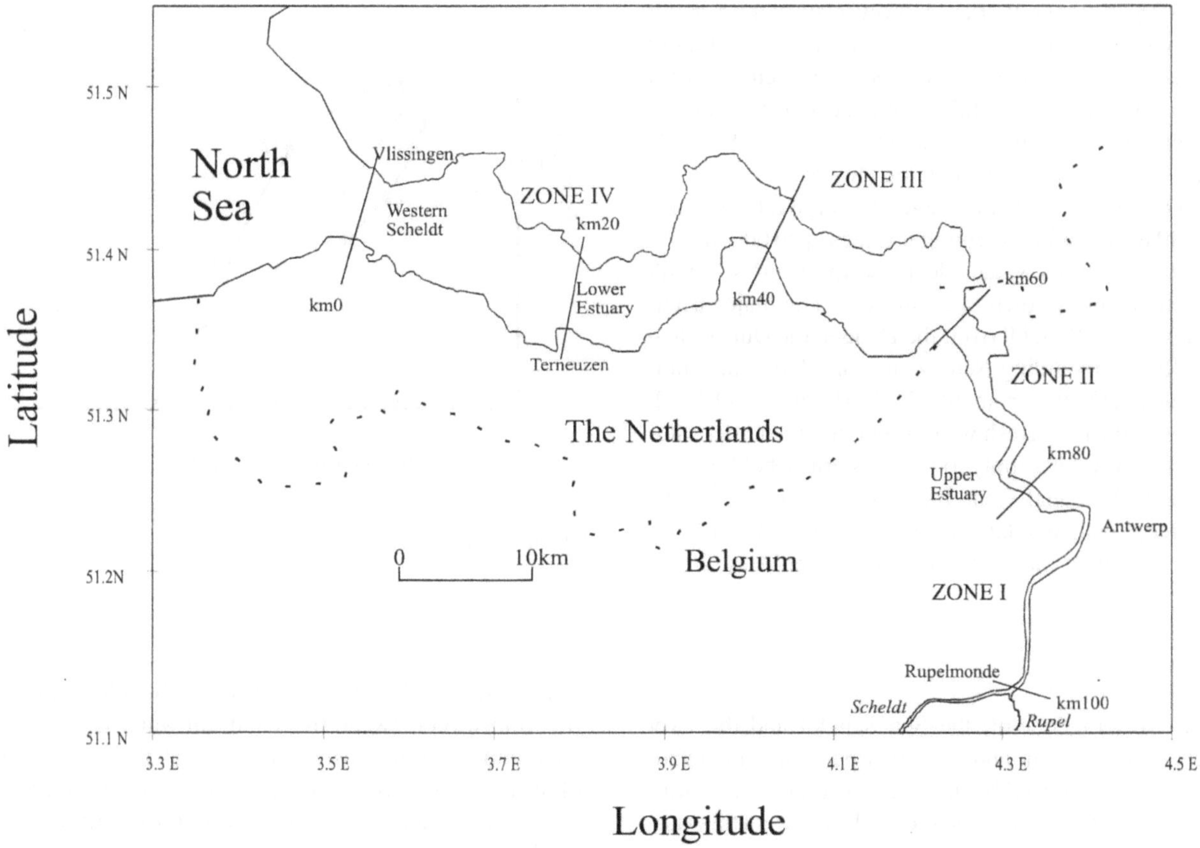

Figure 1. Map of the Scheldt Estuary. Lower estuary (km 0–km 60). Upper estuary (km 60–km 100).

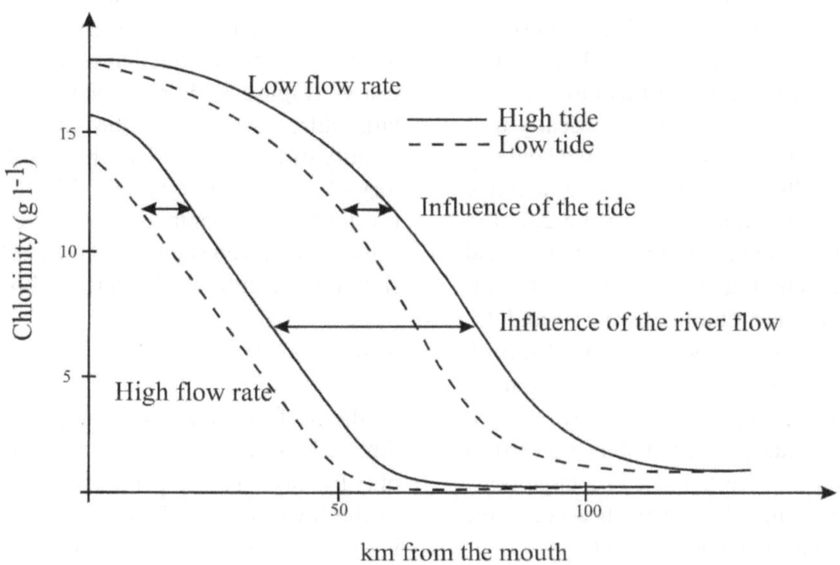

Figure 2. The influence of river flow and tide on the salinity profiles.

s^{-1} for the period 1949–1986 (Pers. commun., The Ministry of Public Works). However, since the Scheldt is a typical rain-river, the actual discharge rate is highly dependent on the season. Over the total catchment area about 30% of the rainfall is drained into the surface water, but in densely populated areas such as Brussels and Antwerp this value increases to 45%.

During winter, the average discharge rate amounts to 180 $m^3 s^{-1}$ with exceptional values up to 600 $m^3 s^{-1}$. Average summer values decrease to 60 $m^3 s^{-1}$ with minimal values down to 20 $m^3 s^{-1}$. The major tributaries are the Scheldt river, the Dender, the Durme, and the Rupel (including Nete, Zenne and Dijle) accounting respectively for about 27, 6, 10 and 56 (17, 12, 27)% of the total fresh water input into the upper estuary. Fresh water supply to the Western Scheldt occurs through the canal Gent-Terneuzen (15 $m^3 s^{-1}$), the discharge-sluice of lake 'Zoommeer' near Bath (11 $m^3 s^{-1}$) and some minor discharges from polders (20 $m^3 s^{-1}$).

Tidal influence

The Western Scheldt, the Sea Scheldt and the river Scheldt up to Gent and the lower parts of some tributaries are influenced by the tide. The tidal range varies throughout the estuary and the tidal rivers, and this variability is paralleled by the volumes of water transported by the tide. The Canter-Cremer number, which is defined as the ratio of the saline water volume flowing up the estuary through a given section during the flood, to the volume of fresh water flowing into the estuary above that section during a complete tidal cycle, is a measure for vertical mixing. When this ratio is large (10–1000) the estuary, or tidal river, is vertically well mixed. A ratio below 10 indicates that vertical density differences occur. In Vlissingen the mean vertical tide is 3.8 m whereas it is 5.0 m in Antwerp and 2.0 m in Gent. The flood volumes are respectively 1000 10^6 and 62 10^6 m^3 per tide in Vlissingen and Antwerp, the corresponding Canter-Cremer numbers are 149 and 12. Hence, the Scheldt estuary is vertically well-mixed. Only in the neighbourhood of Antwerp, occasionally a small vertical stratification may occur. Since the residence time increases quickly with increasing vertical mixing due to dilution of fresh water in a large body of sea water, the residence time of fresh water in the Scheldt estuary is high (two to three months).

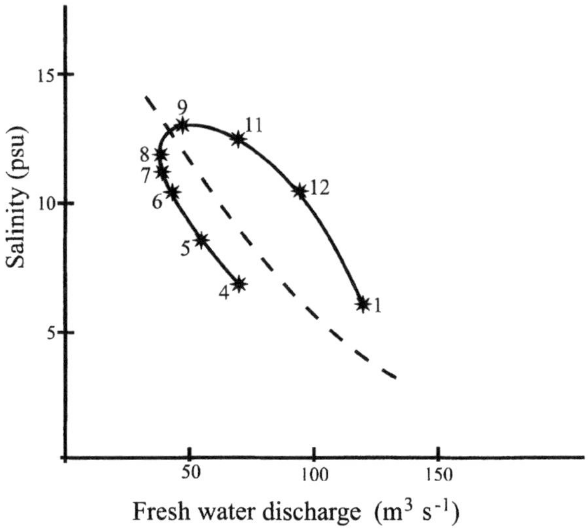

Figure 3. Observed salinities at km 40 as a function of river flow. The broken line represents the computed salinity.

Salinity

The longitudinal salinity profile of the Scheldt estuary (the transition between fresh and salt water is particularly smooth) is primarily determined by the magnitude of the river discharge. The tidal action, on the contrary, contributes to a lesser degree (the salinity shift during a tidal period is much smaller than the salinity shift between low and high river discharge). As an example, the longitudinal chlorinity profiles for different fresh water flow rates are given in Figure 2. The yearly variation of the salinity at a given sampling station and at low tide has been plotted as a function of the river discharge (Figure 3). For the same fresh water flow rate, large differences of salinities are observed. Thus, the salinity distribution reflects here strongly the past history of the hydraulic regime. Ideally, if a steady state is established instantaneously, the relation between the salinity at a given station and the fresh water discharge should not exhibit the hysteresis shown in Figure 3.

Residual current

In the surface layer the residual velocity is always directed towards the sea, but in the bottom layer this is only the case in the upper part of the estuary (Figure 4). In the lower part of the estuary, the residual velocity is directed upstream. As a consequence there is also a region in the estuary where the net residual flow in the bottom layer equals zero. Ronday (1976) calculated the residual circulation in the Scheldt estuary with

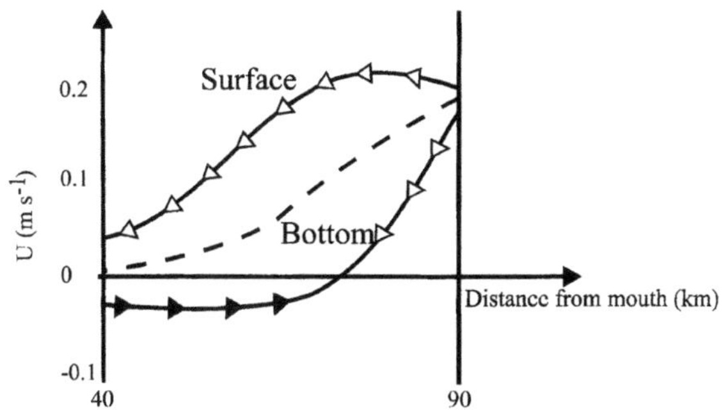

Figure 4. The residual velocity (U) at the surface and at the bottom.

a 2-D mathematical model (x–z) integrated over the width (y). According to a higher or lower fresh water flow rate the line of zero residual current is shifted downstream or upstream. At the mean river flow rate (104 m^3 s^{-1}), the point of zero residual current at the bottom is located in the zone between 2.5 and 5 psu salinity. This means that riverine and marine transported bottom material will accumulate in that area.

Quantitative aspects of suspended sediments (turbidity)

The morphological description of the Scheldt river basin is of particular interest for the transport and sedimentation of pollutants. Various pollutants interact with the finer fraction of the suspended matter, particularly the organic particles. The sediment transport in the different parts of the river basin is controlled by different transport-mechanisms. We will only discuss here the part of the Scheldt estuary between the mouth (km 0) and the Rupel river (km 92). In the Western Scheldt (from the mouth to 55 km inland) spatial turbidity variations are fairly small compared to the variations observed in the upper estuary (see Figures 5a to f). This is due to the much more intense local erosion-sedimentation processes in the upper estuary (see further this paper) and the highly variable suspended matter concentration in the river water end-member over a tidal cycle (Wartel, 1973; 1977; Duinker et al., 1982). On a larger time-scale, meteorological conditions and the river flow will control the turbidity values.

The longitudinal turbidity profiles show a concentration decrease in suspended matter by a factor of 4 to 5 between the upper and downstream estuarine areas. The upper estuary is therefore called the area of maximum turbidity. Compared to several other estuaries (e.g. Gironde, Loire,. . .) it is noted that the maximum turbidity values are relatively low and hence the zone of maximum turbidity less pronounced.

The occurrence of a turbidity maximum has been explained in terms of a hydrodynamical model, involving non-tidal estuarine circulation of water and particles as well as tidal movements (Postma & Kalle, 1955; Ronday, 1976; Wollast & Peters, 1978). Non-tidal estuarine circulation processes account for collection of specific particles in the zone of increased turbidity and deposition in the bottom sediments, while alternating tidal movements account for deposition/erosion and mixing with other water bodies.

To discuss the importance of suspended matter sources in the Scheldt estuary as well as processes affecting the particle dynamics especially related to the turbidity maximum, the area Vlissingen up to Rupelmonde is subdivided in three parts (the schematic diagram of Figure 6 represents the suspended matter transport in each of the 3 zones):

Zone 10–30 psu: the Western Scheldt, extends from the mouth (km 0) up to the brackish marsh of Saeftinge (km 50 to 55). The morphology of the estuarine bed is rather complicated by the formation of several ebb and flood channels. The maximum tidal velocity at the mouth is about 0.9 m s^{-1} and the residual velocity (mean over the cross section) 0.1 cm s^{-1}. The residual bottom current is landward directed. The morphology of the Western Scheldt is also influenced by the dredging of the navigation-channels. The major part of the dredged material, which mainly consists of sand, is again dumped at various sites in the Western Scheldt maintaining the 'natural' balance of the sediment transport in this area. Only a small part of the dredged material is removed from the system and used for other purposes e.g. for landfill. The input of marine suspended matter into the lower estuary is estimated at 79 10^6 kg y^{-1} (Van Eck, 1991). This material is transported into the estuary up to the upper area as isotope measurements indicate (see the section on 'Qualitative aspects of suspended sediments'). Sedimentation amounts to 10 10^6 kg y^{-1} (Van Eck, 1991), and occurs in the intertidal areas bordering the estuary, in particular in the brackish marsh of Saeftinge along the 50–60 km stretch. According to recent estimates about 50% of the fluvial suspended matter is deposited at Saeftinge, and about 35% reaches the coastal zone (Van Eck, 1991).

The bottom sediments mainly consist of sand (coarse, medium-coarse and medium-fine) except at some outflows, such as the Gent-Terneuzen canal, and on the tidal flats. In zone 10–30 psu the amount of bottom material eroded during a tidal cycle is less impor-

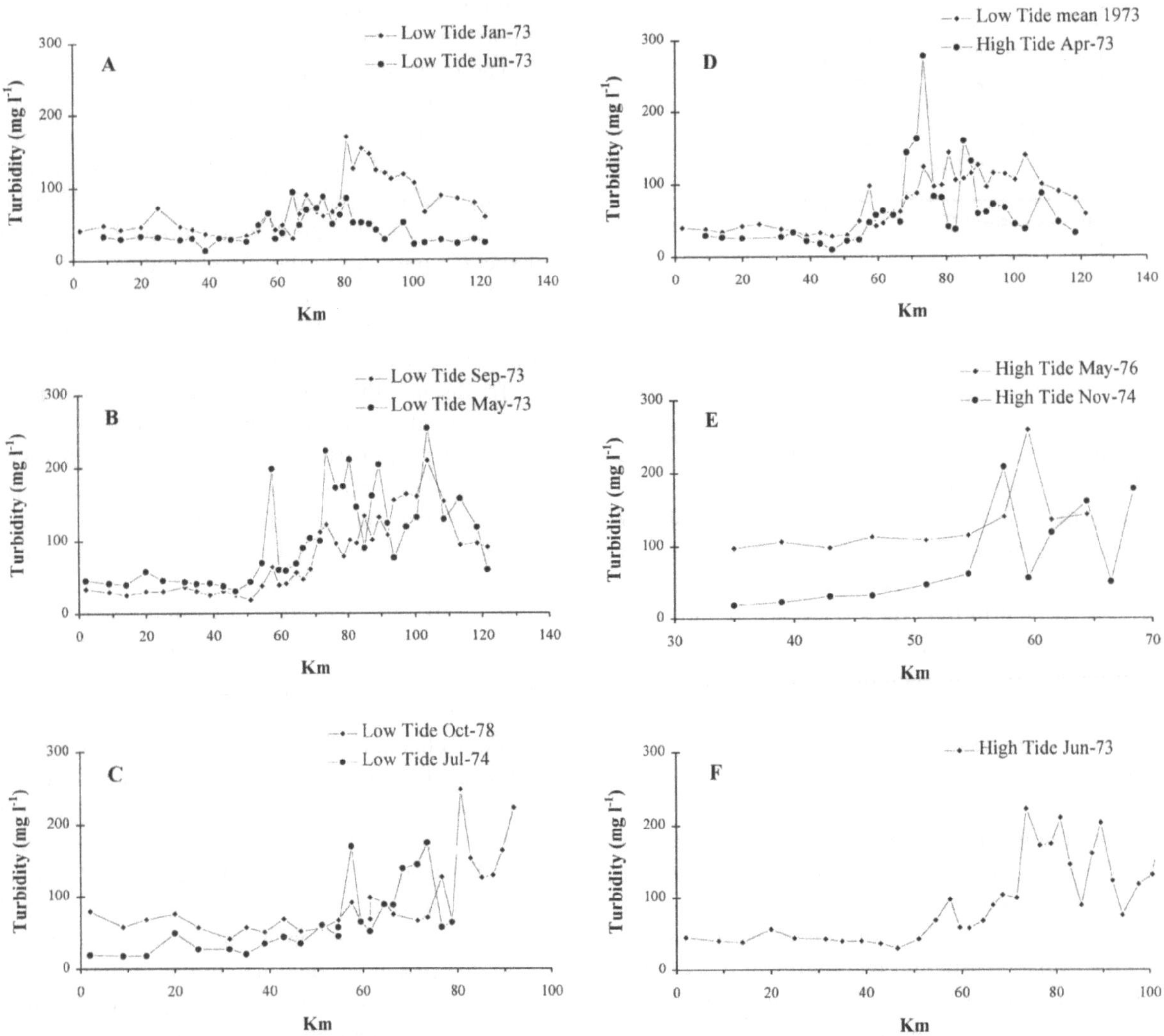

Figure 5. Turbidity profiles at high and at low tide.

tant than in zone 2–10 psu, due to the lower tidal currents and the coarser bottom material. However, at the turn of the current, half the amount of the sediment in suspension is deposited. Accordingly, a thin layer of about 0.25 mm is deposited at the bottom. This layer sedimentates and erodes each tide.

Observations of sand wave movements in the Western Scheldt show a relatively high mixing and exchange rate of the upper sediment layer; twice a year the upper 0.5 m of the sediment is brought in suspension. This process is important regarding the release of pollutants to (or uptake from) the water column.

Zone 2–10 psu: the second salinity zone starts at the brackish marsh of Saeftinghe and continues up to Antwerp (km 78). The morphology is much simpler since the estuarine bed is restricted to one ebb/flood channel. The maximum tidal velocity is 1.1 m s^{-1} and the residual current (mean over the cross section) 1 cm s^{-1}. In this zone there is a point where the residual current at the bottom is zero (Ronday, 1976). Bottom currents upstream (marine) and downstream (riverine) converge there (Figure 4). The sedimentation in this zone is estimated at about 280 10^6 kg y^{-1} (Van Eck, 1991), and mainly occurs in the brackish marsh of Saeftinge.

Many sources contribute to the suspended matter composition in this zone. River-borne suspended matter and rolling bottom sediments are transported down-

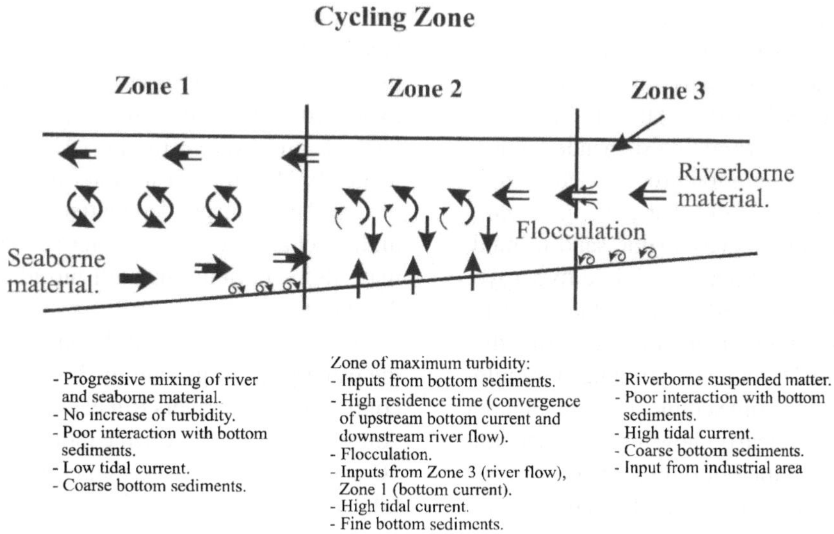

Figure 6. The qualitative suspended matter transport in the 3 zones of the Scheldt estuary.

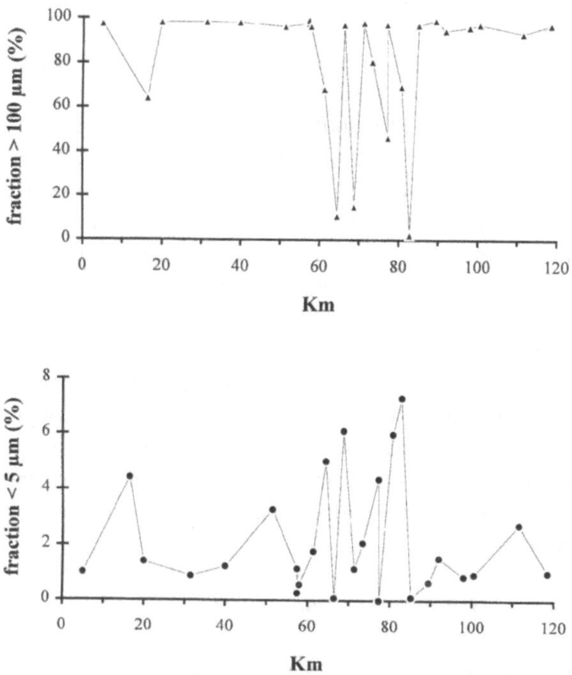

Figure 7. Grain size fractions in bottom sediments.

stream from zone 0–2 psu. Sea-borne material is transported from zone 10–30 psu by the upstream directed saline bottom current and tidal action. At some places in the estuary one finds peaty sediments, mostly there where the riverbanks also contain such material, while at other locations, the riverbanks consist of fine sands and mud. In zone 2–10 psu, part of the continuously suspended matter is influenced by the flocculation process. However, according to Meade (1972), the evidence that the effects of salt flocculation on the concentration of suspended matter can be observed in estuaries, is very limited. Fine continentally derived suspended material coagulates during early mixing (initial raise of salinity) and will more easily settle down. This zone is an exclusive site in an estuary where low density particles can be collected and trapped in suspension and in sediments despite the occurrence of strong currents. The dominant mechanism involved might well be the incorporation of fine sand particles in the aggregates under local conditions of increased concentrations of particles, thus increasing their settling velocities (Wellershaus, 1981). According to Eisma (1990), particle dynamics includes agglomeration of particles as well as deflocculation. In these processes, the particle concentration (Hwang, 1989), the presence of organics and the presence of discrete particle flocs (Wollast & Duinker, 1982) may play an important role. An other important factor contributing to the sedimentation of finer particles in zone 2–10 psu is related to the particular hydrodynamic conditions. In this zone, the saline upstream directed bottom current converges with the downstream directed river flow, increasing the residence time of the particles and enhancing their sedimentation. In addition, a fraction of the finer suspended matter which is exported towards zone 10–30 psu by the net seaward current sedimentates when entering that zone, because there a strong decrease in tidal current velocity occurs (the estuarine cross section strongly increases); these particles are reintroduced in zone 2–10 psu by the upstream directed bottom current, thus also showing an enhanced residence time in that area. The combination of three factors, (1) favourable hydrodynamic conditions, (2) several fine suspended matter sources, and (3) the flocculation process, led in zone 2–10 psu to a bottom sediment that contains locally a high percentage of fine material (fine sand to mud, sometimes even a non-compacted, mobile hyperpycnal or fluid mud layer; according to Metha (1990) a fluid mud layer shows horizontal mobility, highly dissipative kinematics and turbulence damping). Granulometric analyses in the bottom sediments of the Scheldt estuary (Laurent, 1971) clearly show higher concentrations of fine particles in the bottom sediments of zone 2–10 psu (between km 80 and km 55 from the mouth) compared to the other zones (Figure 7). Although Van Eck (1991) suggests a high sedimentation rate for zone 2–10 psu, it does not mean that during a tidal cycle a large amount of the suspended matter is definitively trapped at the bottom. Indeed, with the typical parameter values for zone 2–10 psu (a suspended matter concentration of 200 mg l^{-1}, a depth of 10 m and a bottom surface of 20 km^2) it is sufficient that 0.7% of this material is trapped at the bottom per tidal cycle, to yield an accumulation rate of 280 10^6 kg y^{-1}.

High turbidities in zone 2–10 psu correspond with high tidal current velocities (stirring up of bottom sediments); in fact the denser suspended matter fraction derived from the bottom dominates in suspension practically during the entire tidal cycle, while around the turn of the current, the lowest turbidities, in fact the continuously suspended fractions, are observed. The erosion of the bottom sediments is, however, not equal during ebb and flood tide. The asymmetry between ebb and flood induces relatively more erosion than sedimentation during flood and the opposite during ebb. Turbidity measurements at low, slack water (Figures 5a to 5d) show values increasing from km 50 (10 psu) to km 140 (upstream zone 0–2 psu). These suspensions represent more or less the continuously suspended matter. At high water (Figures 5d to 5f) very high turbidities are observed essentially in zone 2–10 psu (the turbulence in the water column during flood tide is high enough to erode or stir up bottom sediments, especially the finer bottom sediments). Erosion is also different at spring and neap tide conditions. During spring tide, erosion is very strong and distribution of the eroded particles may develop, with a time lag, over the entire water column. This differs from neap-tide conditions when erosion may be weak and distribution of the eroded particles will involve a more limited layer of water above the bottom. Under such conditions, a dense suspension or even a fluid mud may be locally formed. Well known examples are found in the Gironde and Loire estuaries (Allen et al., 1974; Gallenne, 1974). High fresh water flow rates (high inputs of river-borne suspended matter) and to a lesser extent high discharges of waste water (domestic or industry) may also contribute to high turbidity values in zone 2–10 psu.

Zone 0–2 psu: the third zone or the upper part of the estuary (average salinity 2–0 psu) is bound by the confluence of the rivers Rupel and Scheldt (km 92) and Antwerp (km 78). The same morphology as in zone 2–10 psu exists here. The magnitude of the maximum tidal velocity is 1.2 to 1.3 m s^{-1}, that of the residual current (mean over the cross section) 3 cm s^{-1} (only seawards directed).

The quantity of suspended matter supplied by the rivers Scheldt and Rupel (Wartel, 1973, 1977; Duinker et al., 1982) is most variable (several sources are involved) and is estimated at 320 10^6 kg y^{-1}, using a mean suspended matter concentration of 100 mg l^{-1} (Wartel, 1977; D'Hondt & Jacques, 1982). Downstream the confluence of Rupel and Scheldt, the estuary consists in a narrow channel, incised in the Boom clay, a very hard material. The velocity of the current is high (almost no sedimentation of the river-borne suspended matter occurs) and the erosion of the clay bottom is very slow. Close to km 78 (2 psu), local bottom sediments consist of medium-fine to fine sand. Sediments in zone 0–2 psu are eroded only when the tidal current is high and are also transported by rolling to the downstream zone. Direct discharges of domestic origin (city of Antwerp) and downstream the city by the industrial park contribute especially to the finer suspended matter fraction.

Suspended matter fluxes through the estuary

D'Hondt & Jacques (1982) estimated the total suspended matter load produced by the whole watershed at 508 10^6 kg y^{-1} (753 10^6 kg y^{-1} including the estuarine part). The origin of this suspended matter is as follows: 28% domestic, 27% industrial and 45% natural. Information obtained from water treatment plants learns us that for the first two sources, one third is inorganic and two third organic. Suspended matter from erosion is almost entirely inorganic. A large fraction of the organic material is degraded in the riverine system. D'Hondt & Jacques (1982) estimated the fluvial sus-

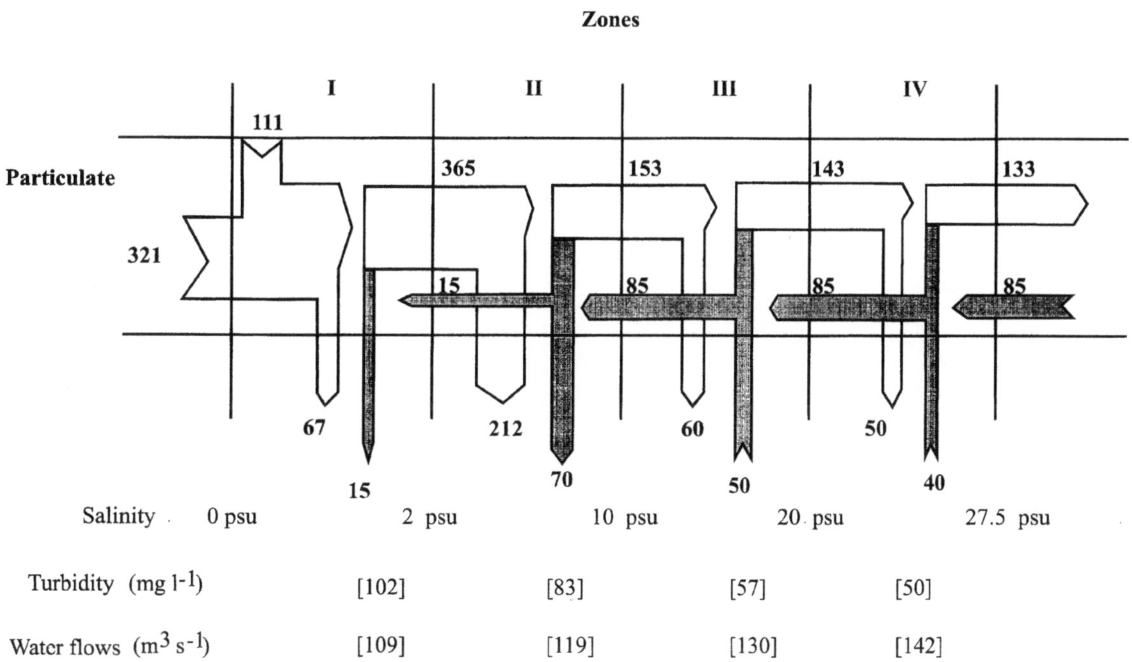

Figure 8. Sediment transport scheme (Baeyens et al., 1996). Flows in 10^6 kg y^{-1}.

pended matter flux at the riverine/estuarine interface (Rupelmonde) at 320 10^6 kg y^{-1}, while an additional amount of 111 10^6 kg y^{-1} is supplied by lateral inputs into the 0 to 2 psu zone. At the estuarine-sea interface an input of 79 10^6 kg y^{-1} of suspended matter into the estuarine system has been estimated by Van Eck (1991). The flux of suspended matter into the sea is estimated at 300 10^6 kg y^{-1} by D'Hondt and Jacques (1982), and 136 10^6 kg y^{-1} by Van Eck (1991). According to the latter author, sedimentation in zones 0–2 psu, 2–10 psu and above 10 psu, respectively amounts to 83 10^6 kg y^{-1}, 282 10^6 kg y^{-1} and 10 10^6 kg y^{-1}. Assuming a net input of fluvial suspended matter of 432 10^6 kg y^{-1} and of marine suspended matter of 79 10^6 kg y^{-1} and a total sedimentation of 375 10^6 kg y^{-1}, the net outflow of suspended matter to the sea should be 136 10^6 kg y^{-1}. Taking into account the particulate metal-mass balances in the Scheldt estuary (Baeyens et al., 1998), the sediment mass balance estimates for the downstream part of the estuary have been updated and refined (Figure 8). Nevertheless, these values are to be considered as orders of magnitude.

Qualitative aspects of suspended sediments

The low river discharge and large tidal influence results in extremely low non-tidal displacements or high residence times of water, and even larger residence times for particles in the estuary. The whole system should thus be in a quasi-equilibrium state as Duinker et al. (1982) suggested but this is only true (1) for time scales much larger than that of a tidal cycle, (2) without significant changes in the inputs from the large industrial site situated at Antwerp, and (3) without significant changes in the inputs from the sediments where the physico-chemical conditions are not necessarily constant nor identical to those in the water column. Sampling at fixed moments of the tidal cycle, at fixed salinities and/or sampling stations and in different seasons of the year is the best approach. This is, however, not always possible in practice and therefore results should be interpreted carefully in relation to a quasi-equilibrium system.

One of the major processes determining the composition of the suspended particles is the estuarine mixing of the river- and sea-derived material. The mineralogy of suspended material has been described by Wollast (1973), Wartel (1977) and Van Alsenoy et al. (1989). The major components are quartz, calcite, clay miner-

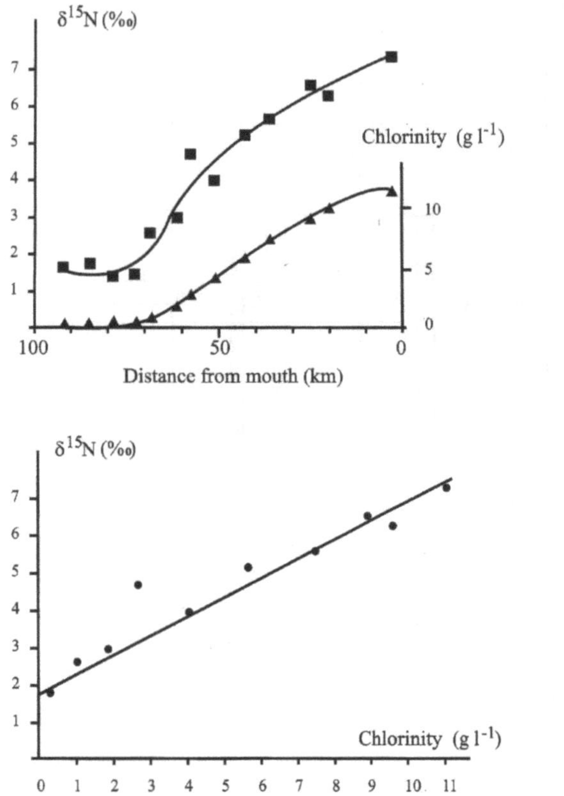

Figure 9. Longitudinal distribution of $\delta^{15}N$ in particulate organic matter (April 1982).

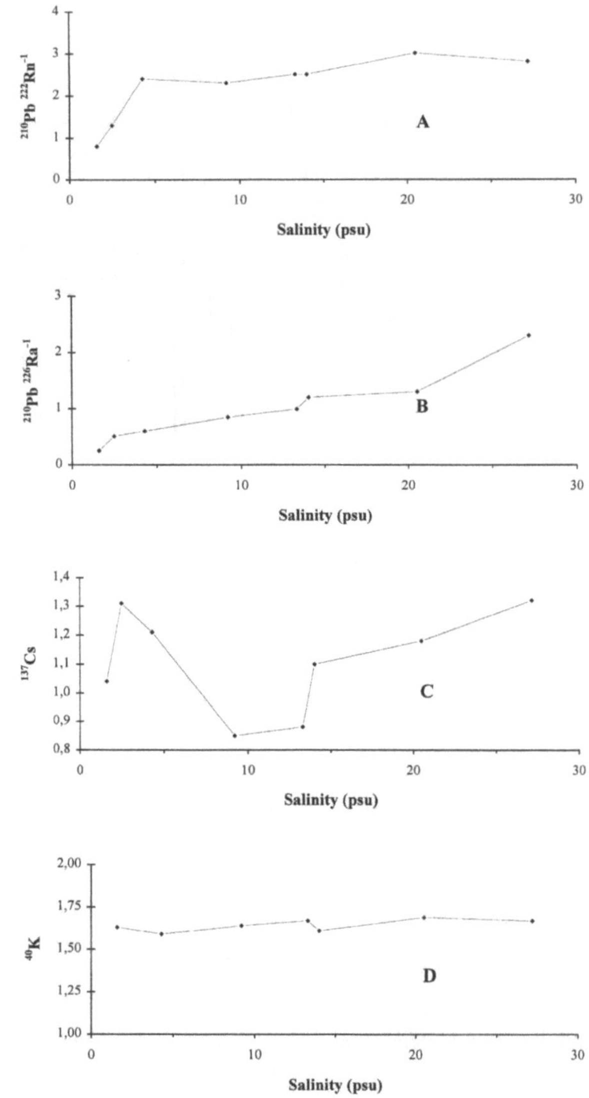

Figure 10. Longitudinal profiles of radioisotopes. Cs and K contents in dpm.g^{-1}

als (illite, montmorillonite, kaolinite and chlorite), and organic matter. Among the minor components are iron oxyhydroxides and phosphates, feldspars, dolomite and aragonite, glauconite, rutile and iron sulphides. Human activities have changed the suspended matter composition significantly (Wollast, 1988). Suspended matter resulting from domestic and agricultural activities has a high organic matter content (40–60%). The industrial suspended load is often strongly contaminated by trace metals and contains large amounts of unusual materials such as gypsum, slag dust and metallic iron spherules. In the bottom sediments the major constituents of the silt and clay fractions are quartz, clay minerals and carbonates (Wartel, 1977).

In a study on the Scheldt estuary Salomons & Eysink (1981) have measured the mixing ratio of marine to fluvial particulate matter by using the stable isotopic composition of carbonate minerals. They found that marine mud is transported into the estuary up to the fresh water area and that the longitudinal particulate profiles of Cs, Hg, Cr comply with a conservative mixing pattern. The isotopic distribution of N in the particulate matter collected in the estuary reveals also some interesting facts (Mariotti, 1982). During April (Figure 9), before the spring phytoplankton bloom, the longitudinal distribution of $\delta^{15}N$ in the particulate organic matter typically corresponds to the mixing of detrital organic matter of continental origin ($\delta^{15}N = 1.5‰$) with plankton of marine origin ($\delta^{15}N = 8.5‰$). The longitudinal profiles of $^{210}Pb/^{222}Rn$, $^{210}Pb/^{226}Ra$, ^{137}Cs and ^{40}K are presented in Figures 10a to 10d. The ratio ^{210}Pb to ^{222}Rn rapidly increases in the zone of early mixing, but tends to a constant value from a salinity of 4 psu on. This ratio can, as a consequence, not be used as a tracer to distinguish between the marine or continental origin of suspended sediments. However, since marine suspended matter is strongly enriched in ^{210}Pb versus ^{226}Ra, the ratio of these radioisotopes gives us an estimate of the marine fraction in that suspended matter. Besides the fact that Figure 10b confirms that marine mud is transported into the upper estuary, it appears that after an initial, rapid mixing of continentally derived material with marine material (up to about 5 psu), a relative more homogenous, well mixed area exists (the area 5 psu to 13 psu which corresponds to the area of maximum turbidity with larger contributions from the bottom layer and a longer residence time of the particles); in the downstream estuary we see again a progressive mixing of the suspended sediments with marine derived material.

The marine suspended matter is also enriched in ^{137}Cs (Figure 10c) but mixing with continentally derived suspended matter results in a decrease of the ^{137}Cs activity in the zone from 30 to about 10 psu. Increased activities are observed in the low salinity area, suggesting a local source: most probably the higher ^{137}Cs values are related to the nuclear power plants located at Doel, 62 km from the mouth. The ^{40}K profile (Figure 10d) shows not only a conservative behaviour, but also since both end members are almost equal, an almost constant value through the estuary.

For a number of other parameters, or for some of the above mentioned ones at other periods of the year (e.g. $\delta^{15}N$), no conservative mixing was observed. Two global transformation zones in the estuary are responsible for the non-conservative behaviour of those variables: a geochemical filter (including microbial processes controlling the redox potential) in the upstream estuary and a biological filter in the lower estuary.

Upstream estuary: the geochemical filter

The organic carbon load of the Scheldt basin is very high (290×10^9 g-C y^{-1}, Wollast, 1982). Most of this organic matter is degraded and respired in the river before the fresh water reaches the estuary. However, the organic matter concentrations entering the estuary are still high. The mean dissolved matter concentration (DOC) is 7 mg-C l^{-1} and the particulate organic matter concentration reaches 15 mg-C l^{-1} whereas in unpolluted rivers these values are respectively estimated at 3 and 2 mg-C l^{-1} (Meybeck, 1982). The high

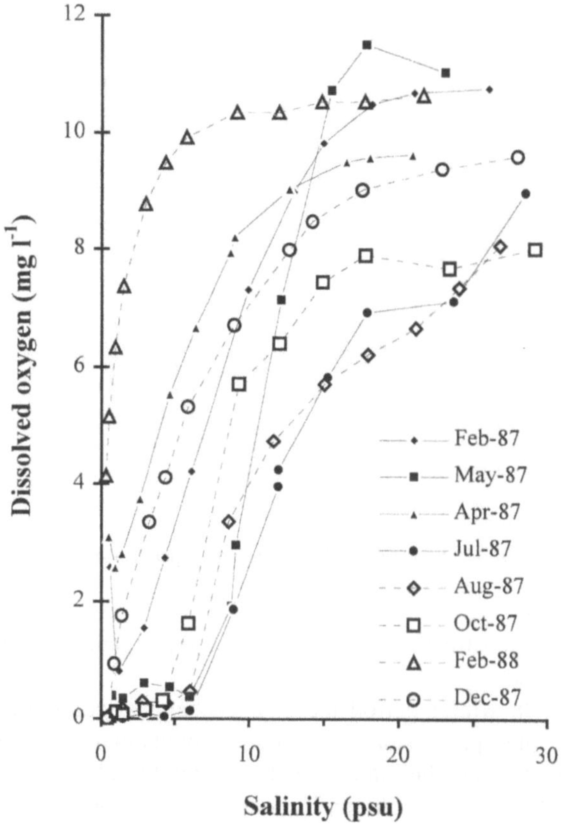

Figure 11. Seasonal O_2 variations.

nutrient concentrations create high productivity in the riverine system with values of 30 mg l^{-1} chlorophyll-*a* (Van Eck et al., 1991) but up to 100 mg l^{-1} in May 1978 (Wollast, 1982). The chlorophyll-*a* values during the latter cruise dropped to almost zero in the zone of turbidity maximum, which in that period of the year is yet strongly depleted in oxygen.

As a consequence of the high organic load entering the estuary, the upper estuary is suboxic in winter, but becomes progressively anoxic when the season proceeds. Simultaneously the suboxic/anoxic area extends. The seasonal variations of the oxygen content at different locations in the estuary in 1982 are shown in Figure 11. In the upper estuary, oxygen may be completely exhausted in the summer months. The longitudinal oxygen profiles as a function of salinity are shown in Figures 12 and 13 for 1982–1983 and 1988 respectively. It is worthwhile to stress that the oxygen profiles presented here are only representative for the years indicated, (1) because the estuarine system is continuously in evolution and (2) the construction

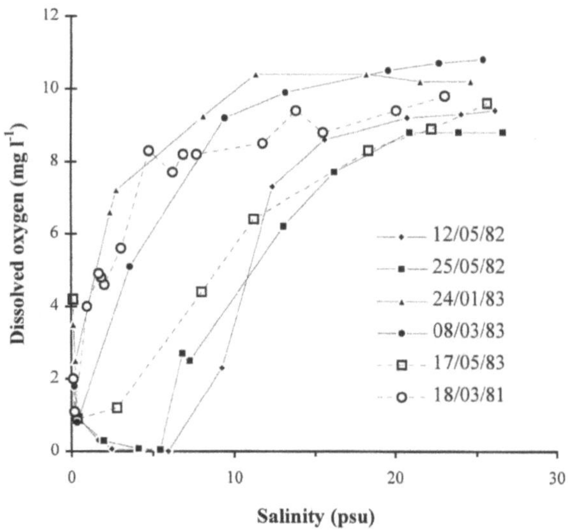

Figure 12. Longitudinal O_2 profiles in the period 1981–1983.

of water treatment plants changed the organic carbon loads.

In winter period the physicochemical conditions in the upper estuary are: high amounts of DOC and nutrients including carbonates, a high turbidity, and a suboxic water column. Under these conditions several processes influence the dissolved/solid distribution of the trace metals:

- adsorption/precipitation: positive deviations from the mixing curve, i.e. precipitation/adsorption processes were found for Fe, Mn and P (and may be valid for trace metals too) in the Scheldt estuary (Salomons & Eysink, 1981).
- Coprecipitation and adsorption on ironhydroxides: a coupling between sedimentary redox processes and enrichment of iron and manganese in North Sea suspended matter has been reported by Dehairs et al. (1989). A similar study was carried out on Scheldt sediments by Panutrakul & Baeyens (1991) and Baeyens et al. (1991), and they observed reprecipitation of iron and manganese oxyhydroxides at the oxic/suboxic interface in the sediments. Zwolsman (1992) confirms the enrichment of iron and manganese in the suspended matter of the high turbidity zone in the Scheldt by typical pore water gradients across the sediment-water interface in that area of 421 mM cm^{-1} for Fe and 45 mM cm^{-1} for Mn.
- precipitation of dissolved organic carbon (DOC): another mechanism is the flocculation process of dissolved organics, removing all associated trace metals from solution (Zwolsman, 1992; Sholkovitz, 1976).
- Oxidation of resuspended metal sulphides: in the reduced sediments in zone 2–10 psu, the oxidisable fraction (metal sulphides and metals bound to the organic matter) constitutes a major fraction according to the general accepted sequential extraction schemes (Panutrakul & Baeyens, 1991; Baeyens et al., 1991). Due to the strong tidal action and the

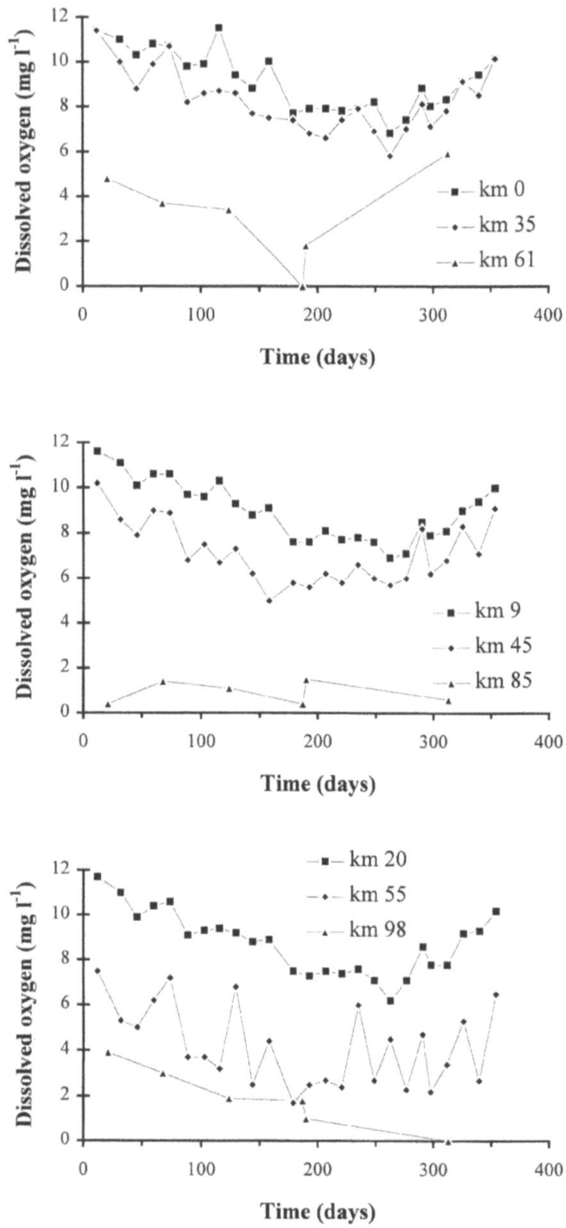

Figure 13. Longitudinal O_2 profiles in the period 1987–1988.

presence of fine muddy sediments, metal sulphides can be resuspended.

The overall result of these processes on the metal distribution is generally a depletion in the dissolved phase and an enrichment in the particulate phase.

Downstream the zone of the turbidity maximum, desorption and oxidation processes may increase the dissolved metal concentrations. The increase of the dissolved cation content when the river water is progressively mixed with sea water affects the composition of the suspended matter, especially with respect to those compounds exhibiting exchange properties. In addition, increased inorganic ligand concentrations such as Cl^- are capable to keep some trace metals such as cadmium and mercury, forming chloro-complexes of a relative high stability (Förstner et al., 1982), in solution. On the other hand, a progressive oxidation of metal sulphides and/or metals bound to particulate organic matter resuspended in zone 2–10 psu, can also enhance the dissolved metal concentrations.

In summer period, the upstream estuary is anoxic. Literature suggests that precipitation of metal sulphides is involved (Duinker et al., 1982; Zwolsman, 1992); the latter authors reported the presence of sulphide in anoxic river water samples of the Scheldt. Sulphide precipitation can present, in addition to the previously mentioned processes, a removal process of dissolved metals in summer in the upper estuary.

Lower estuary: the biological filter

In the anoxic zone of the estuary, the chlorophyll-*a* concentration is high as a result of the nearly complete absence of herbivores (Figure 14). The plankton bloom in the downstream estuary is, however, less pronounced than in the riverine system due to a dilution of the nutrient concentrations. Earlier in this paper, $\delta^{15}N$ in particulate organic matter indicated a conservative mixing of marine and continental derived suspended sediments. This relationship was observed in April, before the plankton bloom. However, during the period of high phytoplankton activity in the downstream estuary, a considerable increase of $\delta^{15}N$ in the particulate organic matter is observed. During the flowering period (April-September) autochtonous phytoplankton contributes up to 50% of the total suspended matter concentration in the lower estuary (Mariotti et al., 1984). This local enrichment perturbs the previously observed conservative mixing pattern of $\delta^{15}N$. According to Wollast (1982), this phenomenon is due

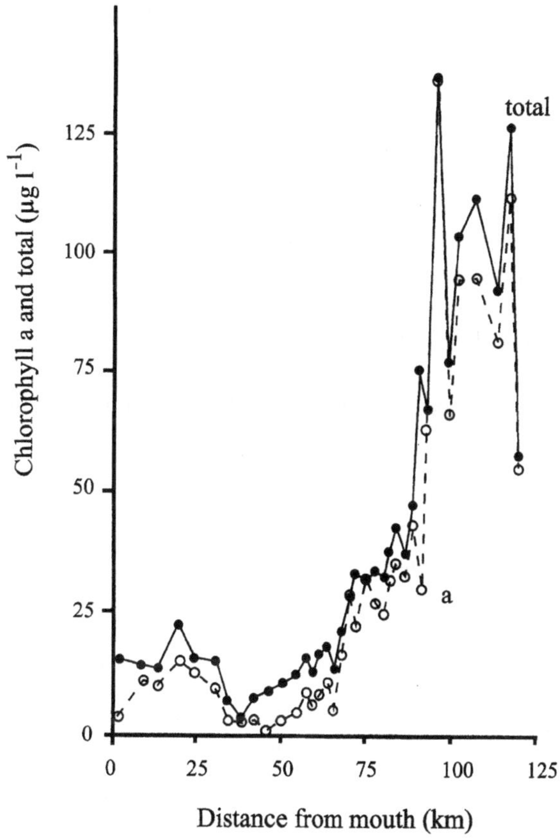

Figure 14. Longitudinal chlorophyll profiles.

to the nitrification process and the progressive enrichment of ammonia in $\delta^{15}N$.

Some metals are actively incorporated during phytoplankton growth. Indeed, Zwolsman (1992) observed a very good correlation between dissolved silica and dissolved Cd and Zn in the lower estuary over the year. Silica is the nutrient which is predominantly assimilated by the phytoplankton (diatoms) in the Scheldt estuary.

Conclusion

The general concepts on hydrology, sediment transport, suspended matter composition and biological activity in the Scheldt described here above, are basic requirements for a better understanding of the trace metal behaviour in that estuary. It appears that he Scheldt estuary is a unique polluted system due to the particular physicochemical conditions in the upstream estuary, especially the high turbidity, high sedimentation, oxygen depletion and the relatively large inputs of metals. The behaviour of Cd, Cu, Hg, Pb and Zn, in relation to these characteristics, is discussed in related papers.

Acknowledgement

The authors are particularly indebted to the Ministry of Science Policy for their continuous and lasting interest regarding oceanographic research, and more specifically for the financial support they offered us during the periods 1970–1982 and 1990–1996. We are also grateful to the Management Unit of the Mathematical Model North Sea and Scheldt Estuary for offering ship-time on the MS Mechelen and the MS Belgica.

References

Allen, G. P., P. Castaing & A. Klingebiel, 1974. Suspended sediment transport and deposition in the Gironde estuary and adjacent shelf. Mem. Inst. Geol. Bassin Aquitaine 7: 27-36.

Baeyens, W., G. Gillain, F. Ronday & F. Dehairs, 1987. Trace metals in the eastern part of the North Sea. II: Flows of Cd, Cu, Hg, Pb and Zn through the coastal area. Oceanol. Acta 10: 301–309.

Baeyens, W., S. Panutrakul, M. Leermakers, J. Navez & M. Elskens, 1991. Geochemical processes in sandy and muddy sediments. Geo-Marine Lett. 11: 188–193.

Baeyens, W., F. Monteny, R. Van Ryssen & M. Leermakers, 1998. A box-model of metal flows through the Scheldt estuary (1981–1983 and 1992–1995). Hydrobiologia 366: 109–128.

D'Hondt, P & T. G. Jacques, 1982. Gesuspendeerde materie in de Schelde. Evaluatie van een onderzoek van de ULB voor het ministerie van Volksgezondheid. Water 4: 133–135.

Dehairs, F., W. Baeyens & D. Van Gansbeke, 1989. Tight coupling between enrichment of iron and manganese in North Sea suspended matter and sedimentary redox processes: evidence for seasonal variability. Estuar. coast. Shelf S. 29: 457–471.

Duinker, J. C., R. F. Nolting & D. Michel, 1982. Effects of salinity, pH and redox conditions on the behaviour of Cd, Zn, Ni and Mn in the Scheldt estuary. Thalassia Jugoslavica 18: 191–202.

Eisma, D., 1990. Particle size analysis of suspended material. In S. Wartel (ed.), Towards a Definition of Mud, Royal Belgian, Institute for National Sciences, Brussels, Belgium, 32.

Förstner, U., W. Calmano & J. Schoer, 1982. Heavy metals in bottom sediments and suspended material from the Elbe, Weser and Ems estuaries and from the German Bight (south eastern North Sea). Thalassia Jugoslavica 18: 97–122.

Gallenne, B., 1974. Study of fine material in suspension in the estuary of the Loire and its dynamic grading. Estuar. coast. Shelf S. 2: 261–272.

Hwang, K.-N., 1989. Erodibility of fine sediment in wave dominated environments. M.S. Thesis, University of Florida, Gainesville, 156 pp.

Laurent, E., 1971. Etude minéralogique de l'envasement de l'Escaut. Rapport N° 1, Ecole Royale Militaire, Bruxelles, 56 pp.

Mariotti, A., 1982. Apports de la geochimie isotopique a la connaissance du cycle de l'azote. These de doctorat, Universite Pierre et Marie Curie, Paris 6.

Mariotti, A., C. Lancelot & G. Billen, 1984. Natural isotopic composition of nitrogen as a tracer of origin for suspended organic matter in the Scheldt estuary. Geochim. Cosmochim. Acta 48: 549–555.

Meade, R. H., 1972. Transport and deposition of sediments in estuaries. Geol. Soc. Mem. 133: 91–120.

Metha, A. J., 1990. Defining fluid mud in a dynamic environment. In S. Wartel (ed.), Towards a Definition of Mud, Royal Belgian Institute for National Sciences, Brussels, Belgium, 58–59.

Meybeck, M., 1982. Carbon, Nitrogen and Phosphorus transport by rivers. Am. J. Sci. 282: 401–450.

Panatrakul, S. & W. Baeyens, 1991. The behaviour of heavy metals in a mud flat of the Scheldt estuary, Belgium. Mar. Pollut. Bull. 22: 128–134.

Postma, H. & K. Kalle, 1955. Die Entstehung von Trübungszonen im Unterlauf der Flüsse, speziell im Hinblick auf die Verhältnisse in der Unterelbe. Deutsche Hydrografische Zeitung 8: 138–144.

Ronday, F., 1976. Modèles hydrodynamiques. In J. C. J. Nihoul (ed.), Projet Mer, Rapport Final, Le Ministère de la Programmation de la Politique Scientifique, Brussels, Vol. 3, 270 pp.

Saeijs, H. L. F., 1977. De Westerschelde een milieu in beweging. De natuurwetenschappelijke waarde van de Westerschelde en de bedreigingen daarvan. Rijkswaterstaat, Middelburg, the Netherlands, 32 pp.

Salomons, W. & W. D. Eysink, 1981. Pathways of mud and particulate trace metals from rivers to the southern North Sea. In S. D. Nio et al. (eds), Holocene Marine Sedimentation in the North Sea Basin. Blackwell, Oxford: 429–450.

Sholkovitz, E. R., 1976. Flocculation of dissolved organic and inorganic matter during the mixing of river water and sea water. Geochim. Cosmochim. Acta 40: 831–845.

Van Alsenoy, V., A. Van Put, P. Bernard & R. Van Grieken, 1989. Chemical characterization of suspensions and sediments in the North Sea and Scheldt estuary. ICES, C.M. 1989/E:31, The Hague, 18 pp.

Van Eck, G. T. M., 1991. De ontwikkeling van een waterkwaliteitsmodel voor het Schelde-estuarium. Water 61: 215–218.

Van Eck, G. T. M., N. De Pauw, M. Van Den Langenbergh & G. Verreet, 1991. Emissies, gehalten, gedrag en effecten van microverontreinigingen in het stroomgebied van de Schelde en het Scheldeestuarium. Water 61: 215–218.

Wartel, S., 1973. Variations in concentration of suspended matter in the Scheldt estuary. Bull. Inst. r. Sci. nat. Belg. 49: 1–11.

Wartel, S., 1977. Composition, transport and origin of sediments in the Scheldt estuary. Geol. Mijnbouw 56: 219–233.

Wellershaus, S., 1981. Turbidity maximum and mud shoaling in the Weser estuary. Arch. Hydrobiol. 29: 161–198.

Wollast, R., 1973. Origine et mecanisme de l'envasement de l'estuaire de l'Escaut. Rapport de synthese, Ministere des Travaux Publics, Borgerhout (Belgium), 140 pp.

Wollast, R., 1982. Behaviour of organic carbon, nitrogen and phosphorus in the Scheldt estuary. Thalassia Jugoslavica 18: 11–34.

Wollast, R., 1988. The Scheldt estuary. In W. Salomons et al. (eds), Pollution of the North Sea: an Assessment. Springer-Verlag, Berlin: 183–193.

Wollast, R. & J. C. Duinker, 1982. General methodology and sampling strategy for studies on the behaviour of chemicals in estuaries. Thalassia Jugoslavica 18: 471–491.

Wollast, R. & J. J. Peters, 1978. Biogeochemical properties of an estuarine ecosystem: the river Scheldt. In E. Goldberg (ed.), Biogeochemistry of Estuarine Sediments, Unesco, Paris: 279–293.

Zwolsman, J. J. G., 1992. Behaviour of trace metals in the Scheldt estuary. Ph.D. Thesis, University of Utrecht.

Hydrobiologia **366:** 15–44, 1998.
W. F. J. Baeyens (ed.), Trace Metals in the Westerschelde Estuary.

Biogeochemical behaviour of Cd, Cu, Pb and Zn in the Scheldt estuary during the period 1981–1983

Willy Baeyens[1], Marc Elskens[1], Gérard Gillain[2] & Leo Goeyens[1]
[1] *Department of Analytical Chemistry, Free University of Brussels, Pleinlaan 2, 1050 Brussels, Belgium*
[2] *Laboratoire d'Océanologie, Université de Liège, Sart-Tilman, 4000-Liège, Belgium*

Key words: Scheldt estuary, trace metals, 1981–1983, dissolved, particulate, biogeochemistry

Abstract

Cu, Zn, Cd and Pb have been analysed in suspended matter and water samples from the Scheldt estuary, collected during five cruises between May 1982 and May 1983. In order to evaluate the overall metal behaviour in the estuary, continuous longitudinal profiles of the total, the particulate and the dissolved metal concentrations are discussed in relation to various specific physico-chemical parameters. By means of correlation and multi-regression techniques, several estuarine processes could be identified which dominate and/or control the behaviour of the individual trace metals in three distinct areas of the Scheldt estuary.

The behaviour of the four selected metals was found to show some common features, but is besides this also characterised by many individual and specific properties. The total metal concentrations seem to co-vary well with the turbidity in the maximum turbidity zone between salinities 2 and 10 psu. In the downstream area a dilution profile is observed for the total metal concentrations of Cu and Cd as well as for the dissolved concentrations of all metals.

Dissolved metal production rates are high for Cu and Cd, but much lower for Zn and Pb, confirming the different longitudinal dissolved metal profiles. For Cu and Cd these dissolved metal profiles are characterised by a broad mobilisation area, covering the whole middle estuary. Pb on the other hand, shows a more confined mobilisation area, and for Zn no such area was found. During winter the dissolved metal production rates seem to be controlled, predominantly, by desorption processes. During summer dissolved metal production is correlated mainly to the dissolved oxygen content, suggesting the domination of redox processes solely or in combination with desorption processes. In some cases, co-existent with nearly zero oxygen concentrations, removal of dissolved metals was observed, possibly reflecting precipitation reactions.

Longitudinal particulate metal profiles all show a negative deviation versus the dilution profile. Cu and Cd provide the higher deviations, characterised by a narrow enrichment zone preceding a broad mobilisation area. Finally, all particulate metal profiles appeared to correlate well with POC, when expressed in volumetric units.

Introduction

The explanation of heavy metal transport in an estuary requires a basic understanding of the physical behaviour of the estuary and of the influence individual parts of the system experience or bring to bear on other parts.

Estuaries are governed by tidal action at the sea side and by river flow. In the mixing zone of salt and fresh water, strong longitudinal gradients in physico-chemical parameters such as salinity, pH, temperature, ... develop. They affect internal processes such as:

- removal of trace elements from the dissolved phase by adsorption, by precipitation and coprecipitation of solutes;
- removal of trace elements associated with fine suspended matter, by flocculation and net sedimentation;

– production of trace elements in the dissolved phase by desorption and by solubilisation of particulate matter, including the associated metals;
– resuspension of sediments and subsequent release of their metals;
– transformation and migration of trace metals species at the sediment/water interface.

These processes as well as external effects, mainly anthropogenic activities such as discharging of pollutants or cooling water, dredging of the channels and in some cases subsequent disposal of the spoil on the shoals or river banks, modify the conservative mixing of river-borne and sea-borne solutes and particulate matter.

The overall effect of internal and external processes is often unexpected and results in complex distribution patterns, which differ from one element to the other and from one estuary to the other.

This paper discusses the behaviour of heavy metals in both the dissolved and particulate phases of the Scheldt estuary. Sampling stations are shown in Figure 1. Concomitantly with the assessment of Cd, Cu, Pb and Zn, related variables such as physico-chemical features, $^{210}Pb/^{226}Ra$ ratios and organic matter content were measured in order (1) to improve our understanding of the role of external and internal processes, (2) to distinguish between the effects of physical mixing and physico-chemical processes and (3) to estimate the heavy metal supply of the river Scheldt to the North Sea.

Methodology

Sampling

Considering the sampling process, marine chemists essentially face two problems: the representativity of the samples and their possible contamination. Besides the sampler, the research vessel itself as well as ships passing by are major causes of pollution, leaving behind a cloud of metals from propellers, zinc anodes, paint, engine exhausts and other possible sources. Airborne material and dust, careless handling by unskilled experimenters, the sampling depth relative to the ships draught, the location of the sampling system on board, the atmospheric conditions, the lack of water mass homogeneity, etc., are many possible sources of error.

Gillain & Brihaye (1985) designed a reliable device for sampling at moderate depth aboard large and small research vessels even with rough sea. The principle of the system is to continuously collect small samples of water from a large volume screened off from atmospheric pollution. A peristaltic pump continuously draws 6 l of estuarine water per minute through a polyvinyl chloride (PVC) tube, previously soaked in HCl (6N) for several days, and rinsed abundantly with ultra-pure water, Milli-Q grade (*Millipore*). During sampling it is necessary to avoid ships discharges, antifouling paints, etc., in close proximity to the sampling point. The ship is best left adrift while the sampling tube is on the lee side to meet water masses which are not polluted by the ship. A second peristaltic pump draws 0.5 l min^{-1} from the main flow, the remaining is returned to the estuary. The unfiltered water is kept into a 5 l polyethylene bottle, equipped with a magnetic stirrer, until filtration.

The representativity is more endangered when particulate matter is sampled. This is especially true with the conventional separation system, where the amount of suspended matter collected on the filter before clogging is generally small. In that case the heavy metal content of particulate matter collected on the filter may not be significantly higher than the filter blank. Consequently, the error introduced during the filtration step only is larger than the bias arising from all other analytical steps.

On the other hand, filtration is a slow process and the characteristics of the filter may change after deposition of a first layer of particulate material. In order to overcome all these problems and to minimise all possible external interferences (dust-free conditions), a shipboard stirred filtration system, pressurised with purified nitrogen, was directly connected to the 5 l polyethylene bottle containing the unfiltered water sample. The type of filtration apparatus used is made from a Pyrex glass tube 120 mm long and 47 mm diameter fitted in between two PVC lids with silicone seals. A magnetic rod is suspended a few millimetres above the filter to accelerate filtration, retard clogging by vigorous stirring and collect the largest possible amount of material. The water drained from the 5 l bottle is replaced by pure nitrogen and filtered estuarine water is continuously collected in a polyethylene flask and stored prior to analysis (Gillain & Brihaye, 1985). There is almost no risk of sample contamination during transfer from the sampler to the storage bottle. For samples containing zoo- and phytoplankton we use a low constant pressure of 0.3 kg cm^{-2}, in order to avoid disintegration of fragile organisms. This may lead to increased heavy metals concentrations in the filtrate and changes in heavy metal speciation.

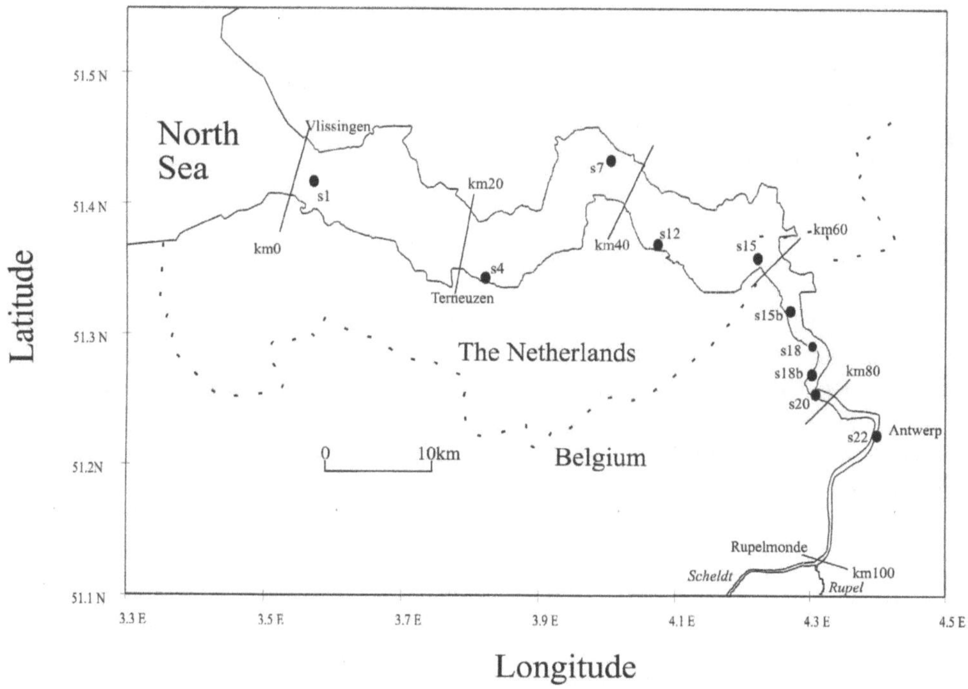

Figure 1. Map of the Scheldt Estuary and position of the sampling stations.

Contamination of the sampling and filtration system is avoided and minimised by cleaning the filters with 100 ml of a 10^{-2} M Diethylenetriaminepenta-acetic acid (DTPA) solution and rinsing them with highly purified water before filtration. Additionally, the sampling and filtration unit is treated for one week with HCl (6N), followed by rinsing with large amounts Milli-Q (*Millipore*) water. In any case, there is little time for possible interaction between the container wall and the sample, since filtration is started immediately after collection and is usually completed within one hour. The first 200 ml of sample are utilised to condition the membrane filter and conducted to the waste.

The performances of this sampling and filtration system and the usually operated systems were compared in the Belgian coastal area. The results (Zn: 0.70–4.00 μg l^{-1}; Cd: 0.020–0.100 μg l^{-1}; Pb: 0.20–0.50 μg l^{-1} and Cu: 0.10–1.0 μg l^{-1}) show that the trace metal concentrations are one order of magnitude lower compared to conventional sampling (Niskin bottle) and filtration methods (open air shipboard suction filtration). Nevertheless, the Cd, Cu, Pb and Zn concentrations in these sampling areas are still high in comparison with most oceanic zones. Two additional tests were carried out.

First, the described sampling and filtration system was used on the Mediterranean Sea (Corsica, Calvi). The trace metal data assessment in this region revealed dissolved levels of Cd in the range from 0.02 to 0.005 μg l^{-1}, of Pb from 0.08 to 0.015 μg l^{-1}, of Cu from 0.10 to 0.04 μg l^{-1} and of Zn from 0.18 to 0.05 μg l^{-1}. All these data are in sharp contrast to those measured in Belgian coastal waters, but compare very well to other uncontaminated areas.

Secondly, the impact of the metal supply by the Scheldt on the trace metals distribution and on the circulation pattern in the Belgian coastal area (Baeyens et al., 1998e), showed that this riverine supply can account for the observed concentration increases in the coastal water masses during their flow through the Belgian coastal zone.

Both results reinforce our confidence in the quality of the sampling and filtration system, specifically designed to minimise contaminations.

Physico-chemical parameters were measured *in situ* by means of a Plessey 9400 automated sensor. Calibrations were performed before, during and after each cruise.

Cleaning

Since results obtained from storage experiments (e.g., Mart, 1979) support the conclusion that PTFE and high density polyethylene bottles show little tendency to adsorb or release heavy metals, new high density polyethylene bottles (1 l) have always been used. These bottles were first rinsed with Milli-Q water (*Millipore*), filled with HCl (6N, analytical reagent grade) and stored for 8 days. Afterwards the containers were rinsed several times with Milli-Q (*Millipore*) water and the acid cleaning procedure was repeated for another 8 days. HCl (1N) (*Merck, suprapur*) was utilised for the final acid treatment. A few days before use the bottles were rinsed with Milli-Q (Millipore) water and conditioned with sea water of very low metal content.

Storage

Zn, Cd, Pb and Cu analyses of filtered sea and brackish water samples, stored over a long period (6 months) in cleaned polyethylene bottles at pH 1 or not-acidified at –20 °C in a freezer showed no significant changes (total metal concentrations). However, significant changes in concentrations occurred after a few days when samples were stored at pH 8 and at room temperature. Similar results, indicating obvious effect of pH and temperature conditions during sample storage, were observed by other investigators (Mart, 1979; Batley & Gardner, 1977).

It is common to stabilise the filtered sample by acidification to pH 1 with either HCl or HNO_3, when only determination of total dissolved metal concentration is required. Since we are highly interested in the metals' speciation in the estuary, the samples were immediately deep-frozen (–20 °C) after collection and filtration, without any additive. With freezing, however, it should be noted that changes in metal speciation may also occur, although the total metal concentrations remain unchanged.

Suspended matter samples, collected on pre-cleaned Millipore membranes for the determination of Cd, Cu, Pb and Zn concentrations, were stored in cleaned Petri dishes and kept frozen at –20 °C till arrival in the home lab.

Analysis

Determination of free and 'very labile' dissolved metals

Ionic and 'very labile' metals were determined in sea and estuarine waters at their natural pH (7 to 8).

Determination of 'moderately labile' dissolved metals

Direct analysis was performed immediately after addition of concentrated HCl (*Merck, suprapur*) to about pH 1. The 'moderately labile' metal concentration is defined as the difference between concentrations measured under acidic (pH 1) and in situ (neutral) conditions.

Determination of total dissolved metal fractions

Total dissolved metal concentrations were determined after exposing acidified solutions (containing 0.05 ml of 30% H_2O_2 per 30 ml) in quartz tubes to UV light with rotational stirring for one night (Gillain et al., 1979). The difference between the total concentration and the concentration measured under acidic (pH 1) condition represents the amount of 'very weak labile' + 'inert' metal.

The differential pulse anodic stripping voltammetry on a hanging mercury drop electrode (DPASV-HMDE) analyses were carried out with a Bruker E 310 modular polarographic analyser and a Hewlett-Packard 7040 A X–Y recorder. A three electrode system was utilised: a Metrohm E 410 hanging drop mercury electrode (HMDE) was used as the working electrode, potentials were measured with reference to a saturated calomel electrode (SCE) and an auxiliary platinum electrode was used to minimise I.R. drop effects across the electrolytic cell (Gillain et al., 1979).

Direct and simultaneous determination of Zn, Cd, Pb and Cu (Gillain et al., 1979)

30 ml of estuarine water, previously filtered through a membrane (0.45 μm), was introduced in the electrolytic cell. The sample was stirred magnetically (450 rpm) and oxygen was removed by bubbling pure nitrogen for 30–40 min. Electrolysis was carried out at a potential of –1.200 V (vs SCE) for 10–20 min, while nitrogen was introduced into the headspace above the solution's surface. One minute before the end of the electrolysis, stirring was stopped and the current-potential curve

was recorded from –1.200 to 0 V with an appropriate current sensitivity, with scan rate 2 mV s^{-1} and pulse 35 mV.

Detection limits (defined as three times the standard deviation of the noise level for a simultaneous determination of the trace metals) and coefficients of variation for Cu, Cd, Pb and Zn were 0.1 $\mu g\ l^{-1}$ and 12% (for a 0.5 $\mu g\ l^{-1}$ Cu solution), 0.005 $\mu g\ l^{-1}$ and 14% (for a 0.05 $\mu g\ l^{-1}$ Cd solution), 0.03 $\mu g\ l^{-1}$ and 13% (for a 0.1$\mu g\ l^{-1}$ Pb solution) and 0.01 $\mu g\ l^{-1}$ and 7% (for a 1 $\mu g\ l^{-1}$ Zn solution), respectively.

The filters containing the collected suspended matter, were stored in acid-cleaned petri-dishes and maintained at –20 °C prior to analyses. To determine the amount of trace metals associated with the particles, filters were dried at 50 °C and inserted in a polytetrafluorideethylene (PTFE) bomb. A mixture of concentrated HNO_3, H_2O_2 (30%) and concentrated HF (4 ml: 4 ml: 1 ml) was added to the solid material. Subsequently, the PTFE-bomb was closed and heated to 160 °C for 6 h. After cooling the solution was evaporated to dryness and an additional ml of concentrated HCl was added. The solution was left overnight and then brought to 100 ml at pH 1, 30 ml of this solution was brought to the polarographic cell for DPASV analysis.

Results

The data of Cd, Cu, Zn and Pb, obtained during five cruises on the Scheldt estuary between May 1982 and May 1983 are presented in the next sections. One additional cruise, characterised by a very elevated fresh water flow rate (390 $m^3\ s^{-1}$) in March 1981 is discussed in relation to some specific processes. All data are collected during the months January to May, and are therefore representative for winter, spring and early summer periods. Zwolsman & Van Eck (1990) measured some of these metals also during summer and autumn, what makes additional information concerning specific seasonal effects also available. As an example the individual data of the 25/5/1982 cruise are presented in Table 1. The mean data are presented in related papers (Baeyens, 1998; Baeyens et al., 1998b). The mean data were calculated at fixed salinities; continuous longitudinal concentration profiles of individual cruises were obtained through an interpolation procedure.

Dissolved metals

Two different approaches were used for the presentation of the overall metal distribution in the Scheldt estuary. Firstly, all individual total dissolved metal concentrations were plotted versus salinity. In order to visualise the upstream-downstream evolution of the concentration, a best fitting third order polynome is calculated (Figure 2). Secondly, the mean dissolved metal (Me_{Diss}) concentrations, as well as the corresponding standard deviations, were compared at fixed salinities as shown in Figure 3.

The third order polynome, representing the average dissolved Cu (Cu_{Diss}) concentration profile in the Scheldt estuary, has a parabolic form starting at the fresh water boundary at about 0.5 $\mu g\ l^{-1}$, reaching a maximum of about 2 $\mu g\ l^{-1}$ at salinity (S) 15 psu and decreasing to below 0.5 $\mu g\ l^{-1}$ at the marine interface. The mean Cu_{Diss} concentration profile calculated at fixed salinities corroborates this trend, although both end member concentrations are somewhat higher. These profiles compare very well to the ones reported by Van Den Berg et al. (1987). Nevertheless, some obvious differences are noticed: (1) Van Den Berg et al. (1987) found higher concentrations (the difference amounts to a factor 3) and (2) our mobilisation/production area is broader, as it covers a salinity range between 5 and 20 psu. The average concentration profile has a convex form, reaching a maximum value of 2.1 $\mu g\ l^{-1}$ at a salinity of 17 psu.

The individual profiles observed during winter and spring also present a broad maximum in the mid-estuary, however based only on few collected data in the very low salinity range. The Cu_{Diss} concentrations observed during winter were significantly higher than those measured in late spring, a pattern that is also observed for other metals (see hereafter) and by other authors.

The average dissolved Cd (Cd_{Diss}) profiles derived by either of the two approaches (the best fitting third order polynome using all data points and the profile interpolating the mean concentrations at fixed salinities) are almost identical. Slight differences were derived for the low salinity zone, similarly to the Cu signature. The estuarine profile is most likely described by a parabola, with a pronounced, very strong mid-estuarine maximum of about 0.15 to 0.2 $\mu g\ l^{-1}$ occurring at the 15 psu region, and with both end members being about four to five times lower. Removal of Cd_{Diss} seems to occur in the low salinity area, where early mixing takes place, as is indicated by the average

Table 1. Individual data of the 25/05/82 cruise

Salinity (psu)	Diss. Oxygen ($mg\ l^{-1}$)	Dissolved metals: Zn Labile ($\mu g\ l^{-1}$)	Zn Total ($\mu g\ l^{-1}$)	Cd Labile ($\mu g\ l^{-1}$)	Cd Total ($\mu g\ l^{-1}$)	Pb Labile ($\mu g\ l^{-1}$)	Pb Total ($\mu g\ l^{-1}$)	Cu Labile ($\mu g\ l^{-1}$)	Cu Total ($\mu g\ l^{-1}$)
27.7	9.99	2.00	6.80	0.020	0.030	0.230	0.320	0.480	0.820
27.2	11.40	2.20	6.50	0.020	0.035	0.200	0.240	0.760	1.550
20.5	10.10	5.40	6.80	0.132	0.153	0.270	0.360	1.430	2.270
15.2	7.03	6.70	13.20	0.185	0.185	0.110	0.370	0.950	2.200
14.0	7.00	5.70	11.10	0.134	0.215	0.240	0.300	0.800	1.940
13.3	6.20	8.20	8.80	0.140	0.170	0.380	0.440	0.900	2.070
9.2	2.30	7.80	13.50	0.063	0.063	0.300		0.700	1.240
5.9	0.01	9.10	20.20	0.046	0.048	0.360	0.460	0.940	0.990
5.2	0.05	6.50	15.90	0.033	0.037	0.340	0.410	0.370	0.500
2.5	0.07	7.00	19.00	0.030	0.050	0.180	0.320	0.500	0.580
1.6	0.32			0.140	0.160	0.160	0.200	0.470	0.610

Salinity (psu)	Turbidity ($mg\ l^{-1}$)	POC ($mg\ g^{-1}$)	Particulate metals: Al ($mg\ g^{-1}$)	Zn ($\mu g\ g^{-1}$)	Cd ($\mu g\ g^{-1}$)	Pb ($\mu g\ g^{-1}$)	Cu ($\mu g\ g^{-1}$)	Hg ($\mu g\ g^{-1}$)
27.2	40.9	33.3	32.4	58	4.4	88	30.6	0.52
20.5	26.4	55.6	28.7	194	5.1	124	60.6	0.45
14.0	39.4	38.2	37.5	212	13	190	123	0.62
13.3	46.6	36.4	40.1	268	13.4	200	126	0.79
9.2	151.0	40.5	68	608	35	290	280	1.3
5.9	136.0	50.5	42.2	580	35	355	314	0.83
5.2	70.4	35.75	35.9	328	19.4	246	240	0.77
2.5	40.9	70.2	44.8	495	27	291	236	1

Cd_{Diss} profile. A similar production process of Cd_{Diss} has been reported by Duinker et al. (1982) and Zwolsman & Van Eck (1991). One should remark that the latter authors mentioned an almost constant low value for the upper estuary. Individual cruise data reveal a seasonal variation, with elevated concentrations in the winter period, however.

The dissolved Pb (Pb_{Diss}) profile is quite different from the Cu_{Diss} and the Cd_{Diss} profile, especially when comparing the shape of the best fitting third order polynome. This curve shows a slight concentration increase (0.30 to 0.41 $\mu g\ l^{-1}$) in the low salinity range (up to 10 psu), followed by a dilution process with the marine water mass. The moderately labile Pb_{Diss} concentration, which is the amount of Pb_{Diss} determined at pH = 1, without UV destruction according to the definition used in the speciation scheme (see 'analysis' section), represents between 64 and 77% of the total Pb_{Diss} burden (pH = 1 and UV destruction). As explained in Baeyens (1998), a significant fraction of the Pb_{Diss} content (sample that passed the 0.45 μm filter) may be associated with colloidal particles. Benoit et al. (1994), for example, found that colloidal Pb accounts for most of the filter passing form of this metal in Galveston Bay, Texas. Hence it is possible that the different analytical techniques used to determine Pb_{Diss} in the Scheldt estuary include these colloidal fractions differently (Baeyens, 1998). Nevertheless, the K_D-values of Pb in the Scheldt estuary, observed in different periods by different authors and with quite different analytical methods (Baeyens, 1998), compare fairly well, if one uses the moderately labile dissolved Pb concentration for the period 1981–1983. This observation needs, however, further consideration in future. The longitudinal profile of moderately labile Pb_{Diss} is similar to that of total Pb_{Diss}, being only about 30% lower in concentration. A slight concentration increase, from 0.23 to 0.28 $\mu g\ l^{-1}$ is also noticed in the salinity range up to 10 psu. For the discussion about the behaviour of Pb in the Scheldt estuary, it does not matter whether total or

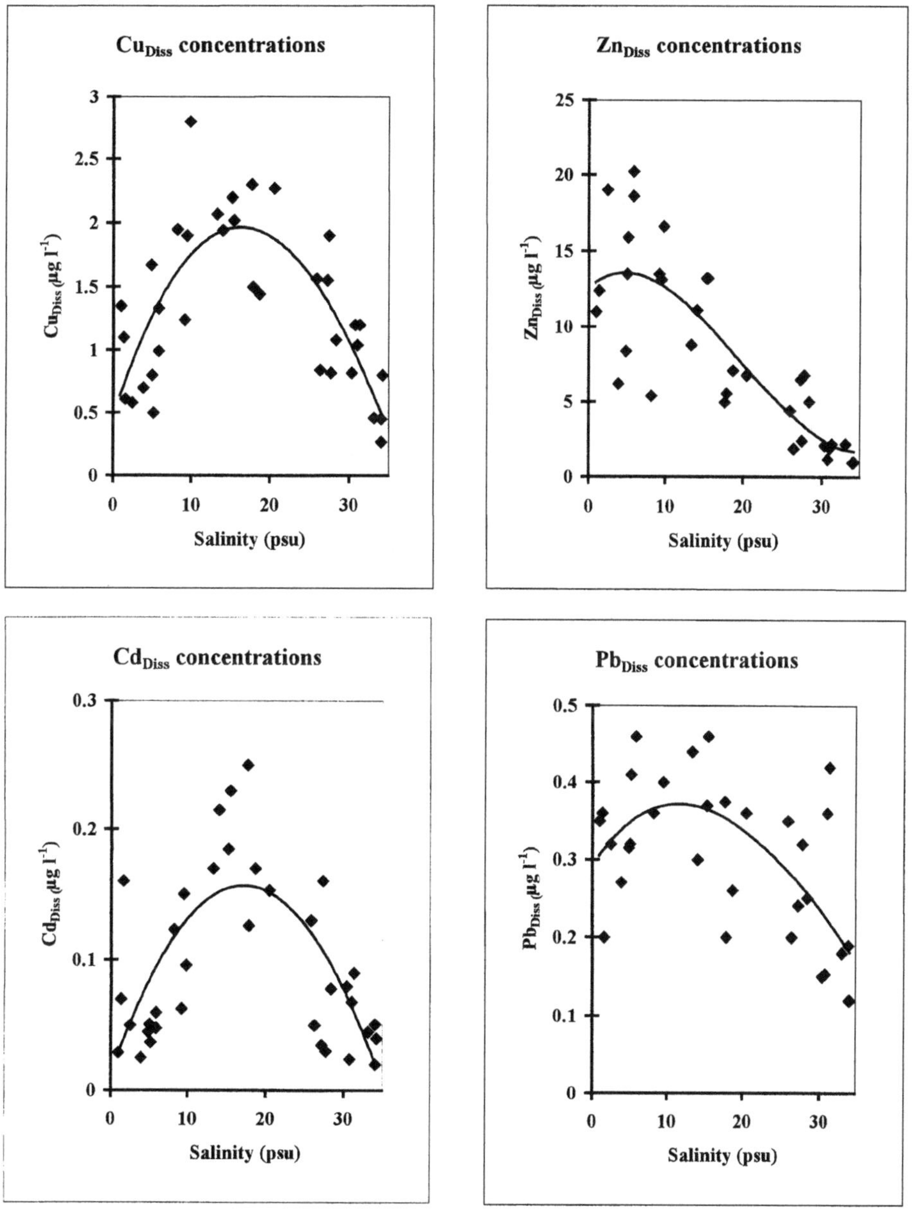

Figure 2. Variability of individual dissolved metal concentrations throughout the Scheldt estuary.

moderately labile Pb_{Diss} concentrations are considered. Unless otherwise stated, total Pb_{Diss} concentrations are presented. However, moderately labile Pb_{Diss} concentrations are used in some related papers dealing with budget calculations, input fluxes or comparisons with other estuarine systems, because the K_D-values based on these results compare with the K_D-values obtained during other periods.

The longitudinal variability of the Pb_{Diss} concentrations is the smallest of the four metals considered in this chapter. Seasonal variations of the Pb_{Diss} concentrations are obvious, with more elevated winter values. Since no other longitudinal Pb_{Diss} profiles are reported for the Scheldt estuary, comparisons are impossible. However, Valenta et al. (1986) measured Pb_{Diss} contents at one fixed station (Hansweert), at a distance of about 35 km from the river mouth and observed similar

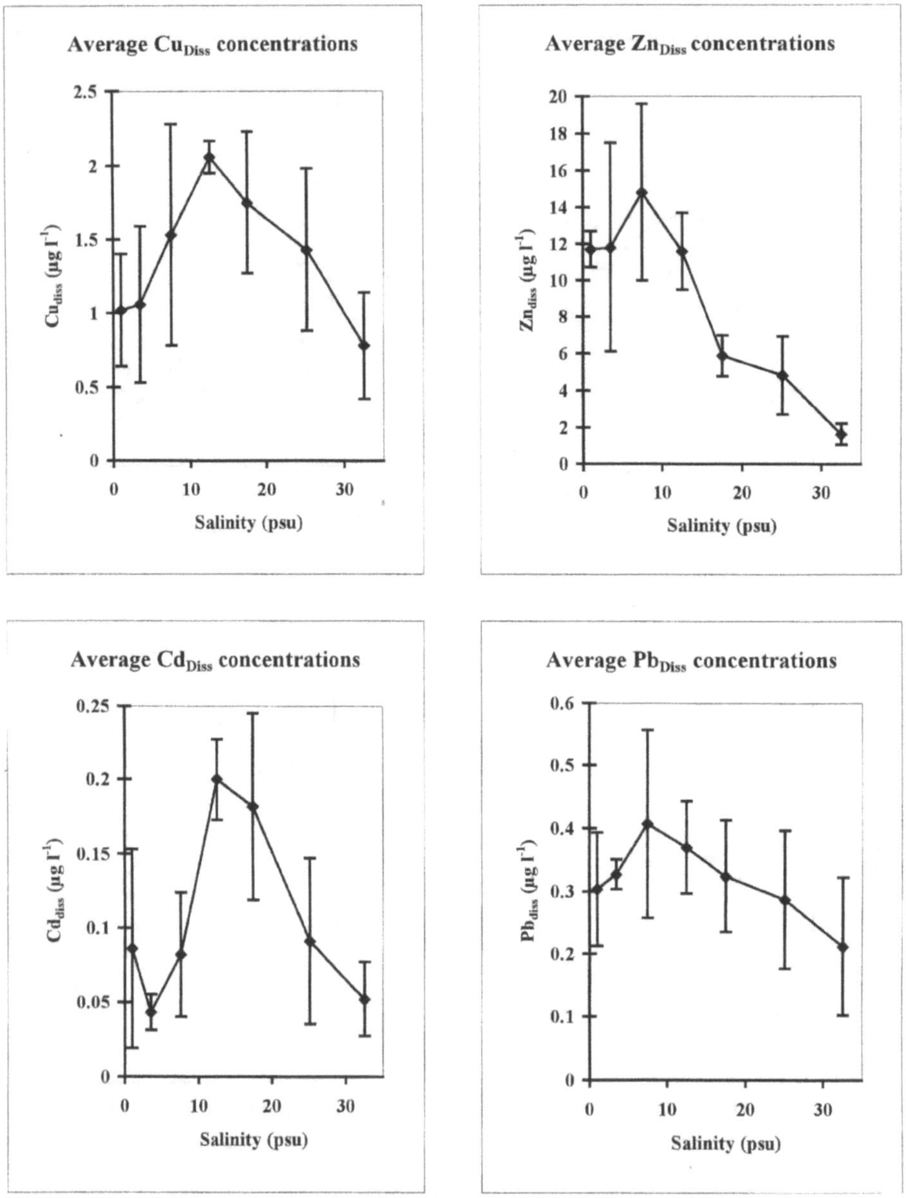

Figure 3. Variability of average dissolved metal concentrations throughout the Scheldt estuary.

seasonal variations with lower values during summer and higher values during winter.

For Zn one observes an almost linear decrease of both the best fitting third order polynome and the average dissolved Zn (Zn_{Diss}) concentration profile towards the river maximum mouth. A mid estuarine maximum can hardly be observed, although within the salinity zone of 5 to 10 psu of the average profile, a slight positive divergence in comparison with the theoreti-

Table 2. Linear dilution equations for total dissolved Zn (Zn_{Diss}), between 10 and 35 psu salinity (S)

Sampling time:	Linear equation:	R2
12/05/82	$[Zn_{Diss}] = 23.199 - 0.656$ S	0.997
25/05/82	$[Zn_{Diss}] = 15.965 - 0.354$ S	0.687
24/01/83	$[Zn_{Diss}] = 7.303 - 0.175$ S	0.941
08/03/83	$[Zn_{Diss}] = 16.836 - 0.470$ S	0.968
18/05/83	$[Zn_{Diss}] = 10.372 - 0.292$ S	0.915

cal dilution line, connecting the two end members, appears. The mean concentrations have such large standard deviations compared to the weak increase of the average curve, that the increase can not be considered as statistically significant. Individual results in the downstream estuarine area (zone 1) reveal a linear decrease for all cruises with correlation coefficients (r) varying between 0.83 and 0.998 (Table 2).

The poorest correlation is observed during the 25th May, 1982, cruise. This aberrant correlation is due to an abrupt increase of the Zn_{Diss} concentration at a salinity of 15 psu.

Seasonal variations, if they ever exist, seem to be limited. Our results are only partly in agreement with literature data. Duinker et al. (1982) reported an almost similar behaviour, while, on the contrary, Zwolsman & Van Eck (1991) reported a pronounced mid estuarine maximum of Zn_{Diss}.

The total Zn_{Diss}-profile is quite different from those of Cu and Cd. The latter ones show a significant mobilisation in the mid- estuary.

We noticed that the March 1981 sampling campaign disturbed significantly the picture of the averaged metal dilution profile, due to the particular hydrological conditions which are only representative for less than 5% of the observations (hydrological variables are daily measured). Hence this curve has been excluded for the calculation of the average situation. The March 1981 cruise was characterised by a very high fresh water discharge rate, 390 $m^3\ s^{-1}$ at Schelle (km 95 from the mouth), compared with the average flow rate of 88 $m^3\ s^{-1}$. The fresh water flow rates are considered to be exceptional, whenever they exceed 350 $m^3\ s^{-1}$ at Schelle. At the Belgian-Dutch border, the salinity was lower than 2 psu, and in this low salinity zone the dissolved metal concentrations (Cd, Pb, Zn, Cu and Hg) were much higher than usually observed in the same salinity zone. A possible explanation for these high concentrations can be that the residence time of these metals in the low salinity zone was, due to the extremely high fresh water flow rate, too small to reach equilibrium between the dissolved and the particulate phases.

Particulate metals

Figures 4 and 5 represent the variability of the particulate metal load ($\mu g\ g^{-1}$) throughout the estuary, the complete particulate metal data set, respectively the mean concentrations, are presented. In the former case we also fitted a third degree polynome through the data points to visualise the global, average behaviour. The longitudinal particulate metal profiles, both for the mean concentrations and the third degree polynomes, show all the same decreasing trend from the riverine end member towards the mouth. We observed that the different metal concentrations were highly variable in the low salinity zone, whereas in the downstream area they were relatively constant at fixed salinities.

The largest concentration differences between end members were observed for Cd and Cu. In addition, all profiles show a negative divergence, compared to the theoretical dilution line, indicating a release of the trace metals associated with the suspended matter during transport towards the sea. A similar trend has been previously described for several temperate estuaries (Jouanneau et al., 1983), for the Rhine and Meuse estuaries by Nolting (1989) and more specifically for the Scheldt estuary by Duinker et al. (1982) and Zwolsman & Van Eck (1991). The observed differences are attributed to mixing processes, different from the conservative one, between end member materials and to mobilisation processes induced by a number of physico-chemical factors (see Discussion).

Figures 6 and 7 represent mean total metal loads (dissolved + particulate, both expressed in $\mu g\ l^{-1}$) throughout the estuary, as well as the partitioning between dissolved and particulate phases, expressed as % of the total metal load. The total concentration maxima were all situated in the area of high turbidity (zone 2), essentially due to a predominant contribution of particulate metals. Downstream the boundary of 10 psu salinity, the total metal concentration drops very quickly. Examination of the dissolved metal fraction throughout the estuary emphasises one particular characteristic for the four metals: the maximum dissolved percentage is consistently situated in the vicinity of the mouth. Cd is the metal which shows the highest dissolved fraction, increasing from about 3% in the low salinity area, to 70% at the mouth. Zn and Cu show fairly similar patterns with a rather constant dissolved fraction at salinities inferior to 10 psu and a subsequent increase to the mouth where both fractions are about equal in size. The evolution is more pronounced for Cu (minimum 10% to maximum 50%) than for Zn (minimum 25% to maximum 50%). The amount of Pb_{Diss} is very small and increases progressively from a few percent in the early mixing zone to less than 10% at the mouth.

Similar distributions for Cd and Pb in the Scheldt estuary have been reported by Valenta et al. (1986). Strong seasonal variations in the partitioning between

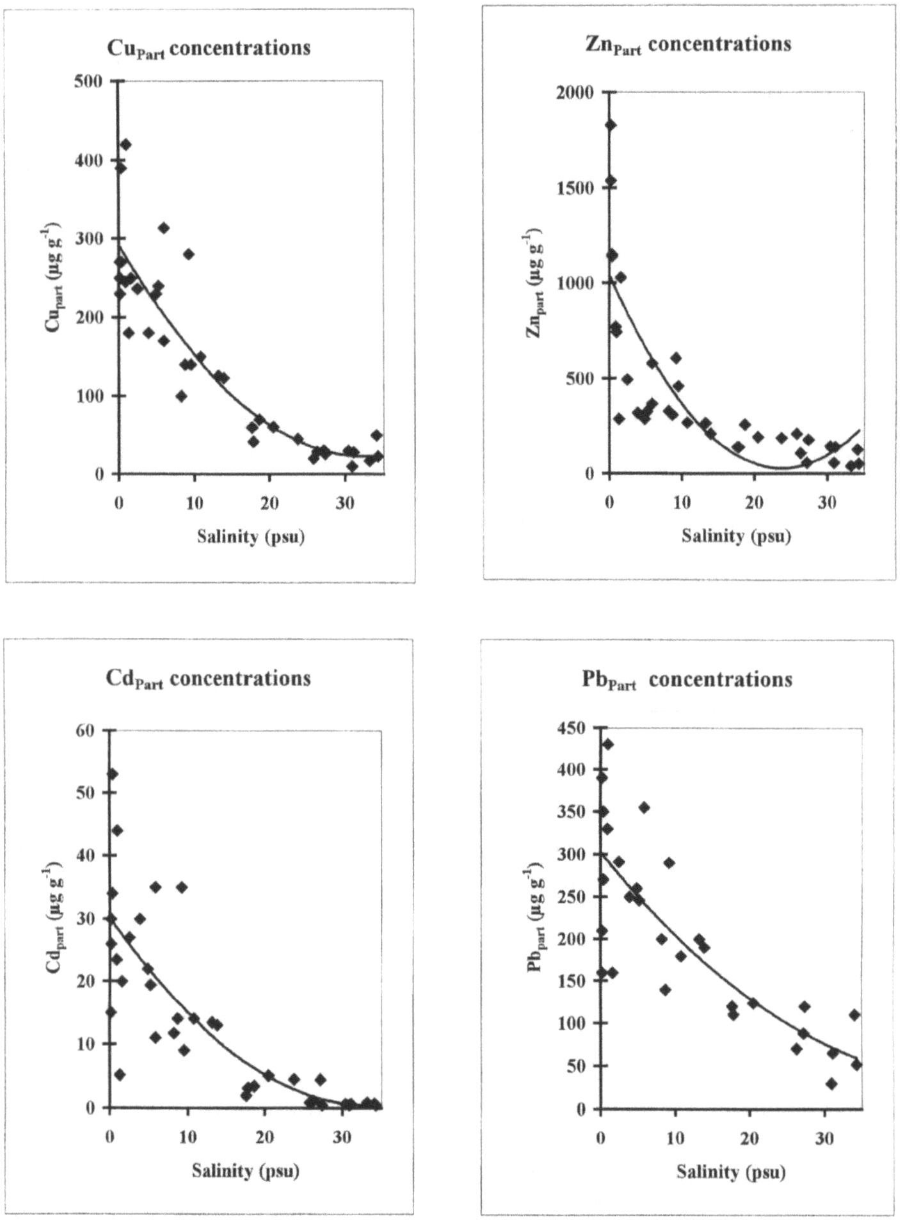

Figure 4. Variability of individual particulate metal concentrations throughout the Scheldt estuary.

both phases being especially observed for Cd, with the dissolved phase being more pronounced during winter than during spring. For the other metals a clear seasonal variation is not observed.

The distribution coefficient (K_D, Figure 8) provides additional information on the partitioning of metals between the dissolved and the particulate phase. K_D values were calculated according to Valenta et al. (1986): the ratio of the particulate metal concentration [μg kg^{-1} suspended matter] over the dissolved metal concentration [μg kg^{-1} water].

K_D^{Cu} decreases in a convex shaped way from an initial value of 250,000 upstream, to 50,000 in the middle estuary, which is also the zone of Cu_{Diss} production, to an almost constant value of approximately 30,000 at higher salinities. The estuarine K_D values vary with

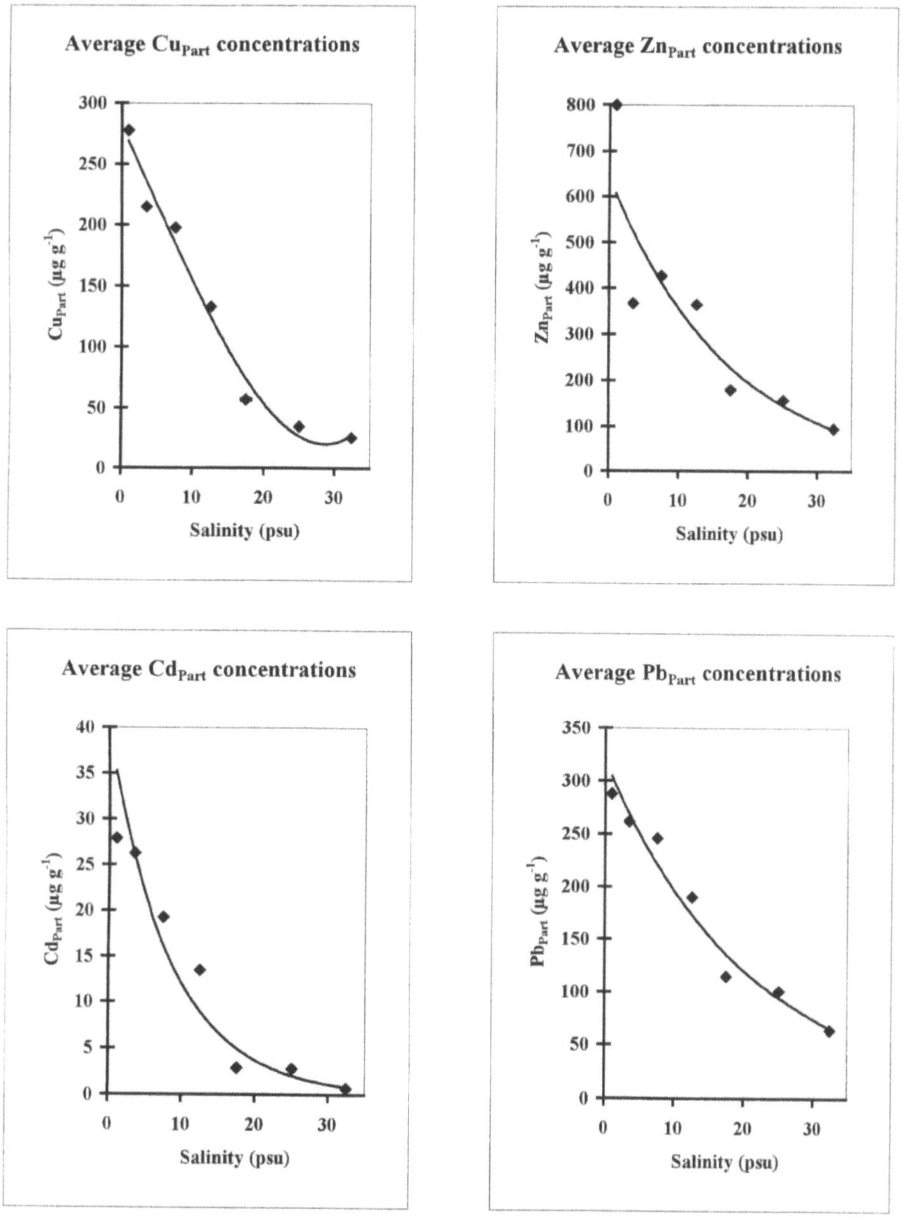

Figure 5. Variability of average particulate metal concentrations throughout the Scheldt estuary.

a factor 8. Balls (1989) reported similar maximal values for environments with an elevated suspended load such as the Dutch coastal waters, and for less loaded waters as in the Forth estuary and in the Baltic Sea. The other K_D^{Cu} values are within the range found in various estuaries and parts of the North Sea.

K_D^{Zn} is high in the upstream area (about 70,000), but decreases rapidly to about 30,000. The difference is mainly due to a decrease of the Zn concentration in the particulate matter. At the mouth of the estuary and in the Belgian coastal waters, the K_D^{Zn} increases again to values of about 50,000. The K_D range in the estuary is fairly small (only a factor 2) when compared to those of some other metals. These K_D values are in the order reported by Balls (1989) for the North Sea and its estuaries.

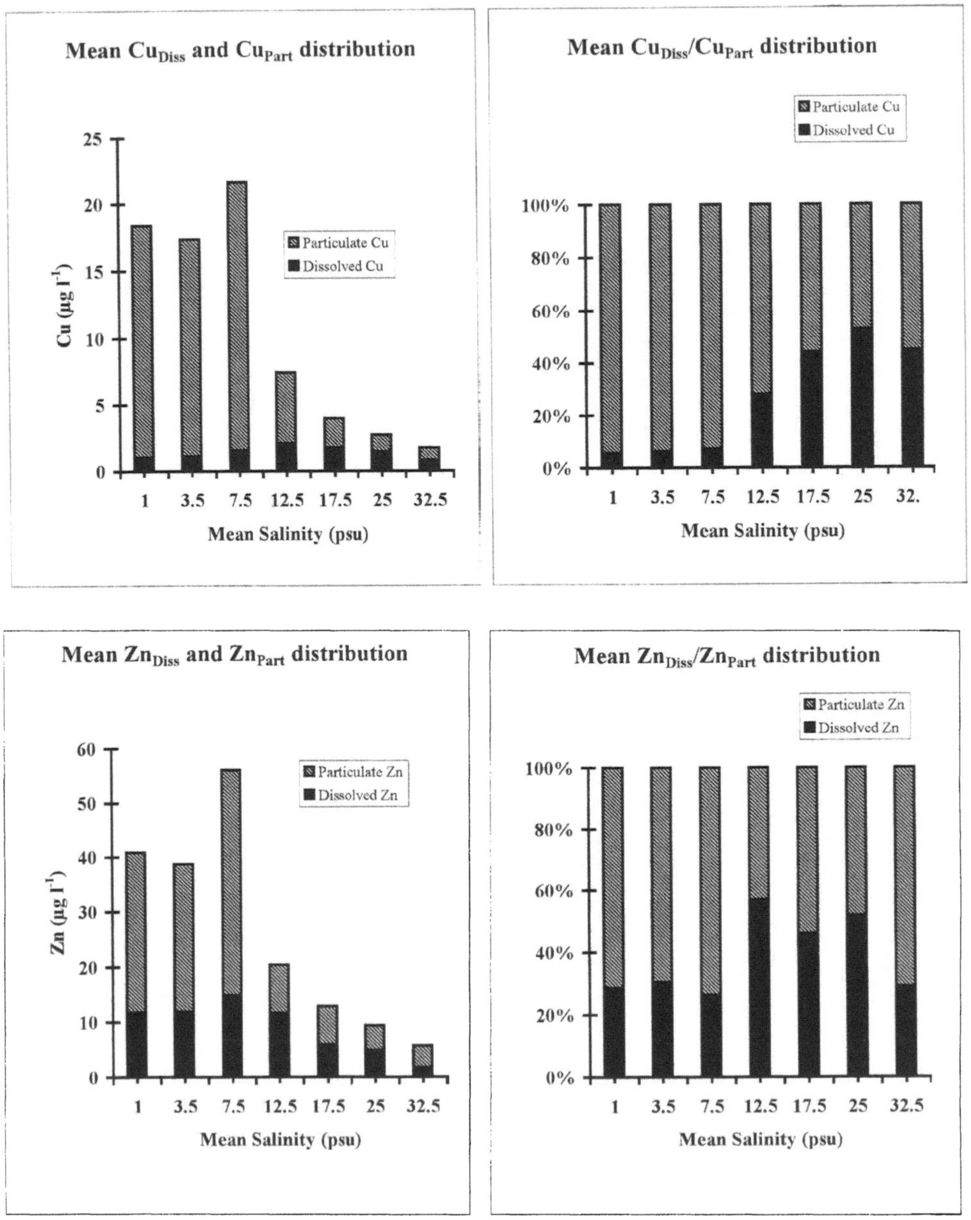

Figure 6. Variability of the mean total Cu and Zn load (dissolved and particulate fractions), and there relative distribution over the dissolved and the particulate fractions, throughout the Scheldt estuary.

The longitudinal K_D^{Pb}-profile shows a tendency to decrease from 900,000 in the upper estuary, to 250,000 in the estuarine plume. For the K_D^{Pb}-profile based on moderately labile Pb_{Diss} concentrations, the upper estuary shows a value of 1,200,000 which decreases to 450,000 in the estuarine plume. This corresponds to the declining Pb concentrations in the suspended matter (Pb_{Part}) (Baeyens & Monteny, 1988). However, a more or less constant K_D value is observed in the mid- and downstream estuarine area. The K_D^{Pb} values do not vary much in the Scheldt, up to a factor 3 only, a variation which is similar to the one for Zn. Our K_D values compare well with the range from 100,000 to 10,000,000 reported by Balls (1989). However, K_D-values above 1,000,000 were only observed in the Firth of Clyde and the North Channel/Irish Sea, where very low Pb_{Diss} concentrations occur. These data clearly indicate that Pb, even more than Hg, is present in the

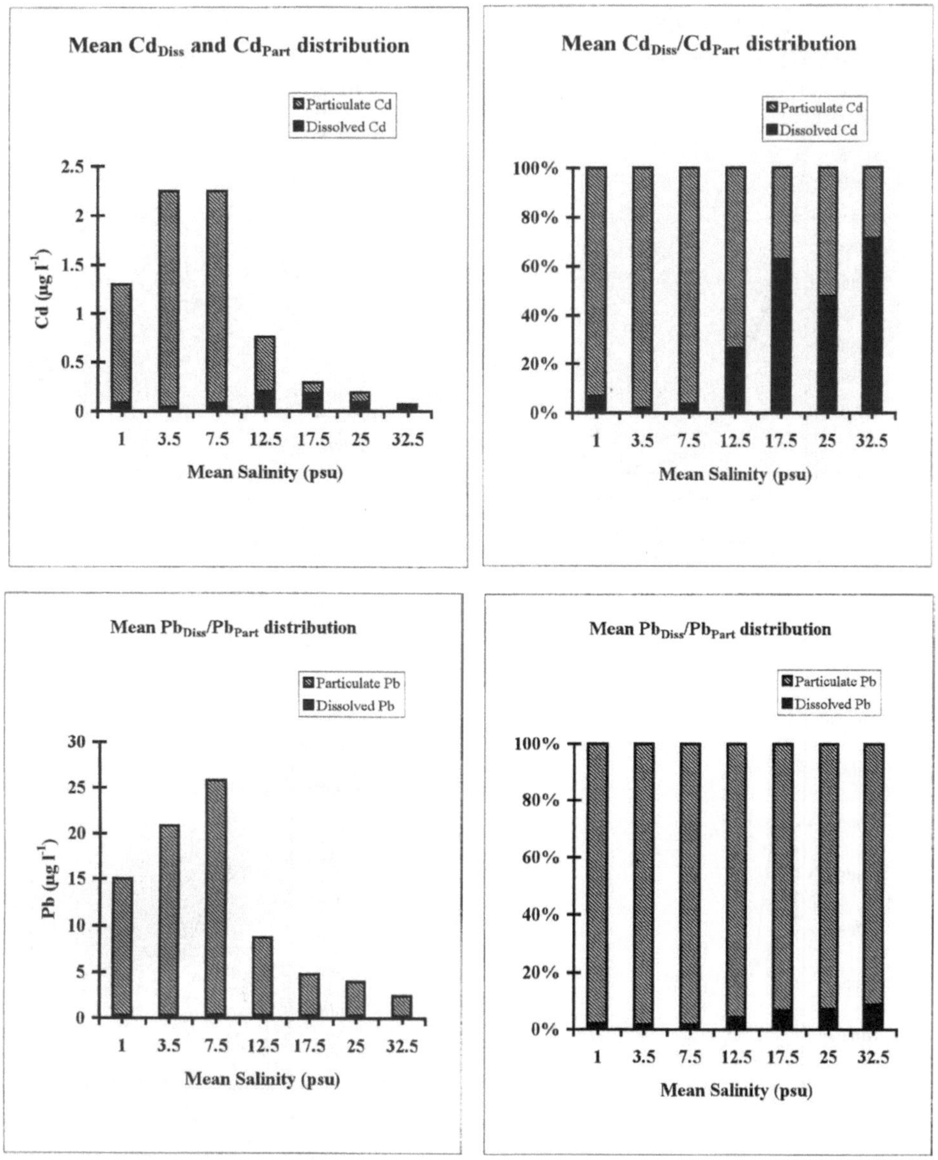

Figure 7. Variability of the mean total Cd and Pb load (dissolved and particulate fractions), and there relative distribution over the dissolved and the particulate fractions, throughout the Scheldt estuary.

particulate phase. For example, at typical K_D^{Pb} and turbidity values in the Scheldt (500,000 and 75 mg l^{-1} respectively) only 3% of the total Pb burden is present in the dissolved state. Complexing Pb ligands are either not competitive enough or not sufficiently abundant to keep Pb in the dissolved state.

The K_D^{Cd} profile initially increases from 300,000 to 600,000 in the low salinity range, due to a lowering of the dissolved fraction in the zone of early mixing, where depletion of oxygen occurs from late spring to autumn. Then, K_D^{Cd} rapidly drops to about 70,000 (S: 12.5 psu) and 20,000 (S: 17.5 psu) in the mid-estuary, where the production of Cd_{Diss} is the highest. In the downstream part of the estuary, the K_D^{Cd} hardly changes (between 20,000 and 25,000). In the upstream area a typical K_D^{Cd} of 500,000 is observed. All values range from 200,000 to 1,100,000, except for the cruise in March 1981, with the exceptionally high fresh water flow rate of 390 m^3 s^{-1}, hence with K_D^{Cd} values in the low salinity zone (S: 0–2 psu) which were much lower

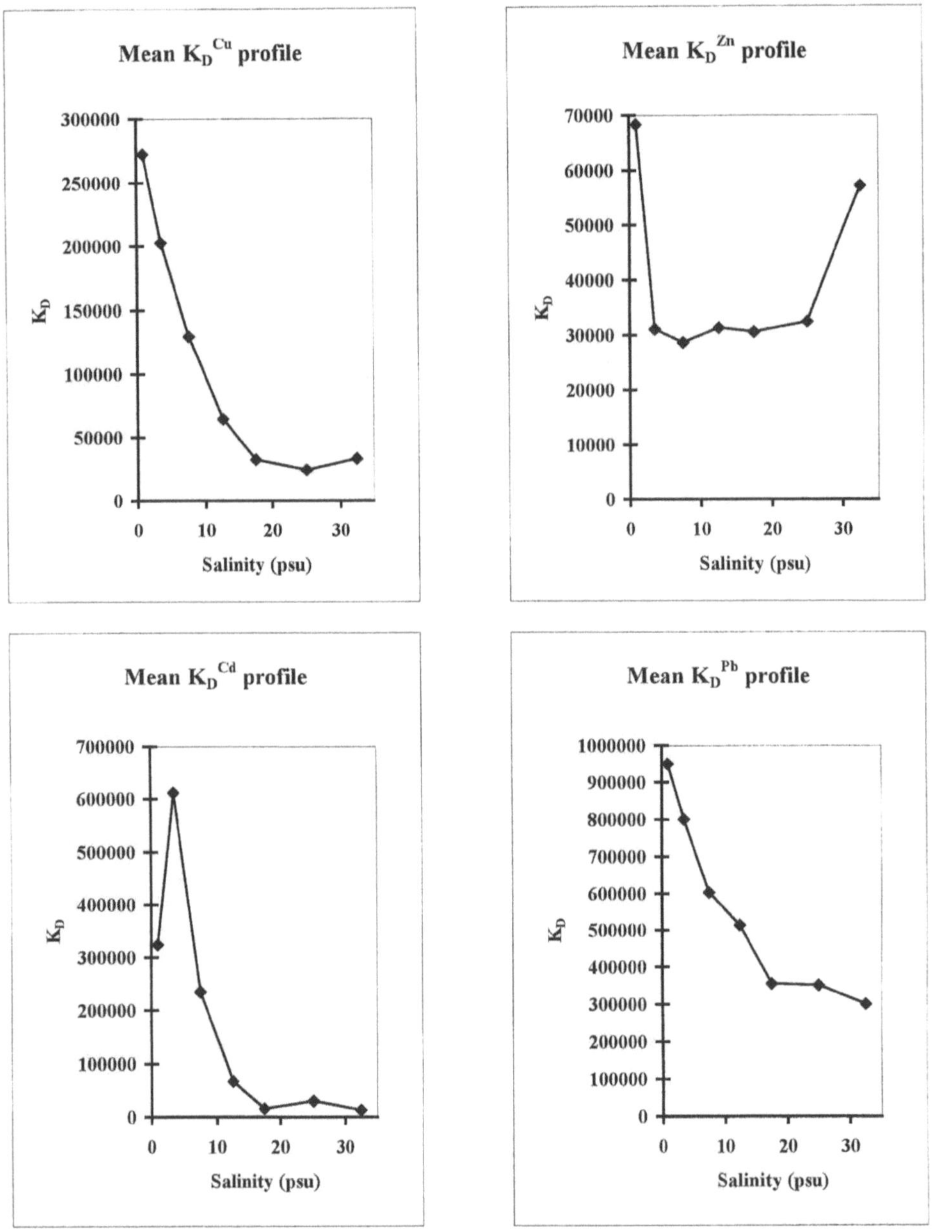

Figure 8. Mean K_D values throughout the Scheldt estuary.

than normally expected. They amounted from 10,000 to 200,000, since Cd_{Diss} levels were much higher.

Considering the values at the riverine (S < 1 psu) and the marine end member (S > 30 psu) of the estuary, two different rankings of K_D values are observed:

upper part: $K_D^{Pb} > K_D^{Cd} > K_D^{Cu} > K_D^{Zn}$

mouth: $K_D^{Pb} > K_D^{Zn} > K_D^{Cu} > K_D^{Cd}$.

The ranking observed at the mouth of the Scheldt estuary is in agreement with the one given by Balls (1989). The elevated K_D^{Pb} values are also in agreement with the observations of other authors (Valenta et al., 1986). Considering the complete estuary, all K_D values for each of the four selected metals were in agreement with the values reported by Valenta et al. (1986). More specifically the values of Cd, Cu and Zn for the upper part of the estuary are within the ranges given by Zwolsman & Van Eck (1991), for the fresh water end member of the estuary.

Discussion

A number of processes control the distribution pattern of the selected trace metals in the Scheldt estuary. For the dissolved, as well as for the particulate phase, two distinct areas can be considered: the downstream estuary with a salinity above 10 psu, and the upstream estuary (S: 0–10 psu). The latter one can sometimes be further subdivided into a low salinity (0–2 psu) and a maximum turbidity (S: 2–10 psu) zone.

The generally observed trend of linearly decreasing metal concentrations in the downstream estuary indicates a dilution process. Neither remobilization from, or adsorption onto, the particulate phase, nor assimilation by phytoplankton seems to be important enough for a perturbation of the dilution pattern. However, it can reasonably be assumed that seasonal processes, such as uptake of metals during a phytoplankton bloom, affect the longitudinal metal concentration profile, but that these effects are only visible during or immediately after that specific period (e.g. Baeyens et al., 1998c).

The interpretation of the data in the upstream estuary is, on the contrary, more complex. For the dissolved phase, three major processes have to be considered.

A first process results from the influence of the redox conditions on the dissolved metal concentrations and distributions. Increases of the dissolved concentration were in different cases noticed during transition from anaerobic to oxygen rich conditions. This observation may suggest that in the anaerobic zone (summer conditions) of the estuary dissolved metals are present as sulphides (Zwolsman & Van Eck, 1991). The origin of metal sulphides may be twofold: (1) metal sulphide formation in the water column or close to the water/sediment interface as suggested by Zwolsman & Van Eck (1991), with sulphide being introduced in the water column through pore water diffusion; (2) metal sulphide formation in the anaerobic muddy sediments of the Scheldt estuary (Panutrakul & Baeyens, 1991), even very close to the surface, and subsequently resuspension of these fine grained sediments containing metal sulphides. When these sulphides are transported downstream to more oxic areas, they are reoxidised, leaching the metal into solution. Progressive shifting of the oxic borderline towards the fresh water end member is observed from fall to winter.

A second process which can explain the non-conservative behaviour of metals in the upstream estuary, is the diffusion of metals from the bottom sediments. In oxic conditions, iron and manganese oxyhydroxides are present in the water column (Benjamin et al., 1981); these compounds possibly coprecipitate, or adsorb, trace metals before, or during, sedimentation and consequently deposit them at the bottom. Since in the maximum turbidity zone muddy sediments are abundantly present, trace metals associated with oxyhydroxydes will be released during their reduction in the reduced upper layers of muddy sediments (Panutrakul & Baeyens, 1991; Baeyens et al., 1991; Elskens et al., 1991). The presence of oxyhydroxide compounds in the water column displays seasonal variability since the upper part of the estuary becomes suboxic to anoxic during summer, while during winter the water remains richer in oxygen.

The third process in view of the non-conservative behaviour in the upstream estuary, is the physical and chemical interaction between the dissolved metal and the particulate suspended material. In the area of the maximum turbidity these adsorption/desorption processes are in favour of the removal of dissolved metals, due to the low salinity of the environment and the high availability of adsorption surfaces (numerous and small sized particles). In addition, marine derived particles with a low content of trace metals are supplied into the upper part of the estuary. Since trace metals show a much higher adsorption affinity than the marine cations (Na, Mg, . . .), displacement on that suspended matter will occur. Downstream, the higher salinity and the reduced amount of suspended matter will favour the desorption process. Organic ligands might play an important role in the adsorption/desorption equilibrium. Their exact role remains, however, still speculative and several points of view have been forwarded in literature. Mantoura et al. (1978) proposed a model to calculate the metal speciation in an estuary. This speciation takes into account the different inorganic ligands, the free ions and the metal fraction associated with humic acids. However, when considering organic speciation in estuarine waters, one should bear in mind that the extent of complexation may vary markedly according to the nature of the organic materials, i.e. concentration and value for the stability constants. For the Scheldt, we calculated the stability constants of organic-metal complexes (Baeyens et al., 1998a) and compared them to those reported by Mantoura et al. (1978) and Van Den Berg et al. (1987).

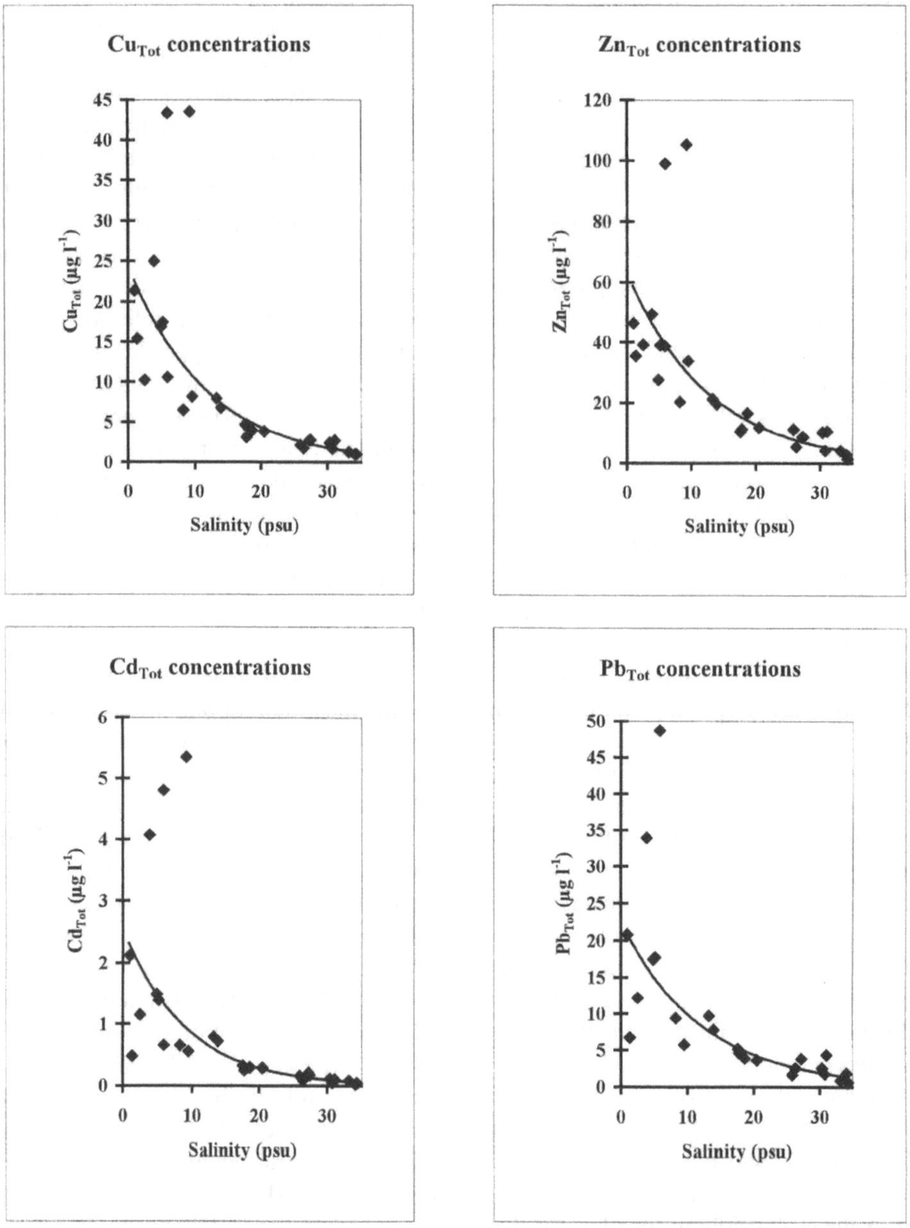

Figure 9. Variability of the total metal concentrations throughout the Scheldt estuary.

The above mentioned adsorption/desorption or precipitation/dissolution processes have only a minor effect on the distribution profile of particulate metals in the upper part of the estuary. There, the presence of the maximum turbidity, the resuspension of bottom material (which is strongly depending on tide and current), and the impact of local industrial discharges (the industrial complex of Antwerp includes a diversity of chemical activities) are the major parameters to consider, resulting in a highly variable particle population in the upper estuary. The selected moment and time of sampling can, therefore, strongly affect the particulate metal profile.

In order to decipher and identify the significance of each of the above mentioned processes on the behav-

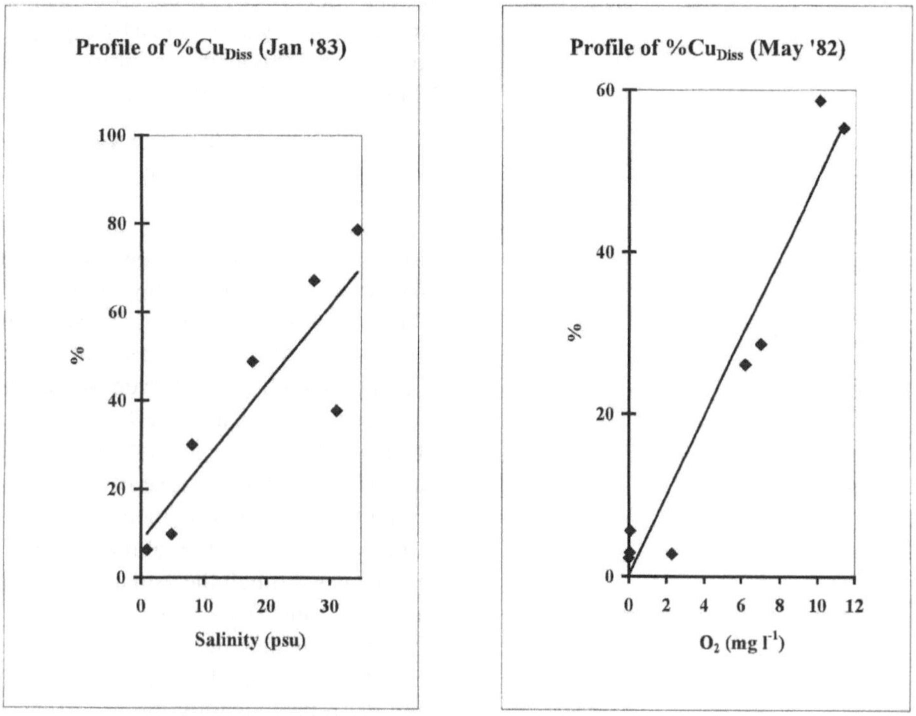

Figure 10. Relation between Cu_{Diss} and salinity and oxygen concentration.

iour of the individual metals in the Scheldt estuary, we will use correlation and multi-regression techniques.

Copper

An interesting analysis of the behaviour of Cu can be made by first considering its total concentration (Cu_{Tot}: dissolved + particulate, both expressed in $\mu g\ l^{-1}$). The variation of the Cu_{Tot} concentration versus salinity for all individual data points, is shown in Figure 9. The correlation with turbidity is good ($r = 0.86$) and when oxygen is added to the variance analysis, the correlation coefficient increases to 0.91.

In the downstream estuary (S: 10–30 psu), the data are relatively homogeneous and the general trend is one of decreasing concentrations with increasing salinity (almost a dilution profile):

$$[Cu_{Tot}] = 9.4 - (0.26 \pm 0.05) \times S \qquad (r = 0.80)$$

When oxygen is added as a second variable, the variance analysis yields a coefficient of $r = 0.87$.

The average Cu concentrations at 23 psu salinity, for each individual cruise, indicate a consistently decreasing trend during the three years of sampling (Table 3).

Table 3. Average total copper concentration in zone 1 at 23 psu salinity

Cruises:	$[Cu_{Tot}]$
18/03/81	$6.90 \pm 2.8\ \mu g\ l^{-1}$
25/05/82	$5.33 \pm 2.4\ \mu g\ l^{-1}$
24/01/83	$2.83 \pm 1.5\ \mu g\ l^{-1}$
08/03/83	$2.32 \pm 1.4\ \mu g\ l^{-1}$
18/05/83	$1.90 \pm 0.9\ \mu g\ l^{-1}$

In the maximum turbidity zone (S: 2–10 psu) the longitudinal profile of the total copper concentration is very high, varying between 43 $\mu g\ l^{-1}$ (cruise of 25/05/82) and 7 $\mu g\ l^{-1}$ (cruise of 24/01/83) with an average value of 20 $\pm$ 13 $\mu g\ l^{-1}$ at a salinity of 6 psu ($n = 10$). The spread of these data seems elevated, but the Cu_{Tot} concentration is well correlated with the turbidity in this zone:

$$[Cu_{Tot}] = -4.1 + (0.30 \pm 0.04)\ T \quad (r = 0.94;\ n = 9)$$

This is also true for the three other metals. The correlations with the turbidity in this zone are very high: r amounts to 0.97, 0.86 and 0.99 respectively for Pb, Zn and Cd respectively. These results suggest that

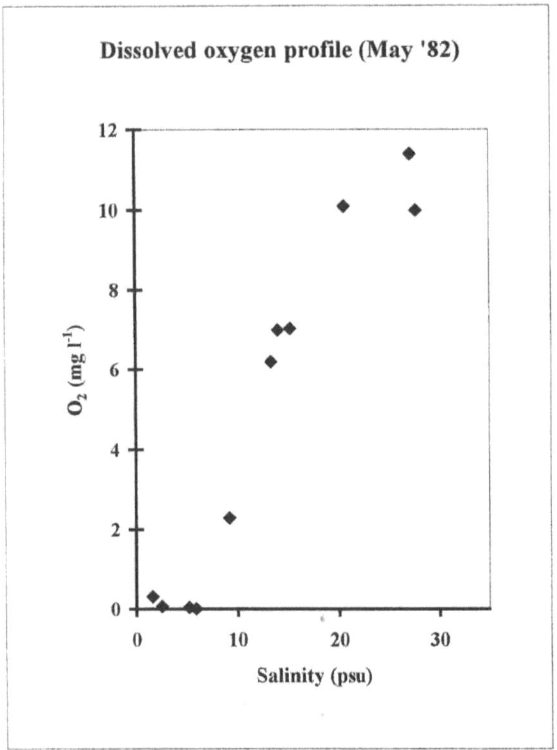

Figure 11. Dissolved oxygen profile throughout the estuary.

the high total metal concentrations in the water column are most probably due to the resuspension of the fine bottom sediment fraction. As already discussed in Baeyens et al. (1991), local domestic and industrial discharges contribute considerably to the formation of fine contaminated material in this zone. Since this fine bottom sediment fraction dominates the suspended matter during almost the entire tidal cycle (except at the turn of the tide), and since the lowest turbidities are observed at the turn of the tide, we conclude that local, time-depending fluctuations cause sawtooth profiles.

The third zone to be considered, is the upper estuary (S: 0–2 psu). The suspended matter load of the rivers Scheldt and Rupel is very variable, but this variability does not affect the amount nor the composition of the suspended matter in this zone. The bottom sediments are eroded only at maximal tidal current, and other sources of suspended matter are less important since large industries and cities do not exist in that area. This explains why the system should be in a quasi-equilibrium state. We observed a slow decrease in total copper concentration, with a mean value of 17 ± 6 μg l^{-1} at a salinity of 0.7 psu ($n=8$). Good agreement between the experimental data and the theoretical results was obtained by applying a multi-linear regression. Introducing salinity and turbidity, as variables, reduced the variance significantly:

$$[Cu_{Tot}] = 7 + (4.0 \pm 3) \times S + (0.15 \pm 0.06) \times T$$
$$(r = 0.84;\ n = 9)$$

Cu_{Diss} production (%) primarily occurs in the downstream estuary, as shown by profiles of the average percentage ratio of Cu_{Diss} to Cu_{Tot}. The average% Cu_{Diss} at low salinities (< 5 psu) equals 6.4% and increases up to 46% at high salinities (> 30 psu). Only Cd shows such high or even higher production percentages, with an average% Cd_{Diss} value of 4.7% in the low salinity zone and 64% at the mouth of the estuary. The absolute increase of the Pb_{Diss} is moderate, going from a very low fraction of 1.6%, in the upper estuary, to 8.7% at high salinities, while% Zn_{Diss} hardly changes through the estuary, showing only a slight increase from 30.0% to 35.9%.

During winter, the relationship between Cu_{Diss} and salinity (for example $r=0.87$ in January 1983, Figure 10) reflects mainly the occurrence of desorption processes due to increased chlorinity and major cation concentrations. On the other hand, the spring values correlate best with the variability of the oxygen concentration ($r=0.97$ for 25/05/82, Figure 10), suggesting a major influence of the redox processes on the mobilisation of copper. In the maximum turbidity zone the copper production (% Cu_{Diss}) correlates well with oxygen ($r=0.84$).

The evolution of the redox potential in the water column is related to the oxygen profile throughout the estuary (Figure 11). During summer the upper part of the estuary becomes suboxic to anoxic, while during winter the water is richer in oxygen. As already stated for Cu, an increase of the dissolved fraction downstream the suboxic region (S: 10–15 psu) is observed. Figure 12 shows that Cu_{Diss} production occurs within a broad range of oxygen (between 0 and 4, respectively, 0 and 8 mg l^{-1} O_2). The concentration increases from about 0.5 to 2.5 μg l^{-1}. The correlation coefficient between Cu_{Diss} and oxygen in zone 2 (the oxygen deficient area) amounts to $r=0.94$ (the addition of the salinity as a variable hardly influences the variance; $r=0.97$). This illustrates most likely the importance of oxidation processes. Indeed, reduction of the bivalent ion Cu^2 to Cu^+ is possible at the pH and Eh values commonly observed in the upper estuary during summer (e.g. pH 6 to 8, Eh 0.1 to –0.4 mV). According to Boulègue (1978), the resulting cuprous ions are able

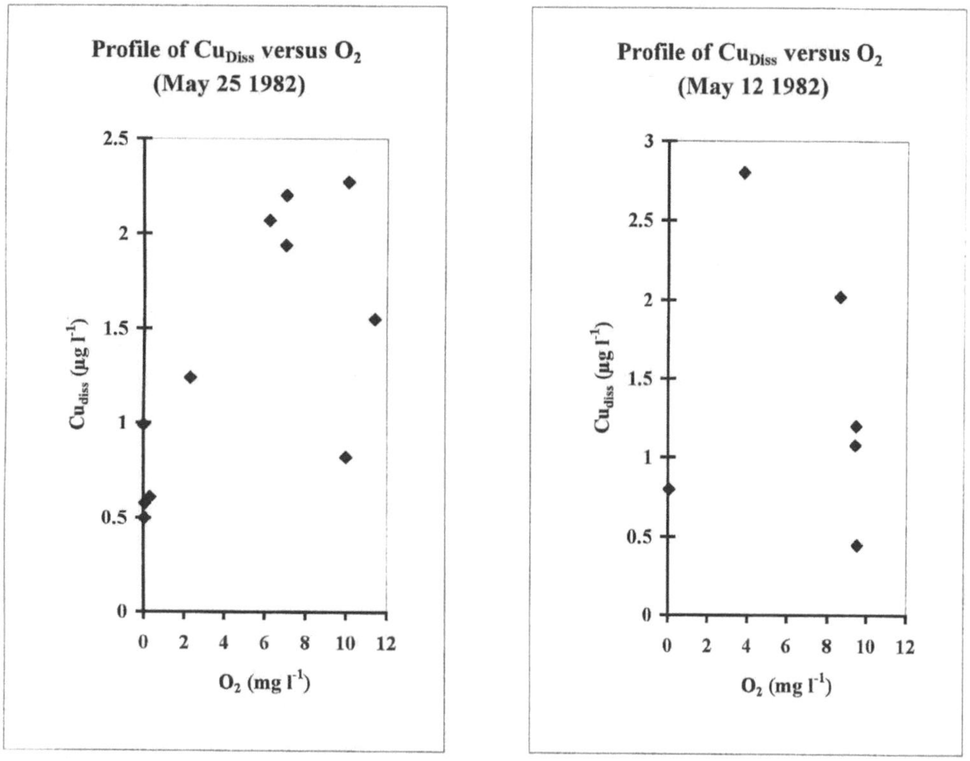

Figure 12. Mobilisation processes for Cu_{Diss}.

to form strongly insoluble complexes with chlorides, disulphides and polysulphides. The latter complexes are of primary importance in complete anoxic conditions, which occur only exceptionally in the Scheldt estuary. Dissolved ionic Cu^{2+} also forms insoluble metal sulphide species (Emerson et al., 1983). Cuprous ion as well as sulphides or polysulphides can then be oxidised in the middle and the lower part of the estuary. Both oxidation as well as desorption are therefore plausible explanations for the broad mid-estuarine Cu_{Diss} maximum in the Scheldt estuary. For zone 1, the correlation coefficient between Cu_{Diss} and oxygen equals 0.43, but adding the salinity improves the correlation ($r = 0.70$).

The longitudinal volumetric (μg l^{-1}) particulate Cu (Cu_{Part}) profiles are not discussed since they are too strongly related to the turbidity profiles. The average absolute (μg g^{-1}) particulate Cu profile (Figure 5) shows a relatively high negative deviation compared to the theoretical dilution line. An exponential function through these data points yields an excellent regression ($r = 0.98$). Such a pattern suggests mobilisation, hence confirming the observed broad Cu_{Diss} production zone in the estuary. During the survey at the end of May 1982, particulate organic matter, Pb^{210} and Ra^{226} were also measured. Assuming that the latter isotopes are in a secular equilibrium (both isotopes disintegrate at the same rate) and that their geochemistry in the Scheldt is quite similar (both isotopes show similar distributions between the dissolved and particulate phases), than their ratio can be used as a conservative parameter for the mixing of suspended matter, in a similar way as salinity for the dissolved phase. The lowest ratio (0.25) was observed at the fresh water end member side (S: 1.6 psu), while the highest ratio (2.3) corresponds to the marine end member side (S: 27.2 psu). Plotting Cu_{Part} concentration versus this parameter (Figure 13) reveals that the Cu_{Part} curve does not follow a dilution pattern. There is a slight increase in the low salinity range, suggesting Cu_{Part} by adsorption, precipitation or resuspension of a highly enriched fluid mud layer (Baeyens et al., 1998d), followed by a broad area of strong release, suggesting mobilisation of Cu_{Part} into the dissolved phase. The decrease of the curve is stronger than the previous enrichment, indicating net

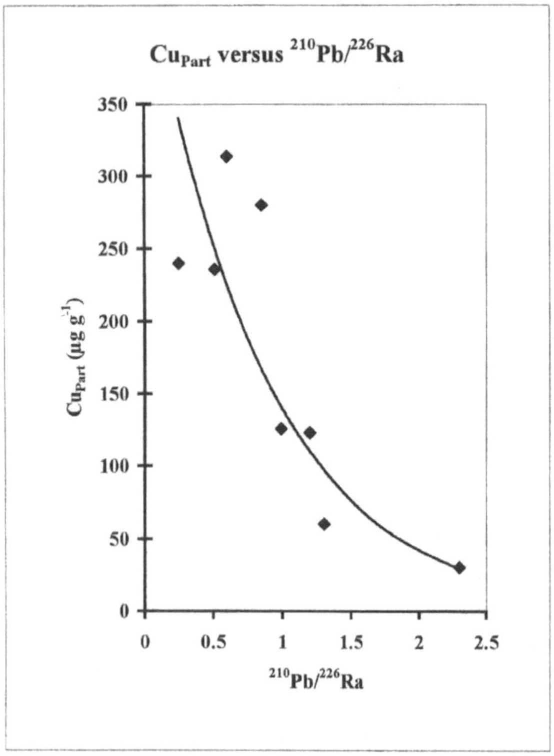

Figure 13. Variability of Cu_{Part} concentrations versus $^{210}Pb/^{226}Ra$.

mobilisation of Cu in the dissolved phase, during this particular cruise.

All particulate metal concentrations correlate very well ($r > 0.95$) with the POC concentration (Figure 14), when both parameters are expressed in molar concentration units, indicating that their transport, sedimentation and resuspension obeys to the same processes.

Zinc

The variation of the total Zn concentration (Zn_{Tot}: dissolved + particulate, both in $\mu g\ l^{-1}$) versus salinity is shown in Figure 6. There is a good correlation with turbidity ($r = 0.89$), but adding other variables does not improve this correlation. In the downstream estuary (S: 10–30 psu), as well as in the maximum turbidity zone (S: 2–10 psu), good correlations, $r = 0.83$ and $r = 0.86$ respectively, are observed with the turbidity only. No clear pattern is obtained in the low salinity zone (0–2 psu).

The Zn_{Diss} production, or the evolution of $Zn_{Diss}\%$, in the estuary is almost negligible; $Zn_{Diss}\%$ is slightly increasing from 30% at low salinity (< 5 psu) to 35.9% at high salinity (> 30 psu). During winter $Zn_{Diss}\%$ shows neither a relation with salinity ($r = 0.4$, Figure 15) nor with oxygen, but it must be stated that low oxygen values do not occur in that period. These correlations improve during late spring ($r = 0.7$, Figure 15). The latter observations suggest that the limited mobilisation of zinc in the low salinity zone during late spring, is influenced by redox processes (either microbial or chemical). Desorption processes controlled by the salinity should act similarly in winter and summer.

The correlation coefficient observed for the Zn_{Diss} concentration versus salinity relation is relatively high for the whole estuary ($r = 0.72$) and also for the downstream estuary ($r = 0.88$). In the low salinity zone (< 2 psu) no correlation with salinity is obtained. In the maximum turbidity area (S: 2–10 psu), the best correlation is observed with oxygen ($r = 0.40$). Mobilisation processes are, only small (a few $\mu g\ l^{-1}$ Zn, Figure 16) and are essentially occurring during spring.

The average absolute particulate Zn (Zn_{Part} in $\mu g\ g^{-1}$) profile (Figure 5) shows a less negative deviation than the Cu_{Part} profile when compared to the theoretical dilution line. The correlation coefficient of the exponential curve through the average data points is $r = 0.92$. The average pattern is largely determined by the very strong initial decrease from about 800 to 400 ppm (S: 0–5 psu) though.

Zn does not show strong mobilisation, with the exception of a small amount in the maximum turbidity zone (S: 2–10 psu), which occurs mainly during late spring. Similarly to the Cu_{Part} pattern, the Zn_{Part} concentration versus the ratio of $^{210}Pb/^{226}Rd$ has been plotted (Figure 17). This curve shows an almost perfect dilution pattern, except in the zone of maximal turbidity where a peak concentration appears (increase by adsorption – precipitation – resuspension), followed by a similar decrease at the next sampling station (desorption – dissolution). Enrichment and loss amounts with respect to the particulate phase are almost equal.

Redox processes as mentioned above, may be involved in this behaviour. If Zn sulphides were eroded from the bottom and added to the prevailing suspended matter, corresponding to an enrichment process, they are transported further downstream, where fast oxidation and dissolution occur, as soon as the dissolved oxygen level increases.

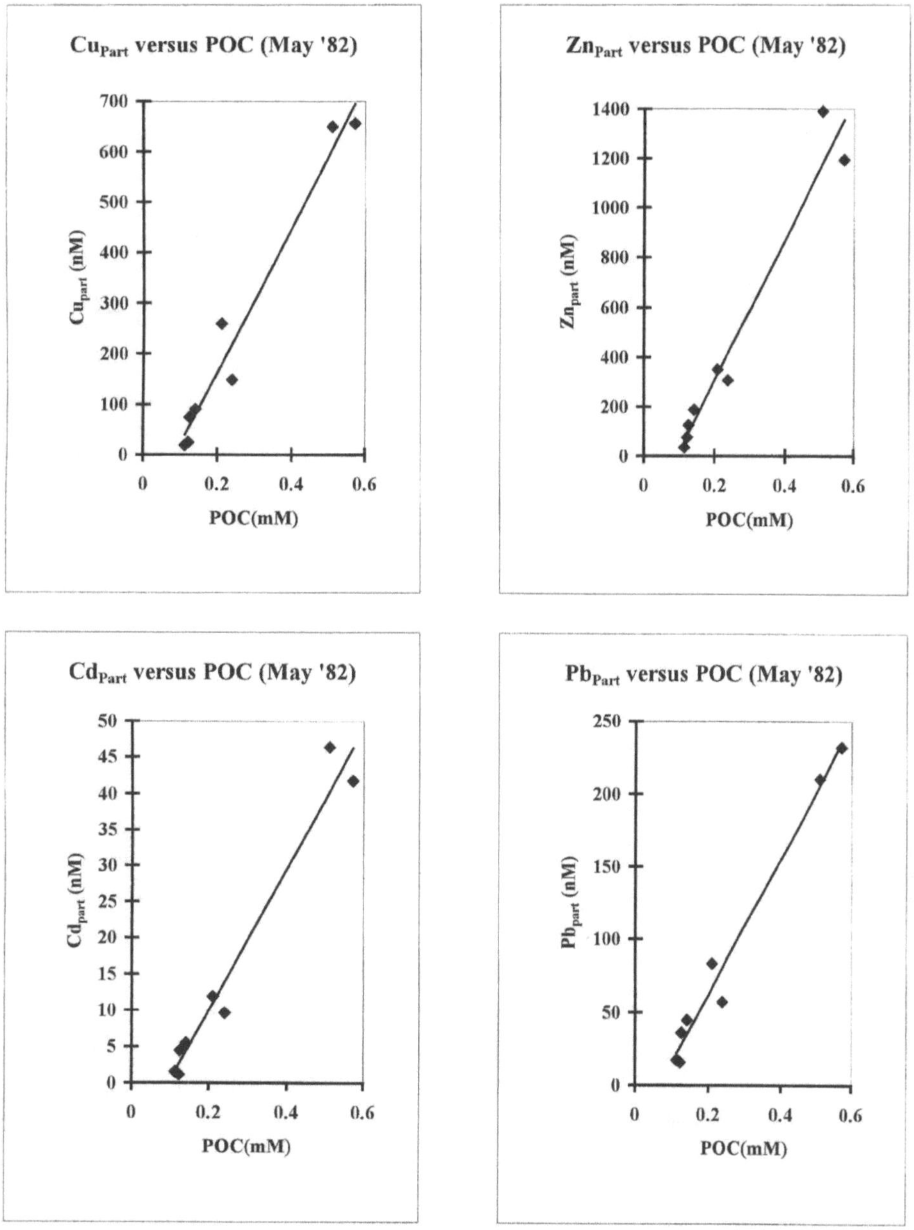

Figure 14. Particulate metal concentrations versus POC values throughout the estuary.

Lead

The variation of the total Pb concentration (Pb_{Tot}, dissolved + particulate, in $\mu g\, l^{-1}$) versus salinity is shown in Figure 7. The best correlation observed is the one with oxygen ($r = 0.68$).

In the downstream estuary (S: 10–30 psu) a fairly good correlation with salinity is observed ($r = 0.83$), this correlation improves slightly when adding oxygen ($r = 0.88$) and turbidity ($r = 0.92$) as variables. The Pb_{Tot} concentrations show a dilution profile like most other metals in the downstream estuary. In the maximum turbidity zone (2–10 psu), Pb_{Tot} is very well related to the turbidity, confirming that the resuspension process primarily controls the prevailing concentration.

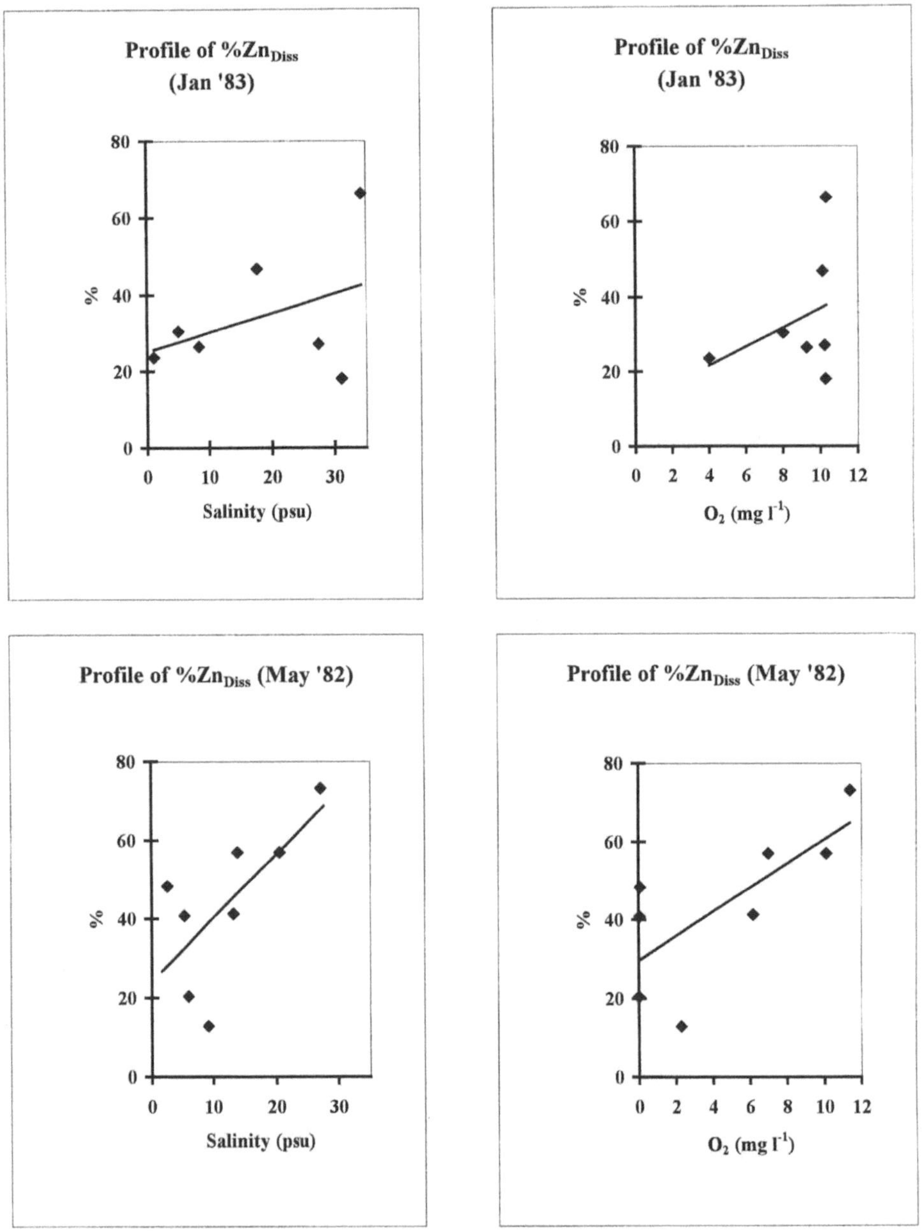

Figure 15. Zn_{Diss} fraction (%) versus salinity and oxygen during winter and late spring.

The Pb_{Diss} production, or the evolution of Pb_{Diss}%, is lower than for cadmium and copper, but still increases with a factor 5 going from low (1.6%) to high salinity (8.7%). However, the ratio of Pb_{Diss} to Pb_{Tot} is by far the lowest of the four metals considered. In winter, a fairly good relationship between Pb_{Diss}% and salinity (Figure 18) is observed ($r = 0.88$), while the linear relation with oxygen is rather weak ($r = 0.62$). Desorption processes due to the salinity increase may thus be involved in a moderate Pb_{Diss} production. The application of an exponential relationship between Pb_{Diss}% and dissolved oxygen (Figure 18) improves the correlation significantly ($r = 0.80$). During late spring the best correlations of Pb_{Diss}% with oxygen are observed: for May 1983 ($r = 0.92$, Figure 18) and for May 1982 ($r = 0.85$, Figure 18). The latter results suggest an influ-

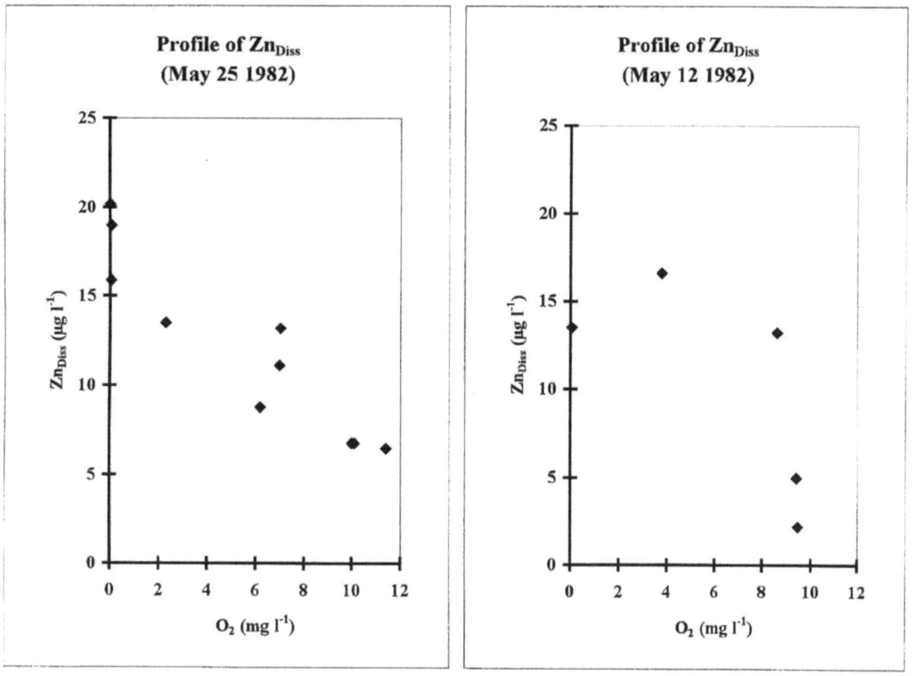

Figure 16. Mobilisation processes for Zn_{Diss}.

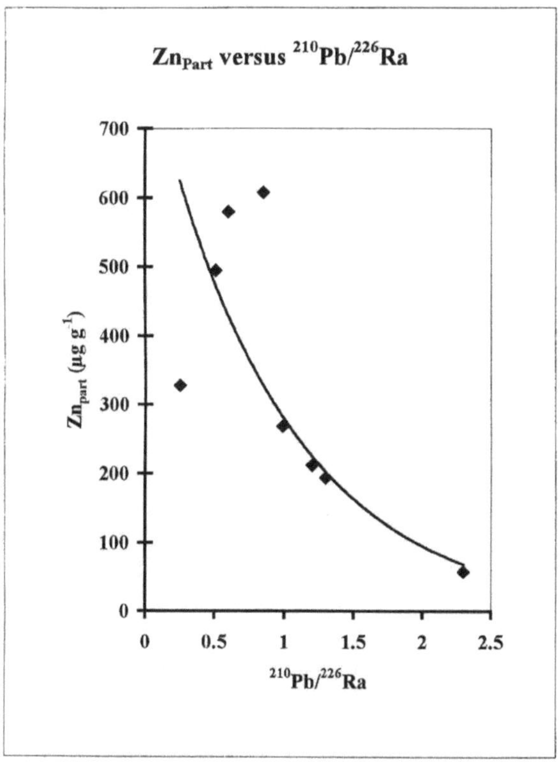

Figure 17. Variability of Zn_{Part} versus $^{210}Pb/^{226}Ra$.

ence of redox processes on the mobilisation of Pb in the middle estuarine zone during late spring and summer, the period when oxygen is lacking in the upstream estuary.

Figure 19 clearly indicates Pb mobilisation in a small oxygen range (from 0 to 2–3 mg l^{-1} O_2). The survey of May 1992 revealed that Pb_{Diss} concentration decreased from 0.45 to 0.20 μg l^{-1} in a zone where oxygen was completely depleted. At the next measuring station, where the dissolved oxygen concentration amounted to 6 mg l^{-1}, an increase in Pb_{Diss} concentration from 0.2 μg l^{-1} to almost 0.45 μg l^{-1}, was observed. Further downstream the Pb_{Diss} pattern corresponded to a dilution profile. The moderately labile Pb_{Diss} concentration profile (Figure 19) shows a similar pattern. The additional data point at the intermediate oxygen level of 2 mg l^{-1}, confirms that mobilisation occurs with increasing oxygen levels.

Salinity is the only variable yielding correlation coefficients higher than 0.5 with Pb_{Diss}. For the whole estuary a correlation coefficient of $r = 0.50$ is observed, whereas r amounts to 0.52 for the downstream estuary (S: 10–30 psu), and to 0.65 for maximum turbidity zone (S: 2–10 psu). Redox processes may reduce (at very low oxygen content) or enhance (in the transition area from low to medium oxygen levels) the Pb_{Diss}

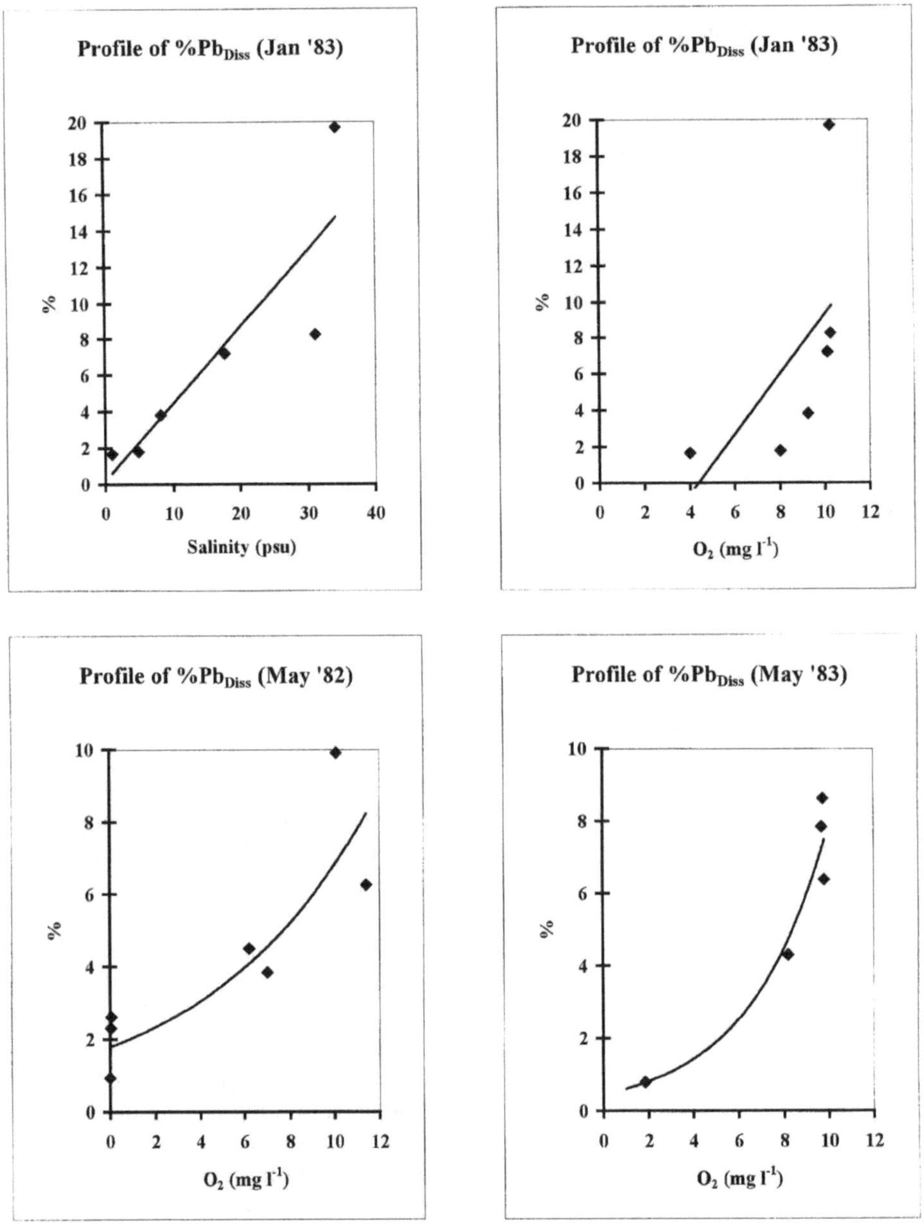

Figure 18. $\%Pb_{Diss}$ versus salinity and oxygen concentration.

content. Therefore, the correlation between Pb_{Diss} and oxygen becomes worse when more redox processes are involved.

The average absolute Pb_{Part} profile ($\mu g\ g^{-1}$, Figure 5) shows the weakest downstream decrease and the smallest negative deviation versus the theoretical dilution line of the four metals considered. On the other hand, the exponential regression fit for Pb, is the best one observed ($r = 0.99$).

The plot of Pb_{Part} versus the ratio of ^{210}Pb to ^{226}Ra, the conservative mixing index for suspended particulate matter (Figure 20), shows neither strong mobilisation, nor enrichment in the estuary. A small zone of Pb_{Part} enrichment, immediately followed by a release

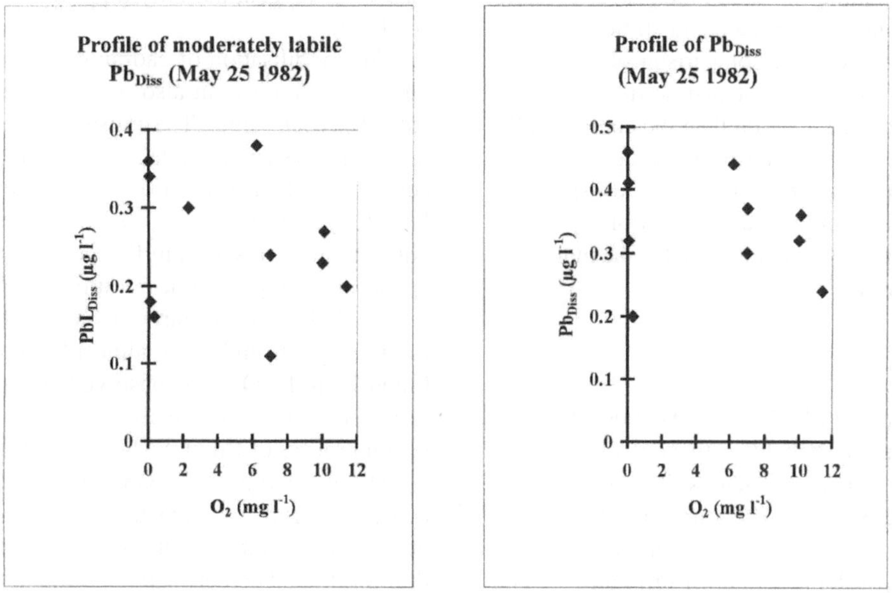

Figure 19. Mobilisation processes for Pb_{Diss}.

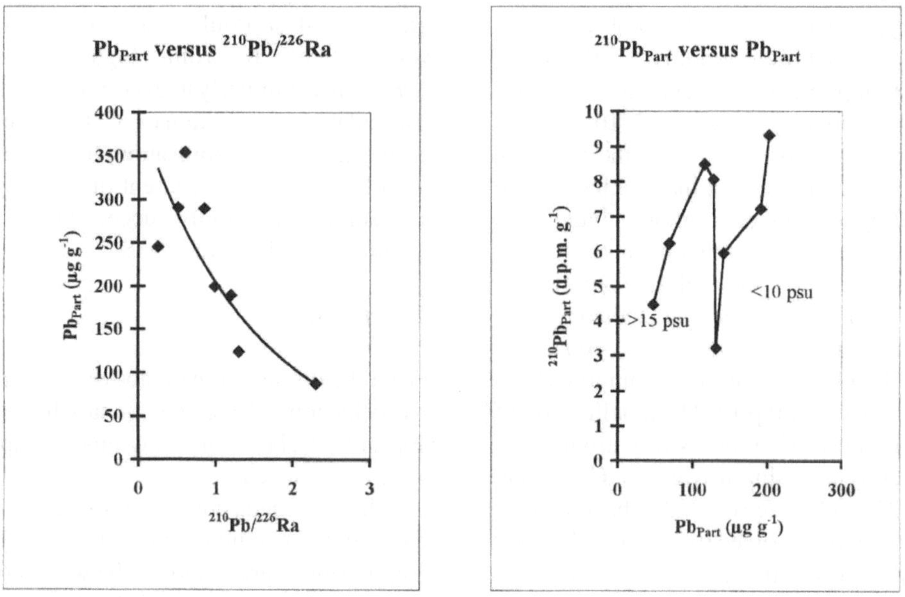

Figure 20. Pb_{Part} versus $^{210}Pb/^{226}Ra$ and $^{210}Pb_{Part}$.

zone, which is similar in strength, is observed in the area of maximum turbidity.

The $^{210}Pb_{Part}$ versus Pb_{Part} values indicate two different groups of particles (Figure 20). In the high salinity area (> 15 psu), this corresponds in Figure 20 to all data points with a Pb_{Part} concentration below 100 μg g^{-1}, a dilution pattern for both the $^{210}Pb_{Part}$ and the Pb_{Part} contents versus the salinity is observed. In the low salinity area (< 10 psu), processes other than only dilution or conservative mixing are apparently involved. Although the concentration range of ^{210}Pb in the upstream area is similar to the range observed in the downstream area (4–10 d.p.m g^{-1}), the related Pb_{Part} concentrations are substantially higher. Both curves

(Figure 20) are parallel to each other, with the one belonging to the upstream area being shifted to higher Pb_{Part} concentrations. A comparison of the ^{210}Pb /Pb ratio clearly indicates that marine derived particles are more enriched in ^{210}Pb than the continentally derived ones. The range of ratios observed in the low salinity water masses can be attributed to resuspension processes in the area of the maximum turbidity and to discharges of particles enriched in lead in the industrial area around Antwerp.

Cadmium

Cadmium is the metal which shows the most alarming concentrations in the Scheldt estuary. Compared to other European rivers and estuaries, its concentration is highest in the Scheldt. The variation of the total Cd concentration (Cd_{Tot}: dissolved + particulate, both in $\mu g\ l^{-1}$) versus salinity is shown in Figure 7. The correlation with the turbidity for the whole estuary is excellent ($r = 0.92$), and the best of all metals.

In the downstream estuary (S: 10–30 psu) the correlation with oxygen appears to be the best ($r = 0.94$), adding turbidity ($r = 0.98$) and salinity ($r = 0.996$) as variables slightly improves the correlation coefficient. In the maximum turbidity zone (S: 2–10 psu), a very good correlation is observed with the turbidity ($r = 0.985$), corroborating the fact that the particulate phase is controlling the total concentration pattern, as was also observed for the three other metals.

The Cd_{Diss} production or the evolution of % Cd_{Diss} in the estuary is the highest of the four studied metals. In the low salinity zone (< 5 psu) % Cd_{Diss} equals 4.7 while it is amplified by a factor of 13.5 up to 64% in the high salinity zone (> 30 psu). The mobilisation of cadmium in the mid-estuarine zone is very strong. During winter a good relationship between % Cd_{Diss} and salinity (Figure 21) is observed ($r = 0.90$), but the exponential regression with dissolved oxygen is still better ($r = 0.91$). Very good ($r = 0.97$; Figure 21) to fairly good ($r = 0.78$; Figure 21) correlation coefficients for the regression between % Cd_{Diss} and dissolved oxygen were observed during late springs 1983 and 1982.

The correlation coefficient between Cd_{Diss} and salinity is relatively good for the whole estuary ($r = 0.72$), and for the zone of maximum turbidity ($r = 0.75$). In the downstream estuary the relationship with the salinity is poor ($r = 0.57$), but adding the variables oxygen ($r = 0.70$) and turbidity ($r = 0.85$) improves the correlation significantly. In the upstream area (S: 0–2 psu), Cd_{Diss} correlates fairly well with dissolved oxygen, but only three results are available ($r = 0.72$).

The mobilisation of cadmium occurs over a wide range of salinities but also over a wide range of dissolved oxygen values. The cruise of May 25, 1992 (Figure 22) shows an initial decrease of the dissolved concentration when oxygen disappears completely, followed by a broad mobilisation area where the Cd concentration raises from 0.05 to 0.2 $\mu g\ l^{-1}$, corresponding to oxygen concentrations ranging from 0 to 7 $mg\ l^{-1}$. Two weeks earlier (May 12, 1982) a similar mobilisation from 0.05 to almost 0.3 $\mu g\ l^{-1}$, between 0 and 7 $mg\ l^{-1} O_2$, was observed (Figure 22). Redox processes may thus actively participate in the mobilisation process of Cd in the Scheldt estuary.

The average absolute particulate Cd (Cd_{Part}) profile ($\mu g\ g^{-1}$; Figure 5) shows the strongest negative deviation of all metals, when compared to the theoretical dilution line. The correlation coefficient of the exponential curve through the average date points is very good ($r = 0.97$).

The strong mobilisation of Cd, reflected by the longitudinal Cd_{Part} profile, is also illustrated by the Cd_{Part} versus $^{210}Pb/^{226}Ra$ profile (Figure 23) during the May 1982 cruise. Similarly to the Cu pattern (Figure 13), the particulate concentration curve lies below the dilution curve, indicating mobilisation. In the area of maximum turbidity a sharp enrichment of the suspended matter by cadmium is rapidly succeeded by a much larger release to the dissolved phase.

Conclusions

From the discussion here above we can conclude that the behaviour of the four metals has some common appearances, but many individual and specific features were also noticed.

The total metal concentrations co-vary with the turbidity in the maximum turbidity zone. The particulate component clearly controls the water column concentration in that area of the Scheldt. In the downstream area a dilution profile for the Cu_{Tot} and Pb_{Tot} concentrations is observed.

The variability of the percentage of the total dissolved metal to the total metal concentration (% Me_{Diss}) throughout the estuary, is a measure for the production or consumption of that dissolved metal in a given area. At the transition between the maximum turbidity zone and the Western Scheldt (downstream estuary), the dissolved metal production is high for Cu and Cd, but much lower for Pb and Zn. These patterns are con-

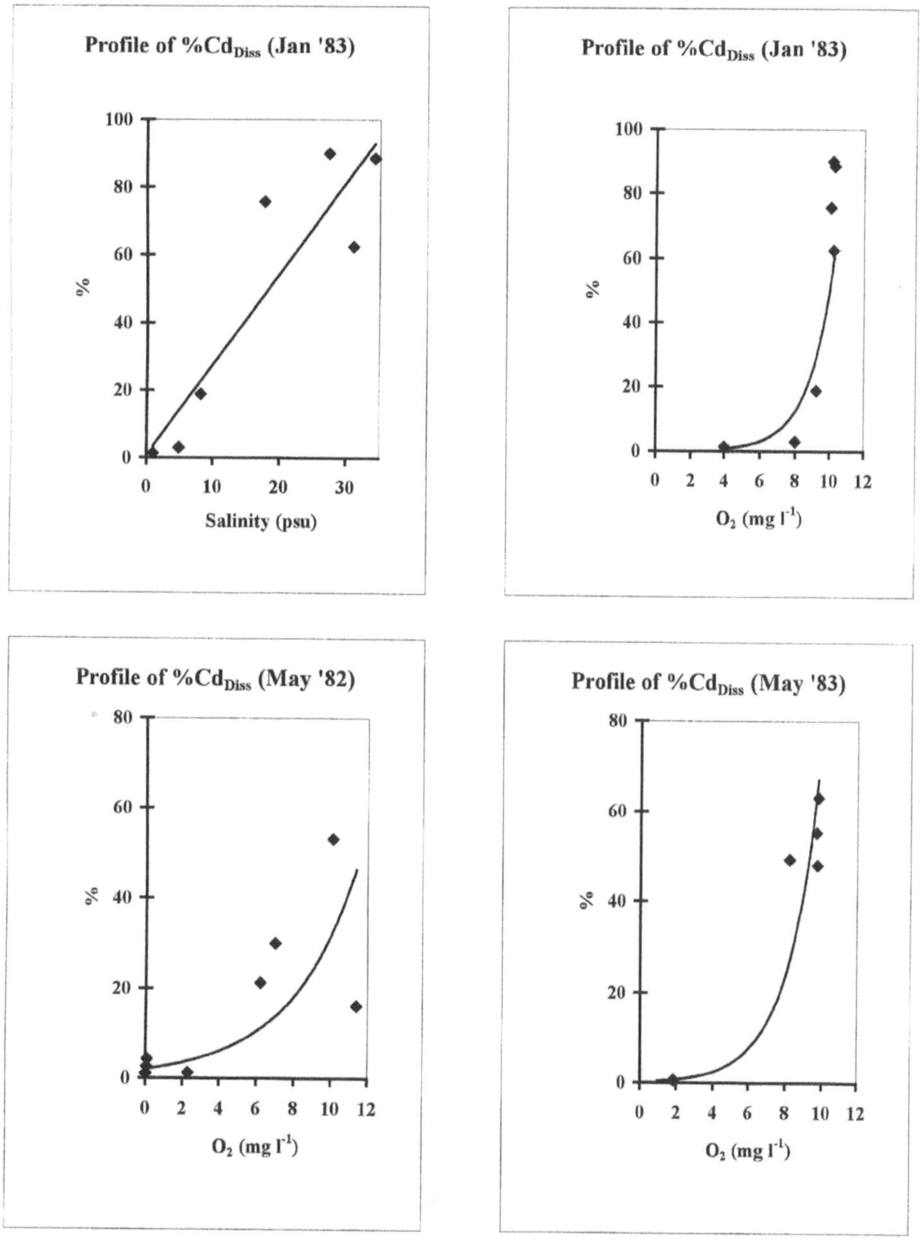

Figure 21. % Cd_{Diss} versus salinity and oxygen concentration.

firmed by the longitudinal total dissolved metal profiles. A broad mobilisation area, covering the whole middle estuary is observed for Cu and Cd, Pb shows a more confined mobilisation area and for Zn no such area was found. In the downstream area a dilution profile for the dissolved concentrations of all metals is observed.

Causes of the dissolved metal production have been investigated by correlating metal concentrations with major parameters, representing typical estuarine processes in different seasons. For all metals but Zn, the dissolved metal production rate, assessed by the evolution of % Me_{Diss}, correlates well with the salinity during winter. Zn is exceptional since it does not correlate with any parameter. Hence, desorption processes

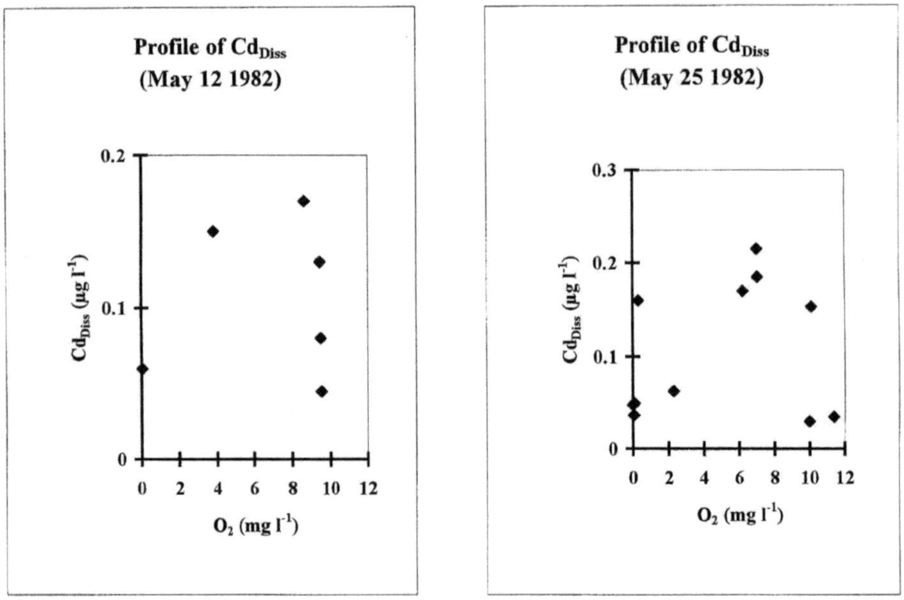

Figure 22. Mobilisation processes for Cd_{Diss}.

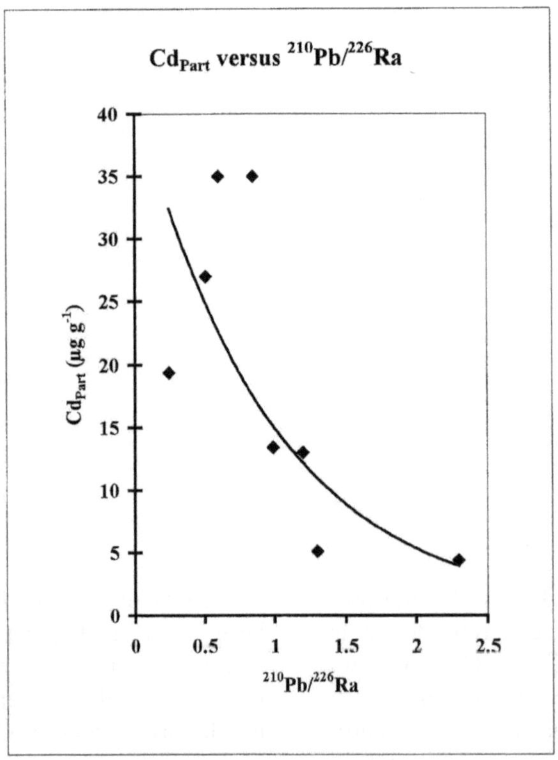

Figure 23. Cd_{Part} versus ^{210}Pb values.

seem to control the dissolved metal production during winter. During summer % Me_{Diss} values for all metals correlate well with dissolved oxygen, suggesting that redox processes solely or in combination with desorption processes, control the dissolved metal production during summer. In some cases removal of the dissolved metal prior to the mobilisation process is observed, probably reflecting precipitation, since it occurs in combination with oxygen depletion.

All longitudinal particulate metal profiles show a negative deviation versus the dilution profile, with Cu and Cd having the strongest deviations. During one cruise, the $^{210}Pb/^{226}Ra$ ratio was measured in order to have a comparison between the particulate metal concentrations and a parameter, which is assumed to behave conservatively, in the suspended matter. For Cd and Cu a narrow enrichment zone preceded a broad mobilisation area, while for Pb and Zn both enrichment and release processes were small in amplitude and in area. Finally, all particulate metal profiles appeared to correlate well with POC, when both are expressed in volumetric units.

Acknowledgement

The authors thank the MUMM (Management Unit for the Mathematical Models of the North Sea and Scheldt estuary) for providing ship time. The authors gratefully acknowledge technical assistance of the captains and crew members aboard RV Belgica. This research was supported by a grant of DWTC to M. Elskens.

References

Baeyens, W. & F. Monteny, 1988. Study of the degree of contamination of dredge in the Scheldt estuary. Synthesis Report. VUB, Brussels, pp 53.

Baeyens, W., 1992. Speciation of mercury in different compartments of the environment. Trends in analytical chemistry, 11: 245–254.

Baeyens, W., S. Panutrakul, M. Elskens, M. Leermakers, J. Navez & F. Monteny, 1991. Geochemical Processes in Muddy and Sandy Tidal Flat Sediments. Geo-Marine Lett. 11: 188–193.

Baeyens, W., 1998. Evolution of trace metal concentrations in the Scheldt estuary (1978–1995). A comparison with other estuaries and background ocean levels. Hydrobiologia 366: 157–167.

Baeyens, W., L. Goeyens, F. Monteny & M. Elskens, 1998a. Effect of organic complexation on the behaviour of dissolved Cd, Cu and Zn in the Scheldt estuary. Hydrobiologia 366: 81–90.

Baeyens, W., F. Monteny, R. Van Ryssen & M. Leermakers, 1998b. A box-model of metal flows through the scheldt estuary (1981–1983 and 1992–1995). Hydrobiologia 366: 109–128.

Baeyens, W., K. Parmentier, L. Goeyens, G. Ducastel, M. De Gieter & M. Leermakers, 1998c. The biogeochemical behaviour of Cd, Cu, Pb and Zn in the Scheldt estuary: results of the 1995 surveys. Hydrobiologia 366: 45–62.

Baeyens, W., G. T. M. Van Eck, C. Lambert, R. Wollast & L. Goeyens, 1998d. General description of the Scheldt estuary. Hydrobiologia 366: 1–14.

Baeyens, W., M. Elskens, R. Van Ryssen & M. Leermakers, 1998e. The impact of the Scheldt input on the trace metal distribution in the Belgian coastal area (results of 1981–1983 and 1995–1996). Hydrobiologia 366: 91–108.

Balls, P. W., 1989. The partition of trace metals between dissolved and particulate phases in European coastal waters: a compilation of field data and comparison with laboratory studies. Neth. J. Sea Res. 23: 7–14.

Batley, G. E. & D. Gardner, 1977. Sampling and storage of natural waters for trace metal analysis. Wat. Res. 1: 745–756.

Benjamin, M. M. & J. O. Leckie, 1981. Multiple-site adsorption of Cd, Cu, Zn and Pb on amorphous iron. O. J. Coll. Interface Sci. 79: 209–221.

Benoit, G., S. D. Oktay-Marshall, A. Cantu, E. M. Hood, C. H. Coleman, M. O. Corapeioghi & P. H. Santschi, 1994. Partitioning of Cu, Pb, Ag, Zn, Fe, Al and Mn between filter-retained particles, colloids and solution in six Texas estuaries. Mar. Chem. 45: 307–336.

Boulègue, J., 1978. Metastable sulfur species and trace metals in hot brines from the French Dogger. Am. J. Sci. 278: 1394.

Duinker, J. C., R. F. Nolting & D. Michel, 1982. Effects of salinity, pH and redox conditions on the behaviour of Cd, Zn, Ni an Mn in the Scheldt Estuary. Thalassia Yugoslavia 18: 191–202.

Elskens, M., M. Leermakers, S. Panatrakul, F. Monteny & W. Baeyens, 1991. Microbial Activity in Sandy and Muddy Estuarine Sediments. Geo-Marine Lett. 11: 194–198.

Emerson, S., L. Jacobs & B. Lebo, 1983. The behaviour of heavy metals in marine anoxic water: Solubilities at the oxygen-hydrogen sulfide interface. In C. S. Wong, E. Boyle, K. W. Bruland, J. D. Eurton & E. D. Goldberg (eds), Trace Metals in Sea Water, Plenum Press, New York, 579–608.

Gillain, G., G. Duyckaerts & A. Distèche., 1979. The determination of trace metals in sea water and suspended matter by classical anodic stripping (Zn, Cd, Pb and Cu) or differential pulse anodic stripping voltammetry with a hanging mercury drop electrode (Zn, Cd, Pb, Cu, Sb and Bi) – An approach to speciation. In Geconcerteerde Onderzoeksacties Oceanologie, Ministerie voor de Programmatie van het Wetenschapsbeleid, Brussels: 123–131.

Gillain, G. & C. Brihaye, 1985. A system for taking a representative sample form a large volume of turbid surface sea water for trace element determinations. Analyt. chim. Acta 167: 387–391.

Jouaneau, J. M., H. Etcheber & C. Latouche, 1983. Impoverishment and decrease of metallic elements associated with suspended matter in the Gironde estuary. In C. S. Wong, E. Boyle, K. W. Bruland, J. D. Eurton & E. D. Goldberg (eds), Trace Metals in Sea Water, Plenum Press, New York: 245–264.

Mantoura, R. F. C., A. Dickson & J. P. Riley, 1978. The complexation of metals with bunic materials in natural waters. Estuar. coast. mar. Sci. 6: 387–408.

Mart, L., 1979. Prevention of contamination and other occurring risks in voltammetric trace metal analysis of natural waters. I. Preparatory steps of filtration and storage of water samples. Fresenius Z. Anal. Chem. 296: 350–355.

Nolting, R. F., 1986. Copper, zinc, cadmium, nickel, iron and manganese in the Southern Bight of the North Sea. Mar. Pollut. Bull. 17: 113–117.

Panutrakul, S. & W. Baeyens, 1991. Behaviour of heavy metals in a mud flat of the Scheldt Estuary, Belgium. Mar. Pollut. Bull. 22: 128–134.

Valenta, P., E. K. Duursma, A. G. A. Merks, H. Rutzel & H. W. Nurnberg, 1986. Distribution of Cd, Pb, Cu between the dissolved

phase and the particulate phase in the Eastern Scheldt and the Western Scheldt Estuary. Sci. tot. Envir. 53: 41–76.
Van Den Berg, C. M. G., A. G. A. Merks & E. K. Duursma, 1987. Organic complexation and its control of the dissolved concentrations of Cu and Zn in the Scheldt Estuary. Estuar. coast. Shelf Sci. 24: 785–797.
Zwolsman, J. J. G. & G. T. M. Van Eck, 1991. The behaviour of dissolved Cd, Cu and Zn in the Scheldt Estuary, Estuar. water Qual. Mgmt. Springer-Verlag, Berlin: 413–420.

Hydrobiologia **366:** 45–62, 1998.
W. F. J. Baeyens (ed.), Trace Metals in the Westerschelde Estuary.

The biogeochemical behaviour of Cd, Cu, Pb and Zn in the Scheldt estuary: results of the 1995 surveys

Willy Baeyens, Koen Parmentier, Leo Goeyens, Geert Ducastel, Marjan De Gieter & Martine Leermakers
Department of Analytical Chemistry, Free University of Brussels, Pleinlaan 2, 1050 Brussels, Belgium

Key words: Scheldt estuary, trace metals, 1995, dissolved, particulate, biogeochemistry

Abstract

Cu, Cd, Zn and Pb concentrations in both dissolved and particulate phase were assessed during 1995. Data Quality Assurance was an integral part of this study and involved all major steps of the analysis procedure such as sampling, sample handling, preconcentration and determination.

Desorption and redox processes clearly control the dissolved Cu and Cd profiles. Mobilisation of dissolved Zn is small and essentially restricted to the low salinity area and the late spring survey. Dissolved Pb shows the clearest dilution pattern. The parameters representative of these processes (dissolved oxygen and salinity) also correlate very well with the particulate metal profiles.

In addition, the plankton activity (expressed by the chlorophyll-*a* levels) may influence the dissolved and particulate metal profiles. In combination with desorption and redox processes, seasonal variations were induced.

Introduction

International regulations (North Sea Conferences (The Hague, 1990), Paris Convention (1992)) aiming at the reduction of the marine pollution, resulted in measures decreasing the input of pollutants into the aquatic environment. This study deals specifically with the contamination due to, and the biogeochemical behaviour of heavy metals in the Scheldt estuary, and repeats in that sense our former investigation performed in the period 1981–83 and discussed in a related paper (Baeyens et al., 1998). Sampling cruises on board of the R.V. Belgica were performed in March, May and October 1995, making use of the newly installed clean lab container. Samples are analysed for trace metals in both dissolved and particulate phase. In addition, salinity, dissolved oxygen and chlorophyll-*a* were measured at the same station.

Materials and methods

Sampling and pre-treatment

Since sampling is one of the most crucial steps in order to obtain reliable results, a maximum of precautions were taken to avoid contamination. All materials in contact with the samples during the sampling process are thoroughly cleaned and stored following a strict standard procedure.

Sampling bottles, separatory funnels and filter holders are submitted to the following decontamination procedure: soaking 1 day in a soap solution (Extran); thoroughly rinsing with Milli-Q water; soaking 1 day in HNO_3 (sub-boiled), 50% at 40 °C; thoroughly rinsing with Milli-Q water; soaking 1 day in HNO_3 (sub-boiled), 10% at 40 °C; thoroughly rinsing with Milli-Q water; drying in a laminar flowhood. After the cleaning procedure, the high density polyethylene (HDPE) sampling bottles are filled with sub-boiled HNO_3 (10%) and stored in closed polyethylene (PE) bags.

The polycarbonate and cellulose acetate filters, used to filter the sea water, are leached for 24 hours in 6M HCl (Merck, *Suprapur*), rinsed thoroughly with Milli-Q water, dried in a laminar flowhood and packed in thoroughly rinsed petri-plates.

The whole estuary is sampled during one tidal cycle in all 10 monitoring stations (Figure 1); sampling starts in Vlissingen at high water, and continues up to Antwerp, where the tide is close to low water. Samples are taken by hand wearing elbow-length PE gloves from a rubber boat or Zodiac at a distance of at least 100 meters from the boat. Moreover, during sampling, the Zodiac is navigated gently upstream and if possible against the wind. In every monitoring station, 1 to 2 litres water are collected in HDPE bottles, for the determination of Cu, Zn, Cd and Pb. The bottles are rinsed three times with the estuarine water before sampling and then capped and packed in two closed PE bags.

At the same time, aboard the ship, water temperature, salinity, turbidity and density are measured *in situ* with a Seabird SCTD-model SBE19 (Seacat) by the Management Unit Mathematical Model (MUMM). Dissolved oxygen is measured using three different methods, namely the Winkler method, and with the online electrodes YSI-55 and YSI-57 (Anker-Schmidt). The average value is calculated, but sometimes one of the three methods shows a strong deviation, which is then rejected. This is not unusual for the Winkler method when the water column is supersaturated in oxygen. Chl-a is determined using the Lorenzen method (extraction by acetone (90%) during 18 h, followed by spectrophotometric determination at 665 and 750 nm).

In order to distinguish between dissolved and particulate phase, all the samples are immediately filtered in the Clean Lab Container (class 100) aboard the research vessel to prevent trace metal redistribution. A 47 mm Nuclepore cellulose acetate (0.4 μm pore size) filter is placed in a Teflon filter holder connected to a Teflon separatory funnel. The sample is poured into the separatory funnel and filtered by means of pressure filtration using purified nitrogen gas. Filters are rinsed with Milli-Q to remove salt. After filtration, filters are stored in petri-plates and PE bags and frozen at –20 °C. The filtered water samples are collected in bottles, and immediately acidified to a pH < 2. The acidified samples are finally stored at room temperature and preconcentrated and/or analysed rapidly following collection. Filters are dried in a laminar flow hood and weighed. Turbidity values are calculated from the filter load and the volume filtrated.

Preconcentration and analysis of dissolved metals

Due to the low concentrations of heavy metals dissolved in estuarine water, a preconcentration is required in order to reach the amounts required for a quantitative determination. In this study, a solvent extraction method, slightly modified from Danielsson et al. (1978, 1982), is applied. Ultrapure reagents are purchased and when possible an additional clean-up procedure was made. The methods of reagent preparation and purification, as used in this work, are briefly summarised:

- *Nitric acid*: Ultrapure nitric acid is obtained through distillation of concentrated HNO_3 (Merck, *Pro Analysis*) using a sub-boiling installation conceived in Teflon.
- *Buffer*: An ammonium acetate buffer is prepared by diluting a solution of 22.5 ml ammoniumhydroxide (NH_4OH; Merck, *Suprapur*) and 11.8 ml acetic acid (CH_3COOH; Merck, *Suprapur*) up to 100 ml with Milli-Q.
- *Complexant*: 1.0 g Ammonium Pyrolidine DithioCarbamate (APDC; Merck, *Pro Analysis*) and 1.0 g Diethylammonium Diethyl DithioCarbamate (DDDC; Merck, *Pro Analysis*) are dissolved in 1 ml NH_4OH (*Suprapur*) and diluted to 100 ml with Milli-Q water. The clean-up of this solution involves a three times extraction with 20 ml freon. Every three days, the solution is freshly prepared.
- *Freon*: 400 ml freon (*Pro Analysis*) is washed three times with 5 ml sub-boiled nitric acid in a 1000 ml Teflon separatory funnel.

In a 1000 ml Teflon separatory funnel, a 300 to 500 g sample is weighed and brought to a pH of 4.5 by means of 1 ml ammonium acetate buffer, in order to obtain an optimal complexation. The pH is immediately checked and if required a second ml of buffer is added. To the buffered solution 2 ml complexant (solution of 1% APDC / 1% DDDC) are added. The metal complexes are then extracted in two steps: first, 20 ml of freon are added and the funnel is shaken vigorously for three minutes, the phases are separated for 5 minutes, the freon phase is layered off in a 100 ml Teflon funnel; this extraction is repeated with 10 ml freon, and the freon phase is again collected in the same separatory funnel. To the separatory funnel, 1 ml HNO_3 is added, followed after 10 minutes by 4 ml of Milli-Q water; after 5 minutes of phase separation, the organic (freon) phase is discarded. The aqueous phase, containing the metals, is finally collected in a 10 ml polypropylene test tube and stored at 4 °C.

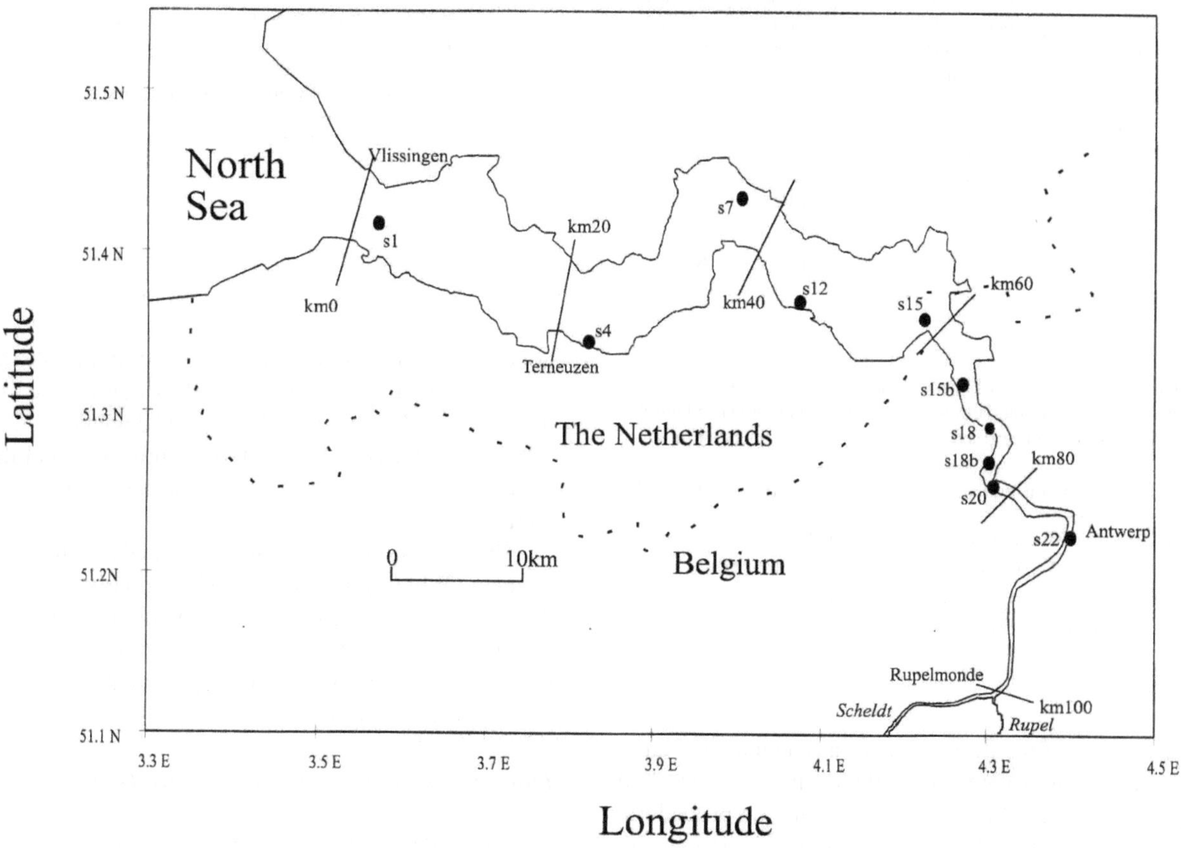

Figure 1. Location of the sampling stations in the Scheldt estuary.

The concentrated solutions are analysed using graphite furnace-atomic absorption spectrometry (GF-AAS) in order to determine Cu, Zn, Cd and Pb. The spectrometer is a Perkin Elmer model 3030 with Zeeman correction, equipped with an AS 60 autosampler. In all analyses pyrolysed graphite tubes with L'vov platform are used. The following matrix-modifiers are added (Hoenig & de Kersabiec, 1989): for the determination of Zn, a solution of 0.1% $mg(NO_3)_2$, for Cd 0.5% $(NH_4)H_2PO_4$ + 0.01% $mg(NO_3)_2$ and for Pb 5% $(NH_4)H_2PO_4$ + 0.1% $mg(NO_3)_2$. Cu is determined without matrix-modifier.

Table 1. Mean blank value (x), standard deviation (s), number of averaged blanks n and detection limit (DL) for analysis with ET-AAS

		Cu (μg l^{-1})	Cd (μg l^{-1})	Zn (μg l^{-1})	Pb (μg l^{-1})
March	x	0.011	0.001	0.19	0.007
	s	0.007	0.001	0.03	0.002
	n	4	8	9	7
	DL (3s)	0.02	0.003	0.09	0.006
May	x	0.027	0.0009	0.22	0.013
	s	0.005	0.0003	0.06	0.003
	n	9	11	9	9
	DL (3s)	0.02	0.0009	0.19	0.009

Quality control on GF-AAS analyses and extraction procedures

The concentrations of the different metals were calculated from calibration curves, based on the peak-surface area of the signals. These calibration curves were obtained from standard solutions, which are prepared daily from the stock solutions, and covering concentration ranges up to 100, 5, 50 and 100 μg l^{-1}, respectively for Cu, Cd, Zn and Pb. Every sample (standard, blank, unknown) was analysed three times and

Table 2. Experimental results (in μg l^{-1}) of the AQUACHECK proficiency testing scheme, reference values and mean values of all participating laboratories, together with % difference between our results and the reference value

AQUACHECK	Cu (μg l^{-1})	Cd (μg l^{-1})	Zn (μg l^{-1})	Pb (μg l^{-1})
Experimental	91.4	3.7	143.5	64.4
Reference	90.0	3.4	155.0	63.0
Mean lab's	91.1	3.2	154.5	60.6
% differ. to refer.	1.6	8.8	–7.5	2.2

Table 3. Experimentally obtained values (in μg l^{-1}) for the NASS-2 seawater reference standard, with standard deviations (s) and number of analyses (n), and certified values with 95% confidence limits

NASS-2	Experimental value $x \pm s$ (n)	Certified value (95% c.l.)
Cu	0.093 ± 0.016 ($n=6$)	0.109 ± 0.011
Cd	0.025 ± 0.003 ($n=7$)	0.029 ± 0.004
Zn	0.196 ± 0.073 ($n=3$)	0.178 ± 0.025
Pb	0.037 ± 0.007 ($n=7$)	0.039 ± 0.006

the average of the three signals was calculated. Detection limits are expressed as three times the standard deviation on the blanks. To determine the procedural blank, 300–500 g Milli-Q was weighed, preconcentrated as described above, and analysed with GF-AAS. Mean blank values for two independent batches (March and May), together with standard deviations and detection limits are summarised in Table 1.

The method's accuracy was tested by participating at the AQUACHECK proficiency testing scheme, organised by the Water Research centre in January 1995. The goal of this program, at which 300 laboratories out of 17 different countries participated, was to check comparability and accuracy of dissolved metals analysis. These samples, however, did not require preconcentration. Table 2 summarises our results, the reference values, the mean values of all participating laboratories, and the standard deviation of our values from the reference values. A good agreement between our results and the reference values for the four metals is observed.

The accuracy of the method as a whole, preconcentration as well as GF-AAS analysis, was evaluated by analysing the commercial seawater reference sample, NASS-2 (National Research Council, Canada). The mean experimental values with their standard deviations, and the certified values with 95% confidence levels are presented in Table 3. For all elements, a good agreement is obtained. It is also remarkable that standard deviations are similar to 95% confidence limits. Only for Zn, a very high standard deviation (37%) was found. It should be mentioned, however, that this value (0.196 μg l^{-1}) is near the detection limit (see Table 1).

Table 4. Relative standard deviation on 6 consecutive analyses of a standard solution with GF-AAS

Trace metal	Relative standard deviation
Cu	12.5%
Cd	1.7%
Zn	1.8%
Pb	4.8%

The precision of the GF-AAS analysis for each element (Table 4) is expressed as the relative standard deviation of 6 replicate measurements on a standard solution. Concentration of the standards of Cu, Cd, Zn and Pb for this purpose were approximately 20, 3, 2 and 20 μg l^{-1}, respectively.

Filter destruction and particulate metals analysis

In order to determine Cu, Zn, Cd and Pb in the particulate phase, an acid mineralisation procedure is carried out. The filters are placed in a 60 ml Teflon bomb, and 4 ml HNO_3 (sub-boiled) and 1 ml HCl (Merck, *Suprapur*) are added. The bombs are tightly closed and oven-heated for 12 hours at 60 °C to obtain complete digestion. After digestion, samples are diluted up to 50 ml with Milli-Q water and transferred in 60 ml PE bottles. Analysis is performed with Inductively Coupled Plasma – Atomic Emission Spectroscopy (ICP-AES) for Cu and Zn and by GF-AAS for Cd and Pb. Calibration curves covering the concentration range of the samples are made with multi-element standard solutions. Concentrations are calculated based on peak surface area of the signals. ICP-AES measurements are performed with a Gobin-Yvon 3848 analyser, GF-AAS with the above mentioned device.

Blank filters are treated like sampling filters, with one difference: instead of sample, 500 ml of Milli-Q were filtered in the clean lab aboard the R.V. Belgica. The filters are digested as described above. Detection limits are expressed as three times the standard deviation on the blanks. Mean blank values, together with standard deviation, number of blanks and detection limits for an analysis batch are summarised in Table 5.

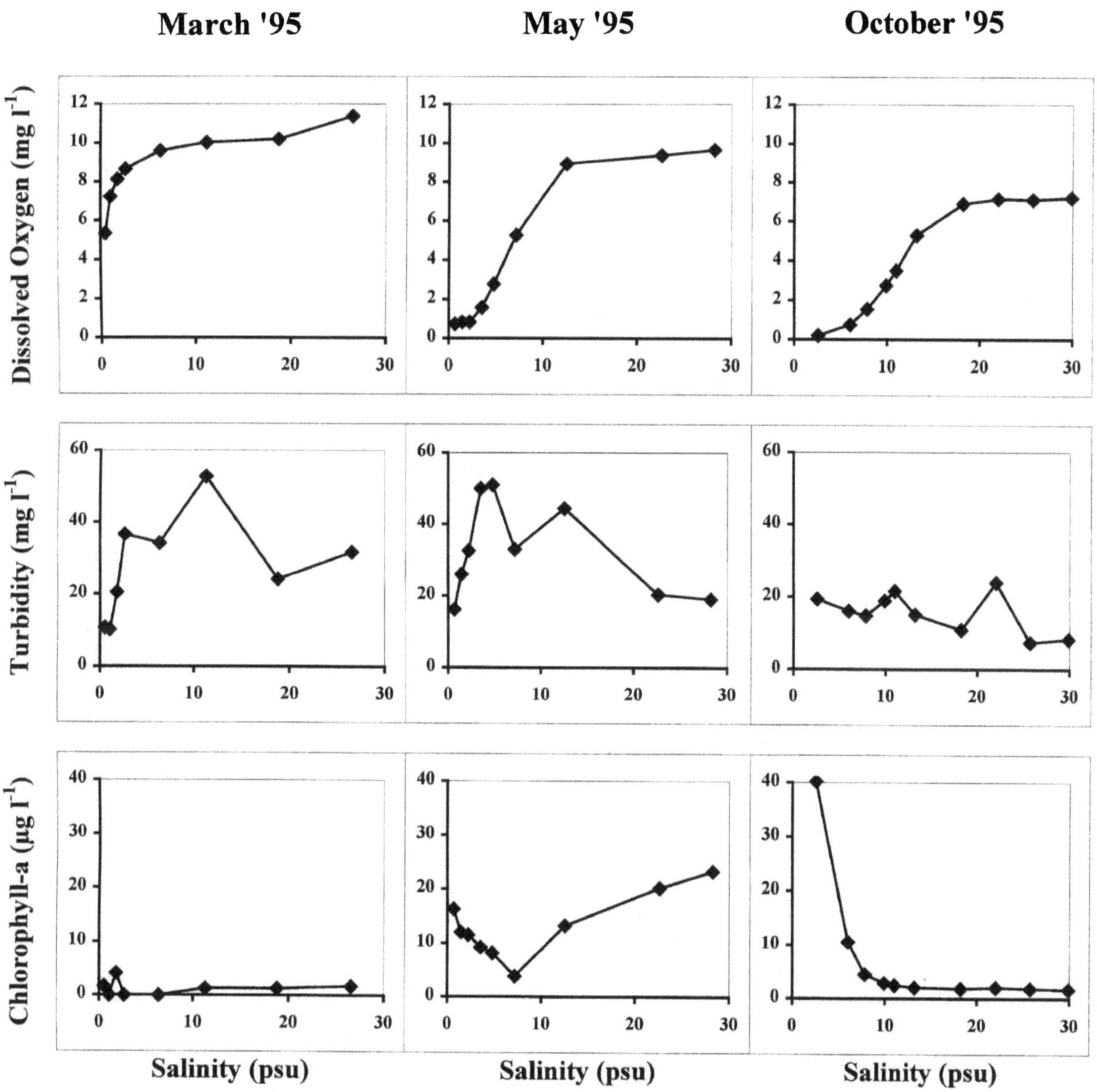

Figure 2. Dissolved oxygen, turbidity and chlorophyll-*a* versus salinity in March, May and October.

Table 5. Mean blank value (x), standard deviation (s), number of averaged blanks (n) and detection limit (DL) for analysis with ICP-AES

March, 1995	Cu ($\mu g\ g^{-1}$)	Cd ($\mu g\ g^{-1}$)	Zn ($\mu g\ g^{-1}$)	Pb ($\mu g\ g^{-1}$)
x	0.162	–0.005	0.55	0.08
s	0.09	0.004	0.39	0.25
n	7	7	7	5
DL (3s)	0.027	0.012	1.17	0.75

Results

The longitudinal profiles of dissolved oxygen, turbidity and chlorophyll-*a* concentrations during the 1995 surveys in March, May and October are shown in Figure 2.

In May and October, the upper estuary is clearly anoxic or at least heavily oxygen depleted while in March the lowest value of dissolved oxygen measured is 5.35 mg l^{-1} at 0.52 psu. In the downstream part of the estuary only in October some depletion is still observable.

Turbidity is highest in zone 2 (salinity 2–10 psu) with values ranging from 40–60 mg l^{-1}, except in October where all turbidity values are low.

The chlorophyll-a values are negligible in March, while in May increased production in the marine and fresh water areas creates a longitudinal V-shape profile of chlorophyll-a, with a minimum at 7 psu (3.8 μg l^{-1} chlorophyll-a). In October high chlorophyll-a concentrations are only observed in the low salinity area (< 1 psu). The occurrence of high chlorophyll-a levels together with low or zero oxygen might seem surprising, but the Scheldt is a very polluted river, and the enormous load of organic material causes the upper estuary not only to be anoxic, but often also depleted in nitrate. Sometimes, even reduction of iron and manganese occurs, leaving the upper estuary in a highly reduced state. All the oxygen produced by the adapted low salinity phytoplankton immediately reacts chemically, explaining why the dissolved oxygen concentrations remain low. Phytoplankton grows slowly (light is the growth determining factor), but the density of grazers and hence grazing is very low due to pollution and low oxygen, explaining why the chlorophyll-a levels can build up (Heip & Herman, 1995; Kromkamp et al., 1995 and references therein; Somville & De Pauw, 1982).

All longitudinal dissolved Cu-profiles have a convex shape (Figure 3). The fresh water end-member ranges from 0.5 to 0.9 μg l^{-1}, the marine end-member from 0.6 to 0.9 μg l^{-1} and the curve maximum from 1.2 to 1.6 μg l^{-1}. The lowest dissolved Cu concentrations are observed during the May cruise.

The shape of the dissolved Cd profiles (Figure 3) is similar to that of the dissolved Cu profiles, although the latter show a broader convexity. The curve maximum fits in a salinity window of 13–18 psu, while the highest dissolved Cd concentration observed amounts to 0.15 μg l^{-1}. This value, however, fluctuates seasonally with much lower concentrations in May than in March (4 times higher) and in October (6 times higher). These findings are comparable to those reported by Zwolsman & Van Eck (1993).

The dissolved Zn and Pb profiles (Figure 3) are completely different from those of Cu and Cd. The dissolved Zn curve shows a dilution profile except in the low salinity area. Particularly in May some mobilisation trend is noticeable: the riverine end-member concentration of 3.7 μg l^{-1} at 0.67 psu first slightly decreases to 3.2 μg l^{-1} at 1.5 psu, then progressively mounts to 6.3 μg l^{-1} at 7.2 psu before finally obeying a dilution pattern. In October, much higher concentrations prevailed (9.4 and 12.5 μg l^{-1}) in the 2 most upstream stations. The seasonal variations are much smaller than those reported by Zwolsman & Van Eck (1993). Their winter values range from 3 to 23 μg l^{-1}, as ours do only from 0.5 to 8 μg l^{-1}.

The dissolved Pb concentrations decrease almost linearly from river to mouth. Seasonal variations are weak: the maximum concentrations which are situated in the upstream estuary decrease from 0.3 μg l^{-1} in March, 0.2 μg l^{-1} in May to 0.1 μg l^{-1} in October.

The longitudinal profiles of the four selected particulate metals show no significant differences (Figure 4). However, some seasonal trend can be noticed. In March, the profiles obey an exponential curve. The negative deviation from the ideal dilution line means that a transfer from the solid to the dissolved phase took place. In October, scavenging of dissolved metals occurred around 10 psu, resulting in the rising of the dilution curve. The May profiles have a kind of intermediate form between the profiles of March and October. The perturbation of the dilution curve in May is less pronounced and occurs at lower salinity than in October.

The total metal concentration (Figure 5) is strongly dominated by dilution, but also by turbidity, which makes its interpretation more complex, because of the high variability of the latter parameter. We observe a substantial increase in the upper estuary, although especially in March and less pronounced in May, the absolute particulate concentration immediately decreases, while the dissolved concentration increase is moderate. The dominating factor in March and May is the strong increase in turbidity in the upper part of the estuary. In October, this increase, observed in a salinity range of up to 10 psu, is tightly related with a reoxygenation of the water column.

The dissolved metal fraction shows a clear evolution throughout the estuary, for Cd and Pb (Figure 6). For these metals the dissolved phase fraction gains importance as salinity increases. It should be noted that the dissolved Pb fraction is very low (2 to 7%) both in the upper and the lower estuary.

The distribution coefficient K_D, see Figures 7 and 8 (K_D = concentration in the solid phase (μg g^{-1})/ concentration in the dissolved phase (μg g^{-1} water)) of Pb is almost constant throughout the estuary and hardly varies over the year (10^6). The K_D profile of Zn changes seasonally: in March it decreases, in October it increases with salinity, while in May higher val-

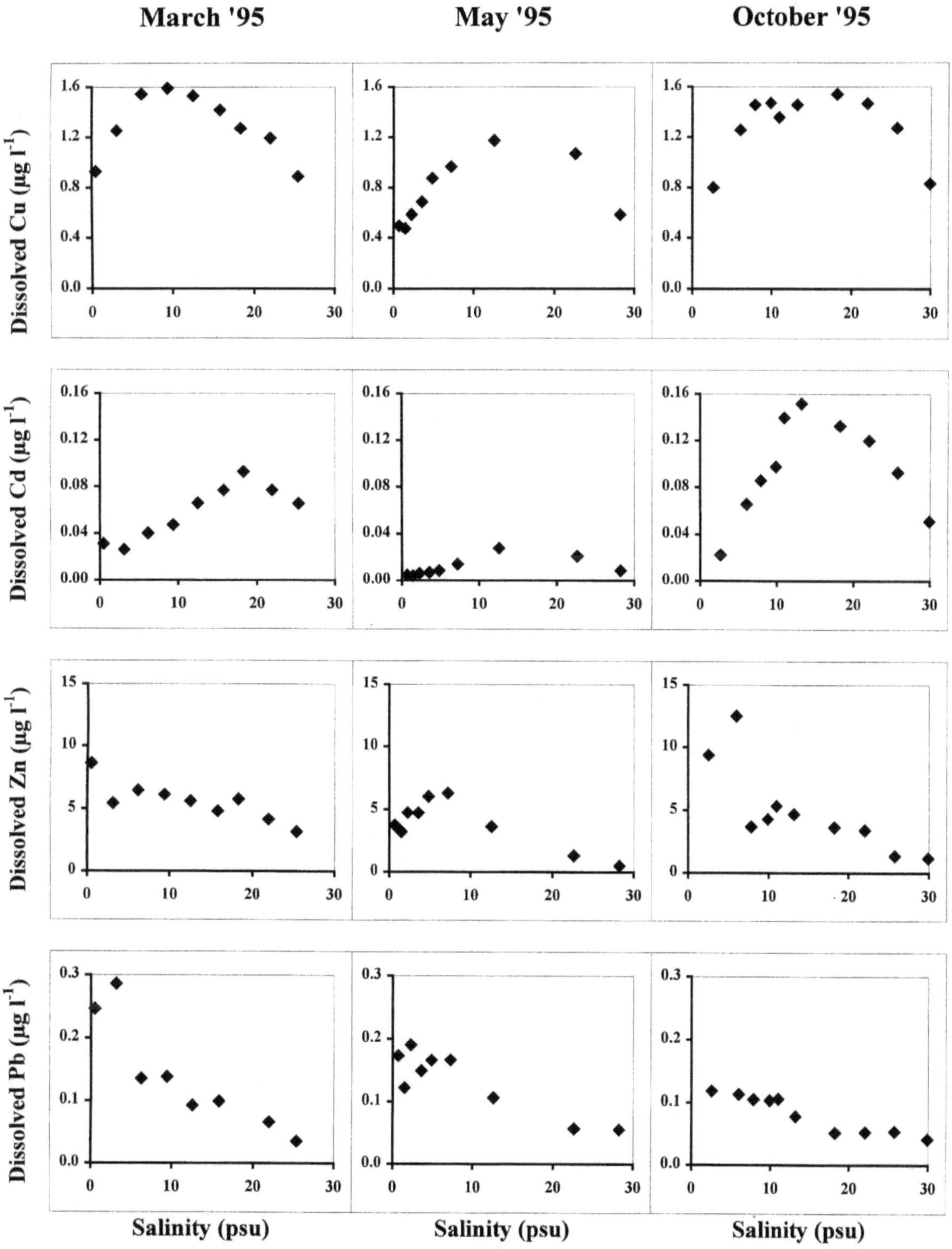

Figure 3. Dissolved Cu, Cd, Zn and Pb versus salinity in March, May and October.

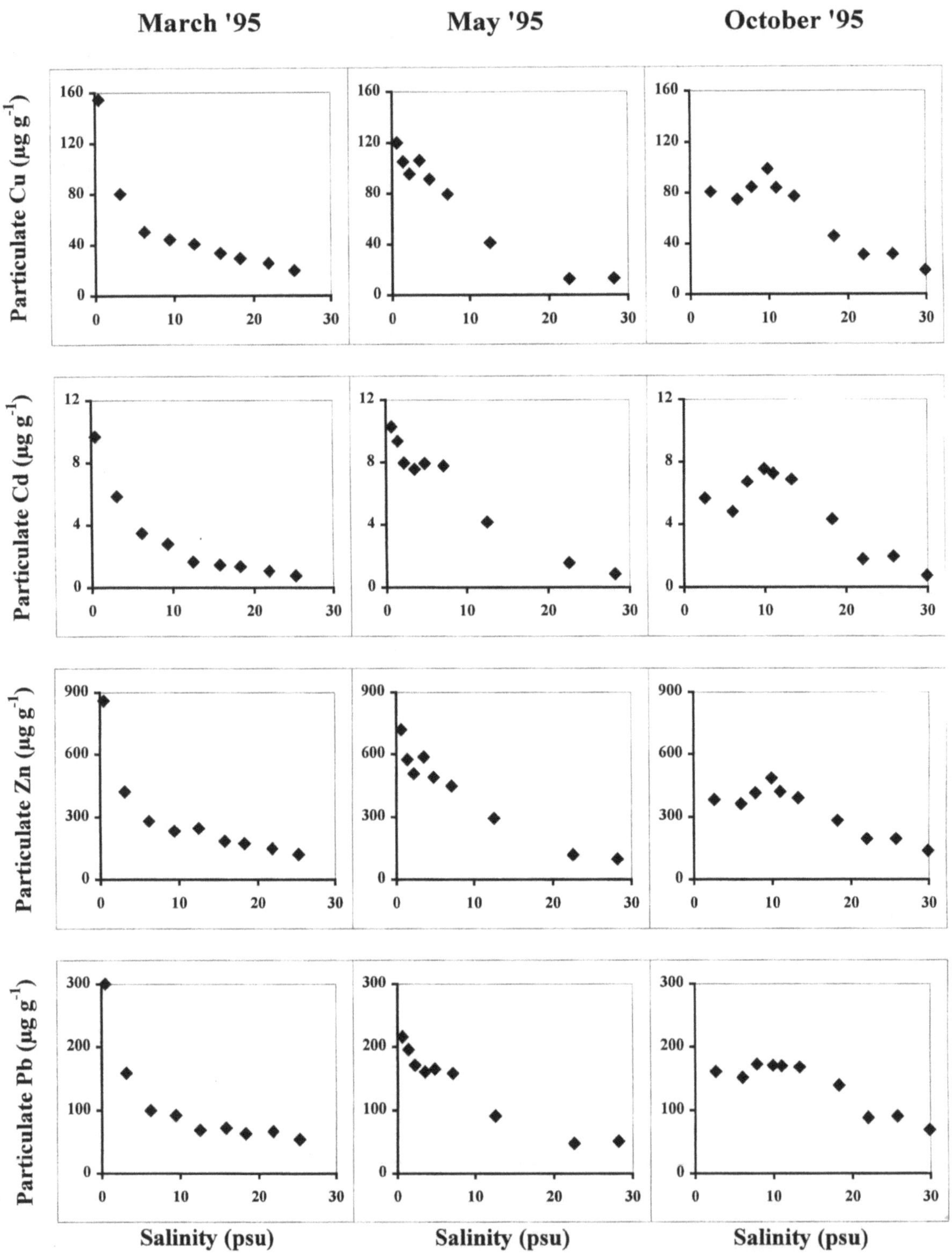

Figure 4. Particulate Cu, Cd, Zn and Pb versus salinity in March, May and October.

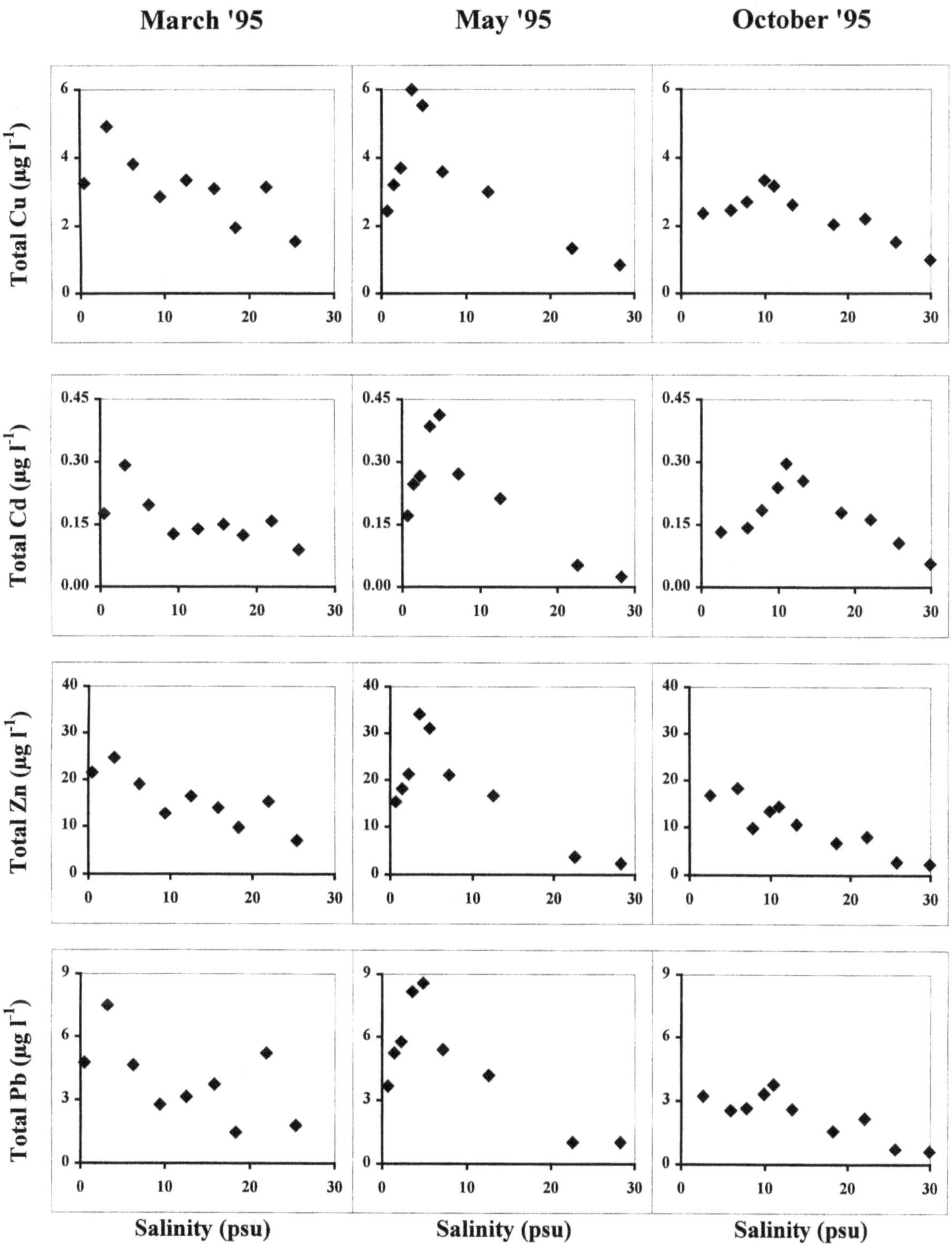

Figure 5. Total Cu, Cd, Zn and Pb versus salinity in March, May and October.

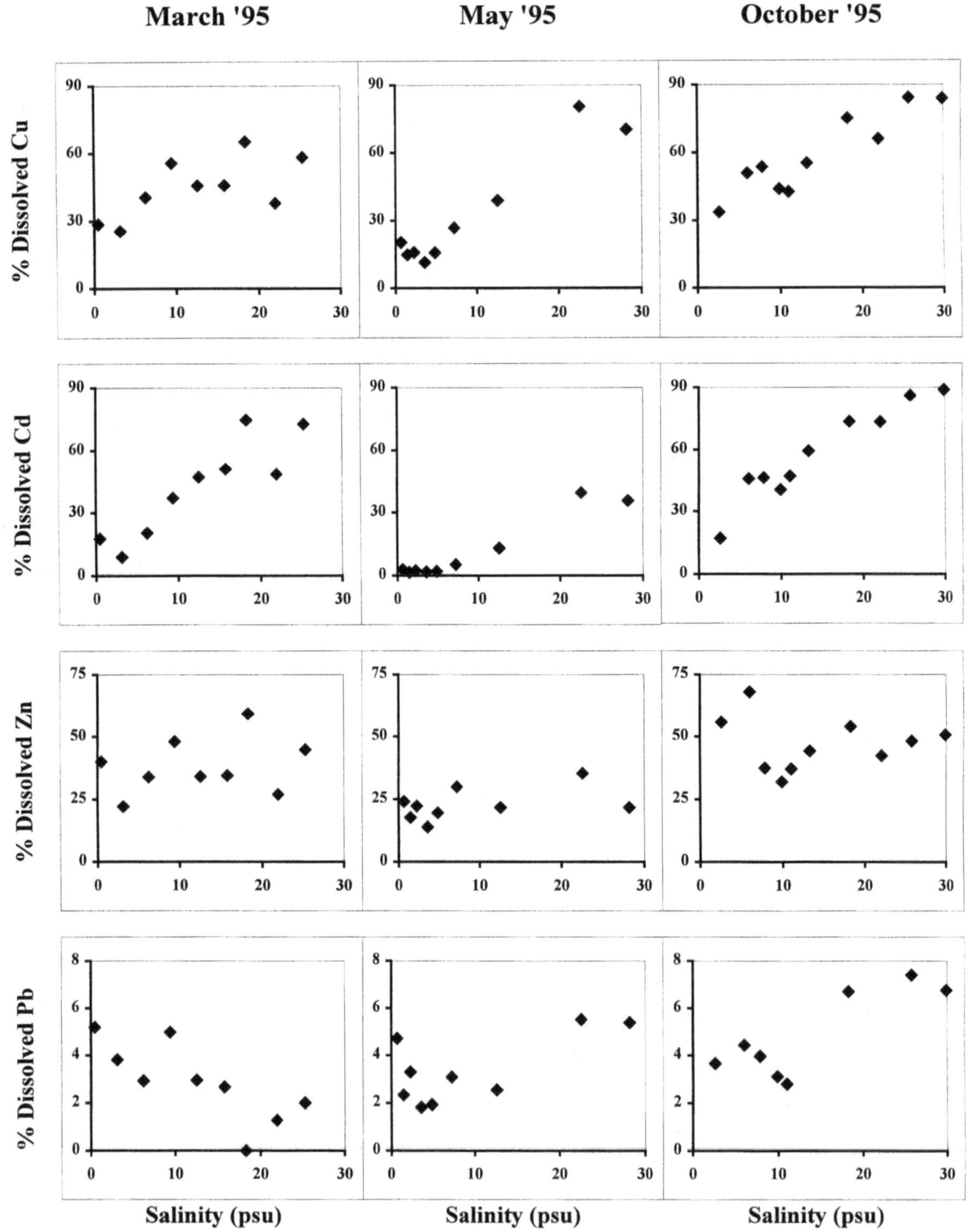

Figure 6. Dissolved fraction of Cu, Cd, Zn and Pb versus salinity in March, May and October.

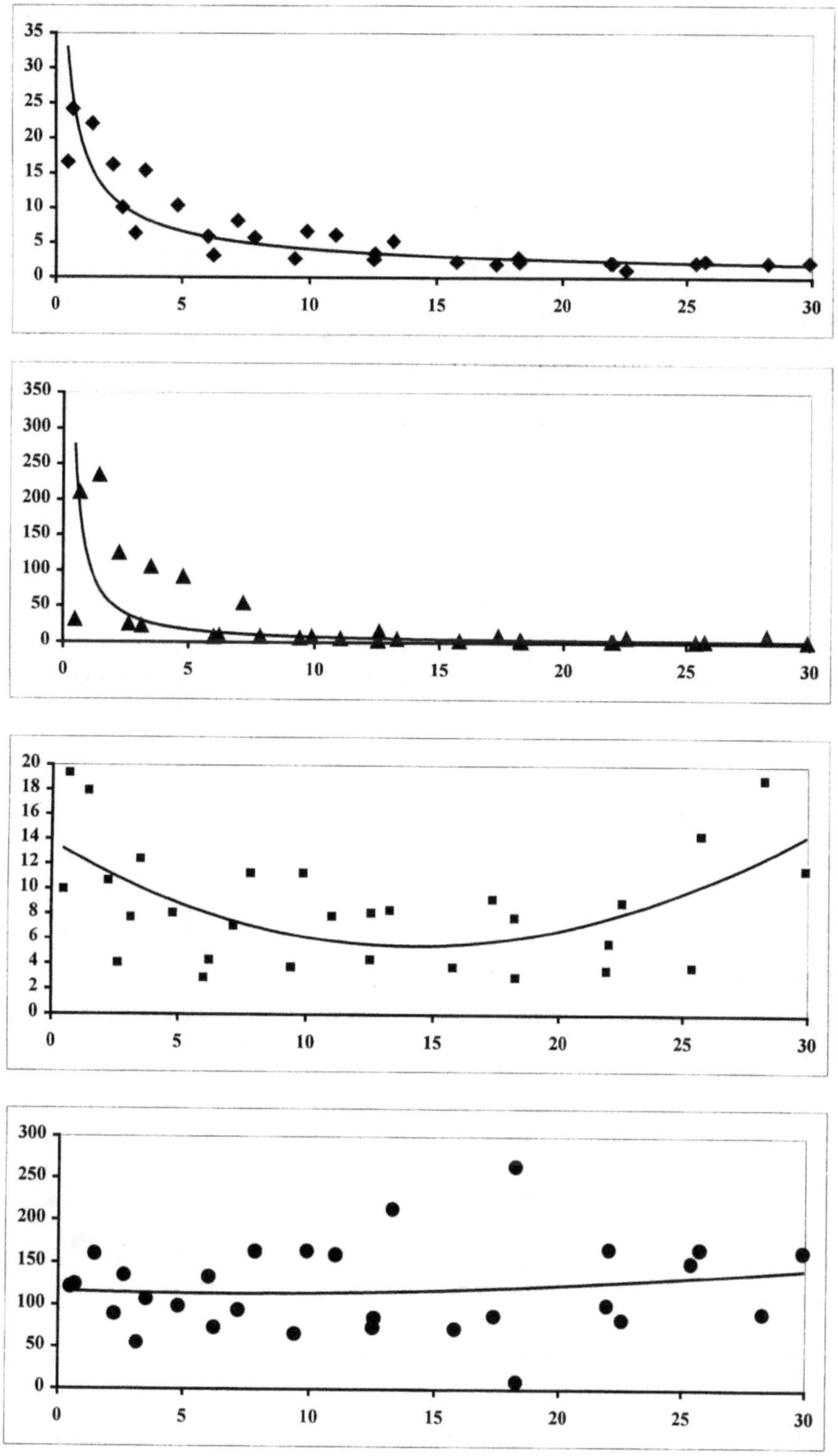

Figure 7. Distribution coefficients (K_D) for Cu, Cd, Zn and Pb versus salinity during all three cruises.

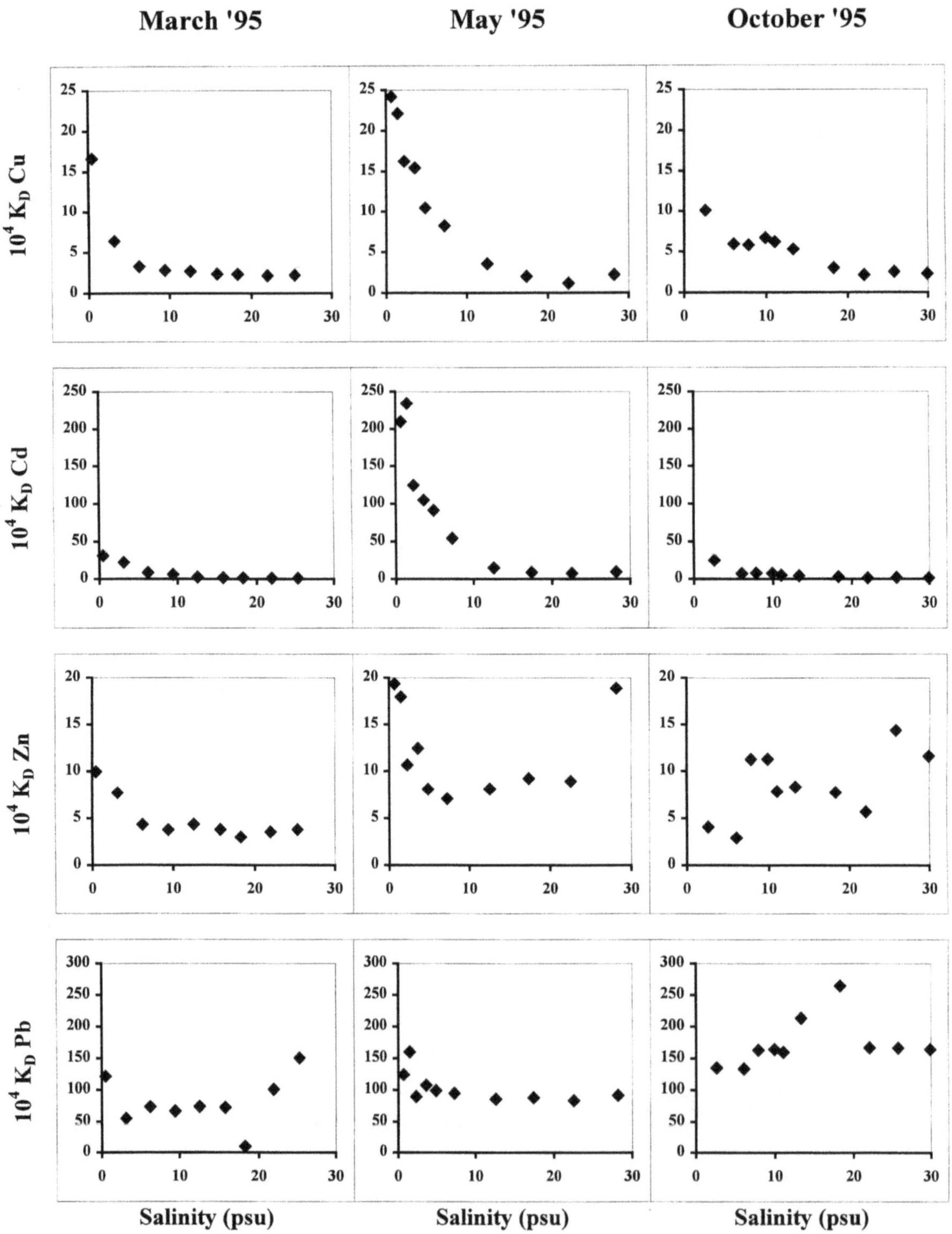

Figure 8. Distribution coefficients (K_D) of Cu, Cd, Zn and Pb versus salinity in March, May and October.

ues are observed at both end-members. The average K_D value of Zn amounts to 10^5. Both K_D values are comparable to the highest K_D values (upstream) in the period 1981–1983 (Baeyens et al., 1998).

The K_D of Cu decreases exponentially with increasing salinity. In the upper estuary, the K_D ranges from 10 to 25 10^4, while downstream this parameter stabilises around 2 10^4. The highest K_D values are observed in May. Such profiles compare well with the profiles observed during the period 1981–1983 (Baeyens et al., 1998).

The K_D profile of Cd also shows an exponential decrease with increasing salinity. However, the maximum K_D values in the upper estuary vary strongly over the year. In May a maximum K_D of 24 10^5 is found while in March and October this value is 8 times lower (3.2 10^5). The upstream K_D of Cd in the period 1981–1983 reaches a value of about 4 10^5 (Baeyens et al., 1998).

Discussion

The different biogeochemical processes governing the behaviour of the dissolved and particulate metal profiles in the estuary are described in Baeyens et al. (1998). A summary of the most important processes includes: (1) redox reactions. In the transition areas from aerobic to anaerobic conditions and vice versa, the dissolved metal content is influenced by sulphide precipitation reactions in strong anoxic conditions, by dissolution of iron and manganese oxyhydroxides, and subsequent release of associated metals in suboxic conditions, by coprecipitation with iron and manganese oxyhydroxides and by dissolution of metal sulphides in oxic conditions. All these redox conditions are met, at least in summer, in the upper estuary (Baeyens et al., 1998). Metal sulphides have been identified in the upper Scheldt estuary (Zwolsman & Van Eck, 1993). In addition, metal sulphides can be resuspended from the bottom sediments (Baeyens et al., 1991); (2) adsorption/desorption reactions. The high turbidity in the salinity zone from 2 to 10 psu favours adsorption, while increasing salinity pairs with increasing desorption; (3) complexation with organic and inorganic ligands. These ligands can compete efficiently with the particulate matter to keep the metals in the dissolved phase. Humic acids are preferential metal scavengers, but so are chlorocomplexes (e.g. with Cd), sulphate complexes (e.g. with Zn) and hydroxide complexes (e.g. with Cu); (4) biological processes. Primary production plays an important role in the removal of dissolved metals in the downstream estuary during the bloom period (Van Eck & De Rooij, 1993).

A detailed discussion of the importance of each of these processes regarding the behaviour of the chosen metals can be found in Baeyens et al. (1998) for the period 1981–1983. Here, only major similarities and discordances with those observations are pointed out.

In the downstream area, dissolved copper shows a dilution pattern. The seasonal variations are small, although the levels in May are somewhat lower (20 to 30%) than in the two other sampling periods. The phytoplankton bloom in May was very important and incorporation of metals by plankton may explain the lower values in that period. This phenomenon has not been observed during the earlier period (1981–1983). In the mid-estuarine area a broad maximum (a cumulative production/consumption pattern) is present. The production zone corresponds to an increase in the dissolved oxygen level (Figure 9): the positive correlation between the dissolved copper and the dissolved oxygen concentrations at low salinity levels (surveys of May and October) is very good, as it was in the period 1981–1983 (Baeyens et al., 1998). Oxidation of reduced copper compounds, probably CuS and/or Cu_2S, occurs as soon as the oxidation conditions are restored. In winter, dissolved oxygen is more abundant in the upper estuary (only suboxic conditions may be present, e.g. the dissolved oxygen profile in March, Figure 2) and adsorption/desorption processes are predominant in the area of high turbidity.

The average particulate copper profile (Figure 4) shows a negative deviation compared to the theoretical dilution line. Such pattern suggests a mobilisation process and confirms the positive convex shape of the dissolved copper profiles. Remarkable is the excellent negative correlation between particulate copper and dissolved oxygen (Figure 10). The best correlation coefficients for all metals are observed in March, but the dissolved oxygen range (6–11 mg l^{-1}) is much smaller than in the other periods. Oxidation processes are thus less important in March. For May, adding chlorophyll-*a* to the variance analysis increases the regression coefficient from 0.93 up to 0.96, in October, the rather poor correlation increases from 0.67 to 0.87 by adding chlorophyll-*a*. These observations show the importance of redox conditions (May and October) and desorption processes (March) on the distribution of metals through the estuary. The relations between dissolved oxygen or salinity and particulate inorganic material may be disturbed by phytoplankton growth

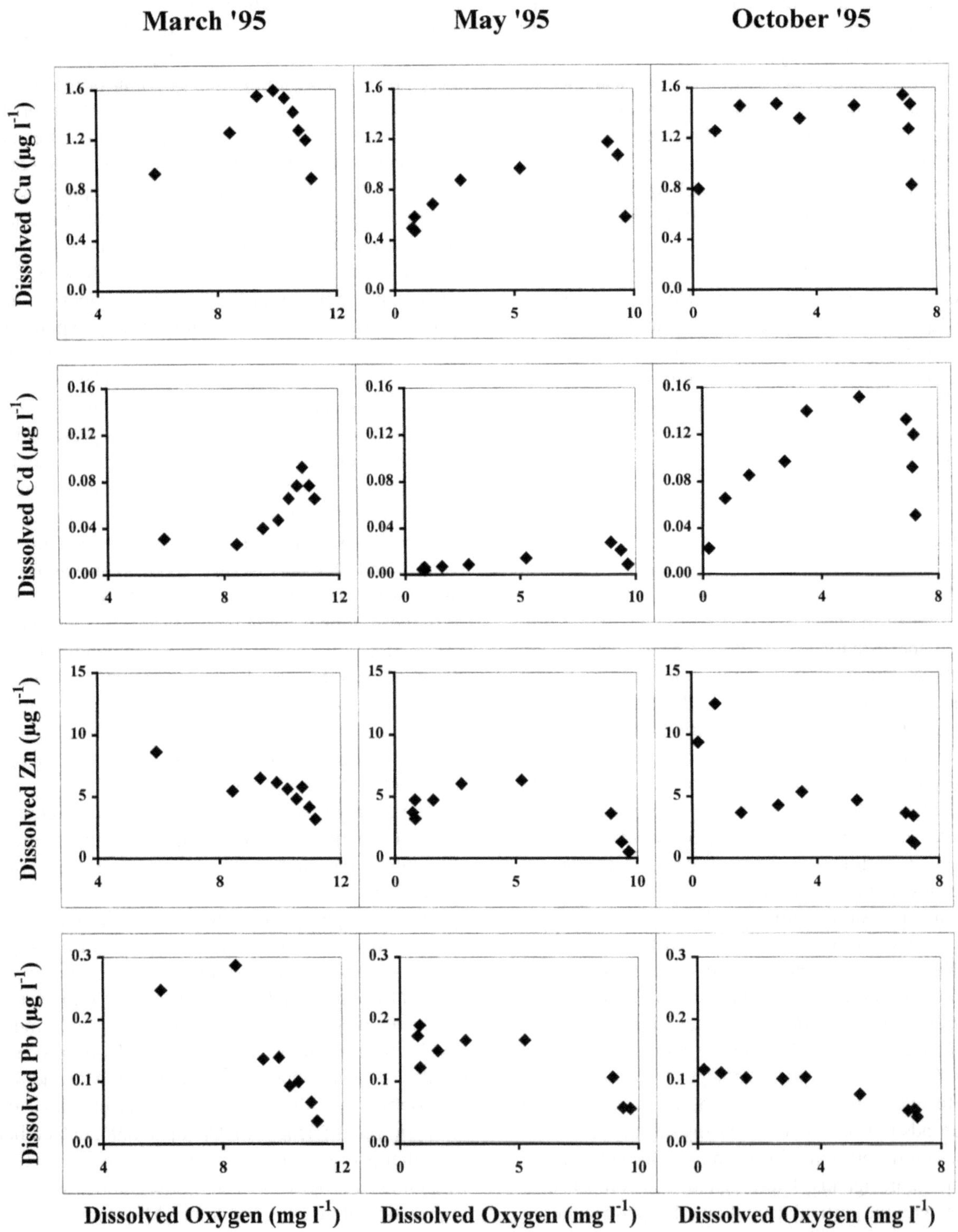

Figure 9. Dissolved Cu, Cd, Zn and Pb versus dissolved oxygen in March, May and October.

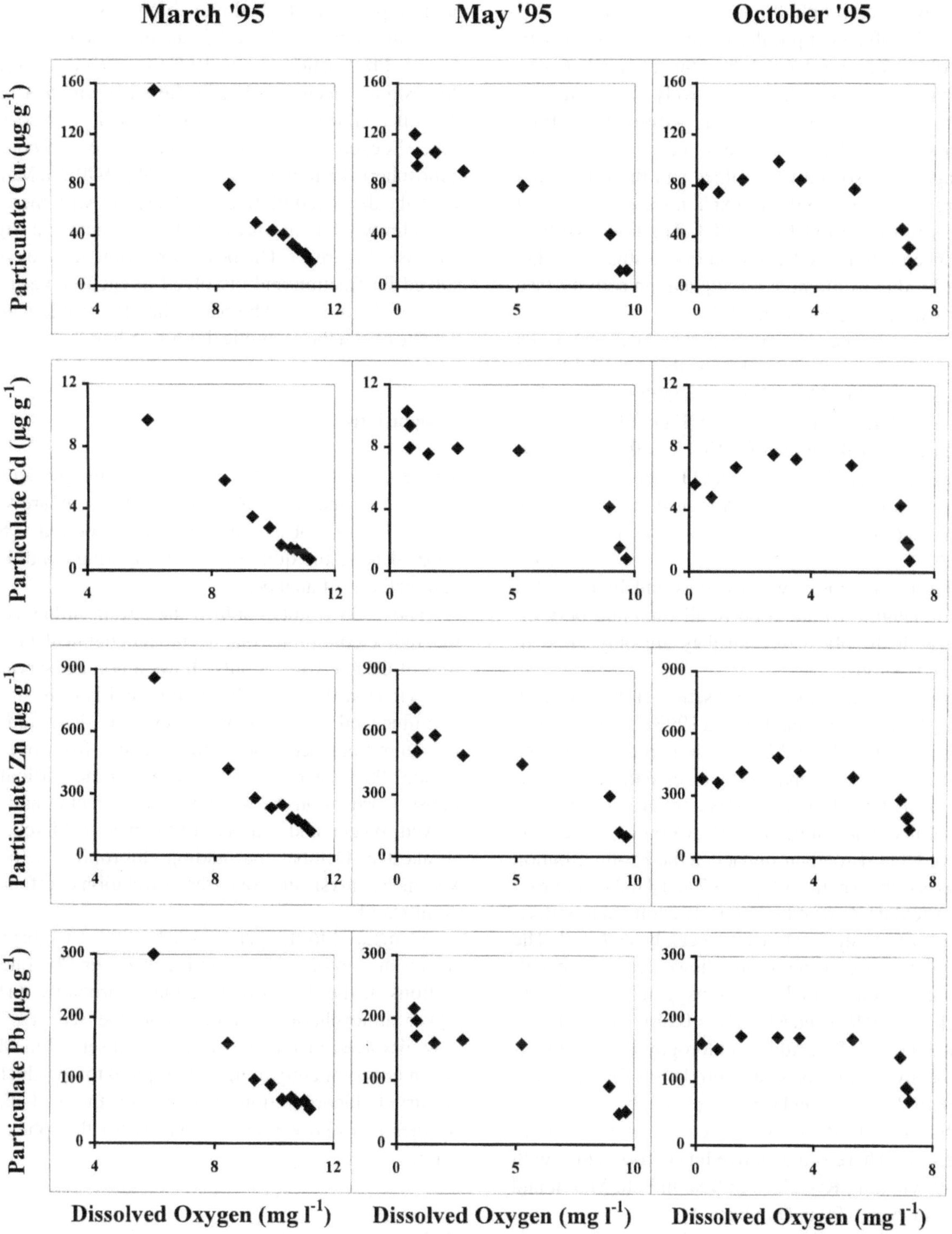

Figure 10. Particulate Cu, Cd, Zn and Pb versus dissolved oxygen in March, May and October.

and death, also occurring in the particulate phase. That would explain that the clearest pattern is observed in March, when chlorophyll-*a* levels are very low.

The dissolved Cd profiles compare very well with those of Cu. Figures 3 and 9 show the relation between dissolved Cd and salinity, respectively dissolved oxygen in the upper estuary. The processes controlling the dissolved Cd profiles are similar to those controlling the dissolved Cu profiles and those reported by Baeyens et al. (1998) to explain the dissolved Cd profiles in the period 1981–1983. The low dissolved Cd concentrations in the mid and low estuary in May might be due to the high phytoplankton activity (see also Zwolsman & Van Eck, 1993).

The particulate Cd profile (Figure 4) also shows a negative deviation compared to the theoretical dilution line, corroborating Cd mobilisation in the estuary. The correlation with dissolved oxygen (Figure 10) in March is even more striking than for Cu ($R^2 = 0.99$), however lower in May ($R^2 = 0.86$) and October ($R^2 = 0.45$). Adding chlorophyll-*a* to the variance analysis yields R^2-values of 0.91, respectively 0.78, for the latter two periods.

The correlations between salinity and dissolved Cd fraction (Figure 6) are good in all sampling periods, while with dissolved oxygen they are only good in May and October.

The difference in the mobilisation of dissolved Zn in the estuary is very small during 1995 (Figure 6), as in the period 1981–1983. Mobilisation processes in both periods (1981–1983 and 1995) are essentially restricted to spring time. Indeed, in May a small increase in the low salinity area is noticeable. Redox processes may be involved, but if so, their impact on the K_D-distribution values is rather small, as Figures 7 and 8 indicate. From the particulate Zn profile (Figure 4), a limited mobilisation activity in the upper estuary can be deduced. The particulate Zn concentrations, however, show excellent correlations with dissolved oxygen (Figure 10) in March ($R^2 = 0.97$) and May ($R^2 = 0.88$), but weaker in October ($R^2 = 0.58$). Adding chlorophyll-*a* to the variance analysis improves the correlation for the latter two periods to 0.89 and 0.83, respectively.

Dissolved lead shows the clearest dilution pattern (Figure 3). There is a negative linear correlation with salinity yielding R^2-values of 0.80 in both March and May and even 0.90 in October.

Pb is mainly present in the particulate phase (about 97%). Again, there is a striking correlation between particulate Pb and dissolved oxygen (Figure 10) in March ($R^2 = 0.96$), a very good one in May ($R^2 = 0.89$), which slightly improved when adding chlorophyll-*a* ($R^2 = 0.91$). In October correlation is poor ($R^2 = 0.55$), but it improves to 0.73 adding chlorophyll-*a*.

In the period 1981–1983, an increase of the dissolved Pb fraction from 1.6% upstream to 8.7% downstream occurred, but in the early eighties high Pb concentrations were still observed in our coastal zone (Baeyens et al., 1987) resulting from important atmospheric Pb inputs (Baeyens et al., 1990). In March 1995 the dissolved Pb fraction decreases with increasing salinity while in May and October there is a slight increase (Figure 6). The best correlation between dissolved Pb fraction and dissolved oxygen is observed in May (Figure 11). The particulate Pb profile hardly deviates from the theoretical dilution line.

Conclusions

The longitudinal profiles of dissolved and particulate Cd, Cu, Pb and Zn in the period 1995 compare very well with those observed earlier in the period 1981–1983, despite the quite different sampling conditions and methods of analysis.

The 1995 results confirm the strong influence of the redox conditions on the biogeochemical behaviour of Cd and Cu. Zn only shows a non-conservative behaviour in the low salinity area and is only slightly influenced by redox processes. The dissolved Pb concentrations decrease almost linearly from river to mouth. Pb appears for 97% present in the particulate phase. Correlations between particulate metals and dissolved oxygen and salinity are the best in March, the weakest in October, but adding chlorophyll-*a* to the variance analysis improves the correlation coefficients significantly.

Compared to the period 1981–1983, two discrepancies are observed: (1) the much lower Pb concentrations in the downstream estuary, probably due to lower atmospheric Pb inputs into the sea; (2) the low dissolved metal concentrations observed in May, when concomitantly a high phytoplankton production occurred. Incorporation of metals by these plankton species is a reasonable explanation for that phenomenon.

Acknowledgements

The authors gratefully acknowledge technical assistance of the captain and crew members aboard R.V.

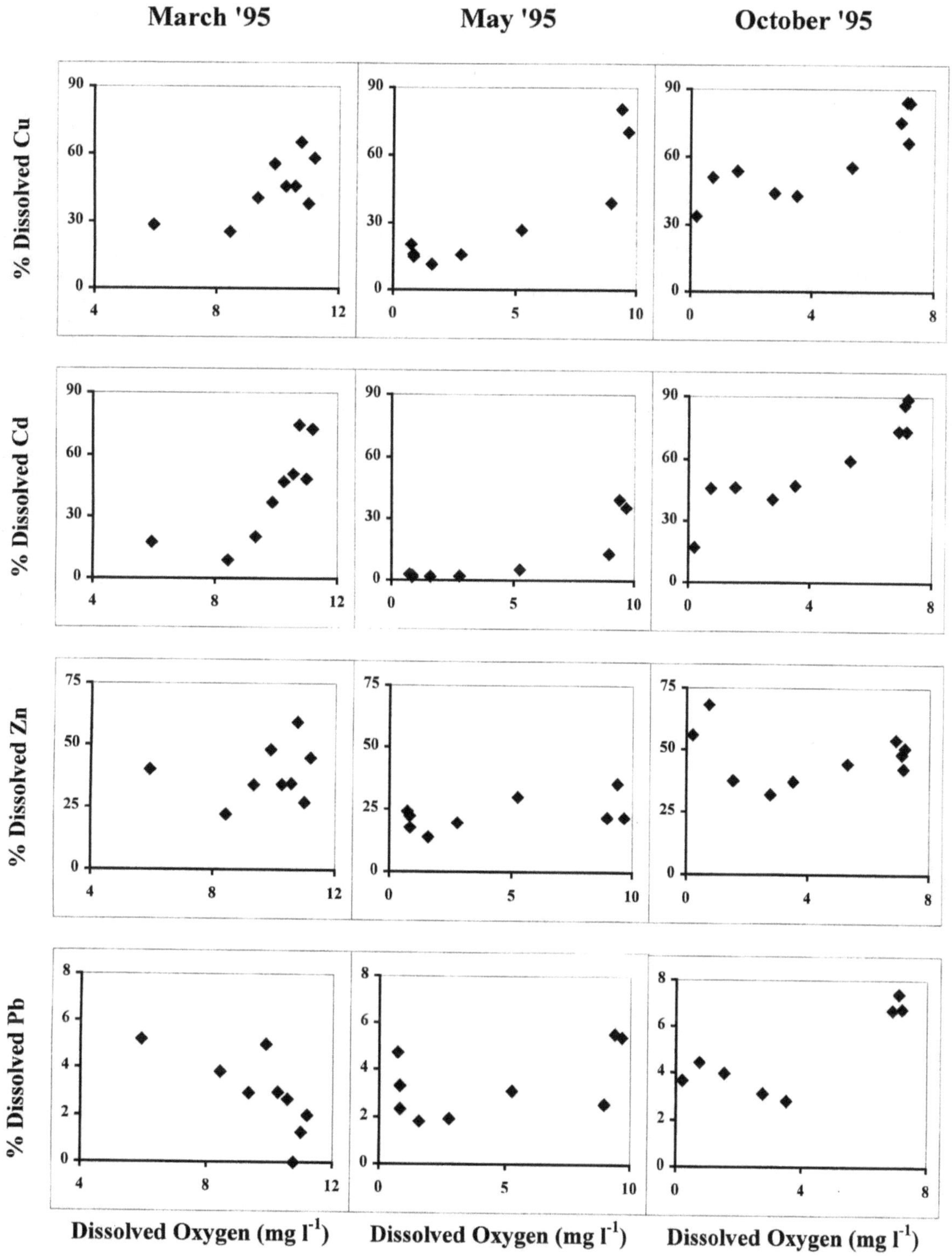

Figure 11. Dissolved fraction of Cu, Cd, Zn and Pb versus dissolved oxygen in March, May and October.

Belgica. They thank the MUMM for providing ship time. The Ministry of Public Health and Environment allocated grants to G. Ducastel and K. Parmentier.

References

Baeyens, W., G. Gillain, G. Decadt & I. Elskens, 1987a. Trace metals in the eastern part of the North Sea I: Analyses and short term distributions. Oceanol. Acta 10: 169–179.

Baeyens, W., G. Gillain, G. Decadt & I. Elskens, 1987b. Trace metals in the eastern part of the North Sea II: Flow of Cd, Cu, Hg, Pb and Zn through the coastal area. Oceanol. Acta 10: 301–309.

Baeyens, W., F. Dehairs & H. Dedeurwaerder, 1990. Wet and dry deposition fluxes above the North Sea. Atmos. Envir. Part A, 24: 1693–1703.

Baeyens, W., S. Panutrakul, M. Elskens, M. Leermakers, J. Navez & F. Monteny, 1991. Geochemical Processes in muddy and sandy tidal flat sediments. Geo-Marine Lett. 11: 188–193.

Baeyens, W., M. Elskens, G. Gillain & L. Goeyens, 1998. Biogeochemical behaviour of Cd, Cu, Pb and Zn in the Scheldt estuary during the period 1981–1983. Hydrobiologica 366: 15–44.

Danielsson, L.-G., B. Magnusson & S. Westerlund, 1978. An improved metal extraction procedure for the determination of trace metals in sea water by atomic absorption spectroscopy with electrothermal atomization. Analyt. chim. Acta 98: 47–57.

Danielsson, L.-G., B. Magnusson, S. Westerlund & K. Zhong, 1982. Trace metal determinations in estuarine waters by electrothermal atomic absorption after extraction of dithiocarbamate complexes into freon. Analyt. chim. Acta 144: 183–188.

Heip, C. & P. M. J. Herman, 1995. Major biological processes in European tidal estuaries: a synthesis of the JEEP-92 Project. Hydrobiologia 311: 1–7.

Hoenig, M. & A.-M. de Kersabiec, 1989. L'atomisation electrothermique en spectrométrie d'absorption atomique. Masson Paris, 296 pp.

Kromkamp, J., J. Peene, P. van Rijswijk, A. Sandee & N. Goosen, 1995. Nutrients, light and primary production by phytoplankton and microphytobenthos in the eutrophic, turbid Westerschelde estuary (The Netherlands). Hydrobiologia 311: 9–19.

Somville, M., & N. De Pauw, 1982. Influence of temperature and river discharge on water quality of the Western Scheldt estuary. Wat. Res. 16: 1349–1356.

Van Eck, G. T. M. & M. De Rooij, 1993. Potential chemical time bombs in the Scheldt Estuary. Land degradation and rehabilitation 4: 317–332.

Zwolsman, J. J. G. & G. T. M. Van Eck, 1993. Dissolved and particulate trace metal geochemistry in the Scheldt Estuary, S.W. Netherlands (Water column and sediments). Neth. J. aquat. ecol. 27: 287–300.

Hydrobiologia **366:** 63–79, 1998.
W. F. J. Baeyens (ed.), Trace Metals in the Westerschelde Estuary.

Behaviour and speciation of mercury in the Scheldt estuary (water, sediments and benthic organisms)

Willy Baeyens, Carine Meuleman, Bamba Muhaya & Martine Leermakers
Department of Analytical Chemistry, Free University of Brussels, Pleinlaan 2, 1050 Brussels, Belgium

Key words: Mercury, methylmercury, biogeochemical cycling, estuaries, benthic organisms

Abstract

Surface waters, sediments and the polychaete *Nereis diversicolor* were sampled in the Scheldt Estuary between 1990 and 1994. In surface waters particulate Hg (Hg_P) concentrations ranged from 350–1610 ng g^{-1}. They are essentially controlled by physical mixing of polluted fluvial particulates with relatively unpolluted marine particulates, but unaffected by seasonal changes. Dissolved Hg species, on the other hand, show large seasonal variations essentially controlled by the redox conditions in the estuary, as well as by bacterial and phytoplankton activity. Total dissolved Hg (Hg_{TD}) concentrations ranged from 0.5 to 3.0 ng l^{-1} with 10 to 90% as reactive Hg. High concentrations of Hg_{TD} are found in the upper estuary in the winter and decrease rapidly with increasing salinity. In summer Hg_{TD} concentrations are low in the anoxic upper estuary and increase as oxygen is restored in the estuary. Significant variations were observed in dissolved Monomethyl Hg (MMHg) concentrations with concentrations ranging from 0.01 to 0.120 ng l^{-1} in the winter and 0.08 to 0.6 ng l^{-1} in summer and autumn. Particulate MMHg ranged from 2 to 6 ng g^{-1} in winter and from 4 to 10 ng g^{-1} in summer and accounted for 20 to 80% of the total MMHg. Hg° concentrations ranged from 0.02 to 0.130 ng l^{-1} and are higher in summer than in winter. In the lower estuary a positive correlation between Hg° and phytoplankton pigments was observed.

Sediments and the polychaete *N. diversicolor* were sampled on the intertidal flat Groot Buitenschoor. Hg_T concentrations in surface sediments ranged from 144 to 1890 ng g^{-1} and MMHg from 0.8 to 6 ng g^{-1} accounting for 0.4 to 0.8% of the total mercury present. Both total Hg (Hg_T) and MMHg concentrations increased with increased organic matter content and anoxic conditions. On the other hand, accumulation of Hg_T and MMHg was higher in *N. diversicolor* living in coarse grain sandy sediments than in muddy sediments. MMHg concentrations in *N. diversicolor* ranged from 2.2 to 20.9 ng g^{-1} accounting for an average of 18% of the Hg_T. Seasonal variations significantly affected Hg speciation in sediments and *N. diversicolor*. Higher Hg_T concentrations were found in the sediments in autumn and winter, whereas MMHg concentrations increased in spring and summer. Likewise, higher MMHg concentrations were also observed in *N. diversicolor* in spring and summer.

Introduction

As heavy metal pollution remains a major threat to coastal and estuarine ecosystems, an understanding of the metal cycling in an estuary and the processes controlling this cycle is essential. Both physico-chemical (water and sediment mixing) as well as biological estuarine processes (primary production, biological degradation of organic matter) play an important role in the behaviour of heavy metals in estuaries. Recent advances in sampling and analytical techniques have enabled accurate measurements of different mercury species at the pg l^{-1} level encountered in aquatic systems (Gill & Fitzgerald, 1987; Bloom & Fitzgerald, 1988; Bloom, 1989) and at the ng g^{-1} level in biota and sediments (Decadt et al., 1985; Horvat et al., 1993). However, compared to the intensive investigations which have been performed in the last years on mercury speciation and bioaccumulation in freshwater systems, currently very little is known about the estuarine behaviour of mercury.

In this paper seasonal profiles of dissolved Hg species (Hg_R, Hg°, $MMHg_D$, Hg_{TD}) and particulate Hg (Hg_P, $MMHg_P$) in estuarine waters as well as total mercury (Hg_T) and MMHg in sediments and the polychaete worm *N. diversicolor* of the intertidal flat Groot Buitenschoor are presented and their relationship with various physicochemical parameters are discussed. Calculations of mercury fluxes from the Scheldt river to the North Sea are presented in a related paper (Baeyens et al., 1996).

Description of the studies site

The river Scheldt passes through France, Belgium and The Netherlands before reaching the North Sea. It has a length of 355 km and a drainage area of 22 000 km^2, mainly on Belgian territory. The zone under investigation (estuarine mixing zone) is situated between Rupelmonde and Vlissingen, covering an area of about 100km (Figure 1). The Groot Buitenschoor is one of the major intertidal mudflats of the Scheldt estuary with a surface area of 200 hectares and is located at the Belgium-Netherlands border (Figure 1). A detailed description of the physico-chemical and biological characteristics of the Scheldt estuary can be found in a related paper (Baeyens et al., 1998).

A number of estuarine aspects, specific for the Scheldt estuary, are of particular importance in the interpretation of the behaviour of trace metals such as mercury: (1) Both the fluvial and the upstream river are heavily polluted as a result of large domestic, industrial and agricultural waste waters discharges resulting in a permanent oxygen undersaturation in this zone and frequent complete depletion of dissolved oxygen in the summer months; (2) Due to the low river discharge (the average water flow is about 100 $m^3\ s^{-1}$ and varies from 20 $m^3\ s^{-1}$ during summer to 400 $m^3\ s^{-1}$ during winter) the Scheldt is a tide governed estuary resulting in long residence times of water masses and pollutants (one to three months); (3) Direct inputs of pollutants occur especially in the upper estuary from the industrial area around Antwerp; (4) The estuary's turbidity maximum, the industrial inputs and anoxic zone are located in the same area complicating the interpretation of the trace metal behaviour.

At the 4th International Conference on the Protection of the North Sea, aquatic inputs and atmospheric emissions of mercury from important industrial sources, domestic waste incinerators, and diffuse inputs from dentistries, Hg-containing products etc. taking place in the whole North Sea draining basin, were presented (Andersen & Niilonen, 1995). For Belgium, the Netherlands and Germany together, these inputs are estimated at 45 tons yr^{-1} for 1995 compared to 69.2 tons yr^{-1} for 1985. Industrial applications of Hg in the Scheldt river basin are essentially the chlor-alkali industry, the non-ferrous industry, PVC production and phosphate industry, located around the harbour of Antwerp or along the tributaries of the Scheldt river (Devolder et al., 1991). Industrial emissions of Hg have, however, decreased significantly in the last ten years (Table 1). The aquatic inputs, discharges and/or losses of mercury to the Scheldt river and its tributaries from Flanders and the Brussels Capital Regions (Table 1), covering the major part of the Scheldt catchment area, have been reported to be 1372 kg yr^{-1} for 1995 (Andersen & Niilonen, 1995). Based on dissolved and particulate Hg concentrations in the Scheldt estuary, Baeyens et al. (1996) calculated that only a small fraction of these emissions (66 kg yr^{-1}) reaches the North Sea.

Methods and materials

Sampling and analysis of estuarine water

Four surveys were carried out on the Scheldt estuary between June 1993 and August 1994 (June 1993, November 1993, February 1994 and August 1994), three on board of the research vessel *R. V. Belgica* and the November 1993 survey on board of the *R.V. Argus* (RIKZ, the Netherlands). Samples were collected from the mouth of the estuary (salinity 25–30 psu) up to the confluence of the Scheldt and the Rupel (salinity 0.3–1 psu). Samples of surface water were collected by hand from a rubber boat by submerging teflon (FEP) and borosilicate glass bottles approximately 20 cm beneath the water surface. Arm length gloves were used during sampling. The rubber boat moved gently against the current during sampling and was positioned approximately 100 m up current of the research vessel. FEP and borosilicate glass bottles had been rigorously cleaned by consecutive washes with Decon-90, boiling concentrated HNO_3 for 6 hours and HCl 1% at 60 °C overnight. After the final rinse with Milli-Q water, the bottles were filled with a 1% HCl solution (Merck, *suprapur*) and stored double bagged until use. The sampling bottles were rinsed twice with the estuarine water before being filled. The bottles were capped while immersed in the water to prevent oxidation of

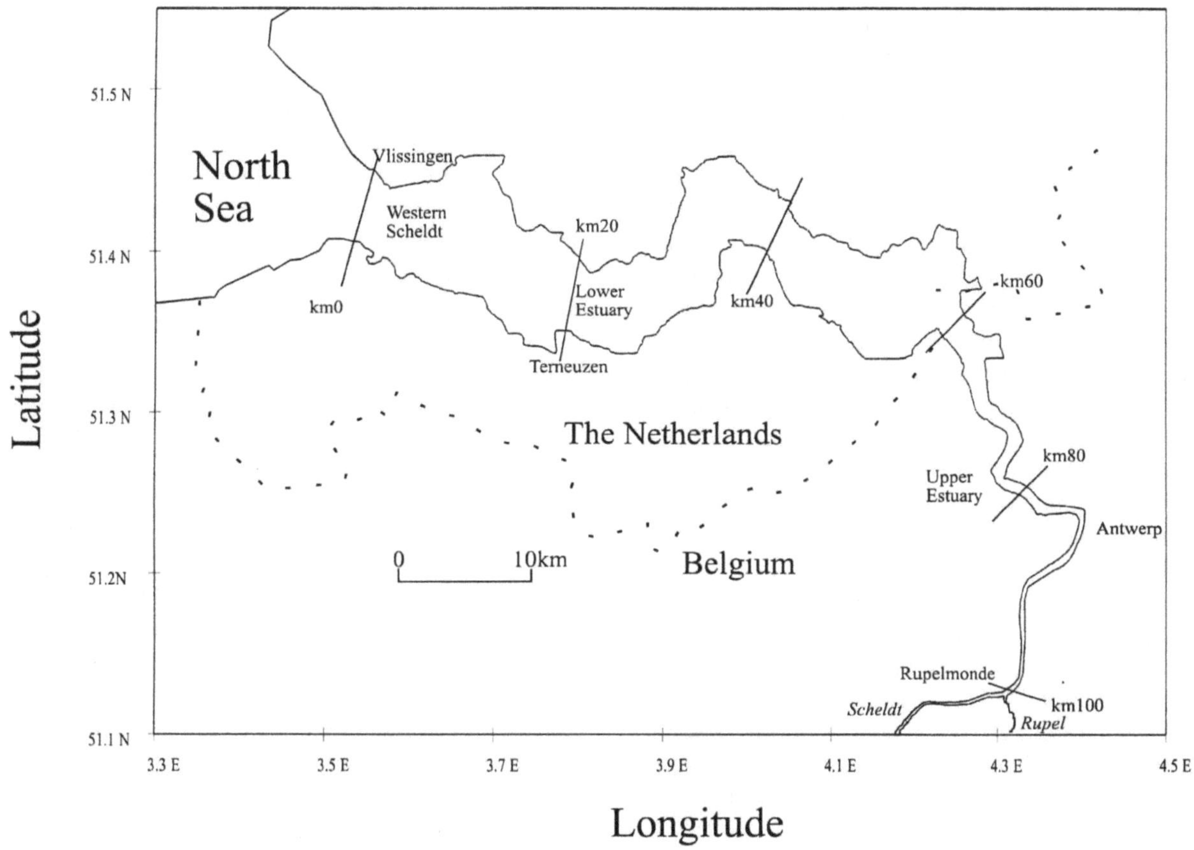

Figure 1. Map of the Scheldt estuary.

Table 1. Atmospheric and aquatic emissions of mercury in Belgium (Source: Andersen & Niilonen, 1995)

Sources	Flanders		Brussels		Belgium	
	1985 (kg yr^{-1})	1995 (kg yr^{-1})	1985 (kg yr^{-1})	1995 (kg yr^{-1})	1985 (kg yr^{-1})	1995 (kg yr^{-1})
Atmospheric						
Waste incineration	3466	1721	1868	943	6334	3484
Chlor-alkali	2435	729	0	0	3685	1492
Energy	650	683	111	104	1186	1234
Iron and steel	413	395	76	0	1145	1004
Electric. on coal	400	382	0	0	912	708
Non-ferrous metal	63	31	–	–	93	31
Total	7427	3941	2055	1047	13325	7953
Aquatic						
Dentistries	1146	1146	198	198	1976	1976
Chlor-alkali	114	17	0	0	224	74
Non-ferrous	150	11	–	–	150	11
Total	1410	1174	198	198	2350	2061
Total	8837	5115	2253	1255	15675	10014

reduced waters in the time between sampling and handling. Once on board of the ship the samples were filtered on 0.45 μm cellulose acetate (Millipore) filters by pressure filtration using a FEP filtration apparatus placed in a laminar flow hood and Ar as inert gas (Baeyens, 1992). Filtered samples were collected in FEP or borosilicate glass bottles. Samples for dissolved reactive Hg (Hg_R) and dissolved total Hg (Hg_{TD}) were acidified with 1% HCl (Merck, suprapur) (Gill & Fitzgerald, 1987). Samples for dissolved methylmercury ($MMHg_D$) analysis were stored deep frozen and unacidified (Leermakers et al., 1990). Suspended particulate matter (spm) collected on 0.45 μm filters were stored deep frozen in Millipore petri dishes for the analysis of particulate Hg (Hg_P) and particulate methylmercury ($MMHg_P$). When measurements of Hg°, Hg_R and Hg_{TD} were carried out analysis was performed on board of the *R.V. Belgica* directly after sampling.

Hg_{TD}, Hg_R and Hg° were determined by cold vapour atomic fluorescence spectrometry (CVAFS) using a Au- amalgamation preconcentration step (Gill & Fitzgerald, 1987). Hg° was measured by purging a 1 l sample with Ar for 1 hour at a flow rate of 400 ml min^{-1} and collection on a gold column. Hg_R was measured using $SnCl_2$ as reducing agent; Hg_{TD} was analysed by BrCl oxidation and reduction with $NH_2OH.HCl$ prior to reduction with $SnCl_2$ (Bloom & Crecelius, 1983). Detection limits are respectively 5 pg l^{-1} for Hg°, 15 pg l^{-1} for Hg_R and 50 pg l^{-1} for Hg_{TD}. Total dissolved MMHg ($MMHg_D$) was analysed by aqueous phase ethylation after separating MMHg from the interfering chloride matrix by an extraction with methylene chloride (Bloom, 1989). Hg species are transformed to their volatile ethylderivates, purged out of solution, collected on Carbotrap columns, subsequently removed at 270 °C and transferred to a cryogenic gaschromatographic trap, held in liquid nitrogen. The species are separated by controlled heating of the GC column from –196 °C to 180 °C and the final measurement is made by CVAFS. The detection limit for $MMHg_D$ is 10 pg l^{-1}. Total particulate Hg was measured by digesting the filters in 5 ml HNO_3/HCl (4:1) at 60 °C for 12 hours in teflon digestion vessels; diluting the sample to 50 ml and analysis of a fraction of the sample by CVAFS as described above. Particulate MMHg ($MMHg_P$) was analysed by the combined distillation-ethylation technique described by Horvat et al. (1993). MMHg was distillated from the filter using a H_2SO_4/NaCl mixture heated to 150 °C while a N_2 gas flow of 150 ml min^{-1} passes through the solution. The distillate is collected in an ice bath. After distillation the Hg compounds are analysed in the same way as $MMHg_D$.

Data of temperature, salinity, pH, oxygen and Chlorophyll *a* (Chl*a*) were obtained from the Université Libre de Bruxelles (ULB), the Management Unit for Mathematical Modelling of the North Sea (MUMM) and the Rijksinstituut voor Kust en Zeewater (RIKZ), the Netherlands.

Mercury methylation experiments in sediments

Incubation experiments were carried out on the sediments of the intertidal flat Groot Buitenschoor in January 1990 and August 1990 in order to determine potential methylation rates of mercury as well as biotic and abiotic factors influencing this transformation. A detailed description of the experimental procedure is described elsewhere (Leermakers et al., 1993). A sediment sample is spiked with either Hg^2, Hg^2 and Na_2MoO_4, Hg^2 and SO_4^{-2} or Hg^2 after autoclaving the sediment. The final concentration of Hg in the sediment was 80 μg g^{-1}. Incubations were performed under anoxic conditions in the dark. Periodically subsamples were taken for MMHg analysis, lyophilised and analysed as described above for suspended matter. The HgS fraction was determined using the method of Revis (1989), which is based on the insolubility of HgS (in contrast to other Hg compounds) in HNO_3. The difference between the total (HCl/HNO_3) and the HNO_3 digestion is used as a measure of the HgS fraction.

Sampling and analysis of sediments and polychaete

Surface sediments (0–5 cm) and the polychaete *N. diversicolor* were collected seasonally from 1992 to 1993 at thirteen stations of the Groot Buitenschoor intertidal flat (Muhaya et al., 1997). The sediments were lyophilised, grinded and homogenised before analysis. The polychaete were put in glass dishes containing diluted sea water and were left for 48–60 hours to clean them from the mud on their skin and empty their guts. The animals were then lyophilised and stored at room temperature until analysis.

Sediment samples were digested with 5 ml of a HNO_3/HCl (4:1) acid mixture using a Prolabo microwave digestion procedure. The polychaete are digested with 5 ml HNO_3 and further oxidised with 1 ml BrCl . After dilution total mercury in the digested samples are analysed by cold vapour atomic absorption spectrometry (Mercury Module and Mercury Monitor 3200, Thermo Separation Products).

MMHg in the sediment and polychaete samples was isolated by a combined distillation-ethylation technique as described above for suspended matter (Horvat et al., 1993).

Results

Dissolved and particulate Hg species in estuarine waters

In Figures 2 and 3 the general water quality parameters, respectively the measured Hg species, are plotted. The summer surveys (June 1993 and August 1994) are characterised by high temperatures, anoxic conditions in the lowest salinity range and an increase in oxygen as salinity rises due to mixing with aerated seawater as well as phytoplankton activity. In June 1993 Chl*a* concentrations were elevated throughout the whole estuary, whereas in August 1994 a pronounced phytoplankton bloom was observed in the lower estuary. This phytoplankton bloom resulted in a sharp increase in oxygen concentrations (up to 11.5 mg l^{-1}) as well as an increase in the pH. In autumn (November 1993) low oxygen concentrations are still observed in the upper estuary, whereas during the winter (February 1994) oxygen concentrations do not decrease below 4 mg l^{-1}. Suspended matter concentrations are generally higher in winter than in the summer months.

In February 1994 relatively high concentrations of Hg_{TD} were found in the upper estuary (up to 3 ng l^{-1}). An exponential decrease was observed in the salinity range 0 to 10 psu, followed by a slow decrease in the lower estuary. Hg_R concentrations were low in the upper estuary (20% of the total dissolved Hg) and increase gradually to over 90% at the river mouth. A totally different picture is obtained during the summer cruises of June 1993 and August 1994. During both these cruises Hg_{TD} concentrations were relatively low at the lowest salinities (1 to 1.3 ng l^{-1}). During the June 1993 cruise the Hg_{TD} concentrations rise rapidly to reach a maximum of 2.8 ng l^{-1} at 6 psu followed by a rapid decrease as salinity rises. A second increase of Hg_{TD}, however less pronounced, was observed at 20 psu. At salinities higher than 20 psu, Hg_{TD} decreases gradually from 1.5 ng l^{-1} to 1.0 ng l^{-1}. Hg_R concentrations follow the same pattern as Hg_{TD}. In the upper estuary the fraction of Hg_R is higher than in the winter cruise (55% compared to 20% in winter) but in the lower estuary this fraction is lower (46% compared to 90% in winter). In August 1994 Hg_{TD} concentrations were low in the upper estuary, reach a mid estuarine maximum at 15 psu followed by a decrease towards the mouth. In November 1993 Hg_{TD} concentrations decrease rapidly from 7 to 10 psu and remain almost constant in mid- and lower estuary. The low salinity range was not sampled on this cruise.

Particulate mercury concentrations (Hg_P) do not show any significant seasonal trend. Hg_P concentrations decrease from an upper estuary maximum of 1.6 μg g^{-1} to a seaward end member of 0.4 μg g^{-1}. The concentration decrease is generally not linear; decreasing slowly from 0 to 5 psu salinity, then dropping almost linearly between 5 and 20 psu and again decreasing slowly in the lower estuary. Particulate Hg is the major species in the estuary; Hg_{TD} only accounts for 1 to 6% of the Hg_T present in the water.

Hg° concentrations were generally lower in winter than in summer (average 0.05 ng l^{-1} compared to 0.10 ng l^{-1} in the summer). Hg° accounts for 2 to 18% of the Hg_R in winter and 4 to 46% in summer. A maximum concentration is found at 10 psu in the winter whereas in the summer Hg° concentrations also appear to be related to the phytoplankton pigments. In June 1993 the Chl*a* and Hg° concentrations remained fairly constant over the estuary and increased at the mouth, whereas in August 1994 a first maximum in Hg° concentrations is observed at 12 psu followed by a second maximum at salinities higher than 20 psu, in the region where elevated Chl*a* concentrations were observed. Using all the results of salinities higher than 15 psu, a positive correlation is found between Chl*a* and Hg° ($r = 0.62$).

In winter $MMHg_D$ concentrations were much lower than in summer and autumn (0.011–0.127 ng l^{-1} in winter compared to 0.08–0.6 ng l^{-1} in summer). In winter maximum concentrations are found in the upper estuary (salinities 1–5 psu) and remain relatively constant in the lower estuary. The percentage of $MMHg_D$ ranged from 0.7 to 8% of the Hg_{TD} concentrations. Both the concentrations of $MMHg_D$, as well as the variations with the salinity, are very similar for the June 1993, the August 1994 and the November 1993 cruise. $MMHg_D$ concentrations ranged from 0.08 to 0.6 ng l^{-1} accounting for 7 to 60% of the Hg_{TD}. Comparable to the variation in Hg_{TD}, dissolved MMHg concentrations are low at the lowest salinities and increase at salinities ranging from 8 to 12 psu, followed by a decrease in the mid estuary and a second peak in the lower estuary at 25 to 30 psu. A good inverse correlation between

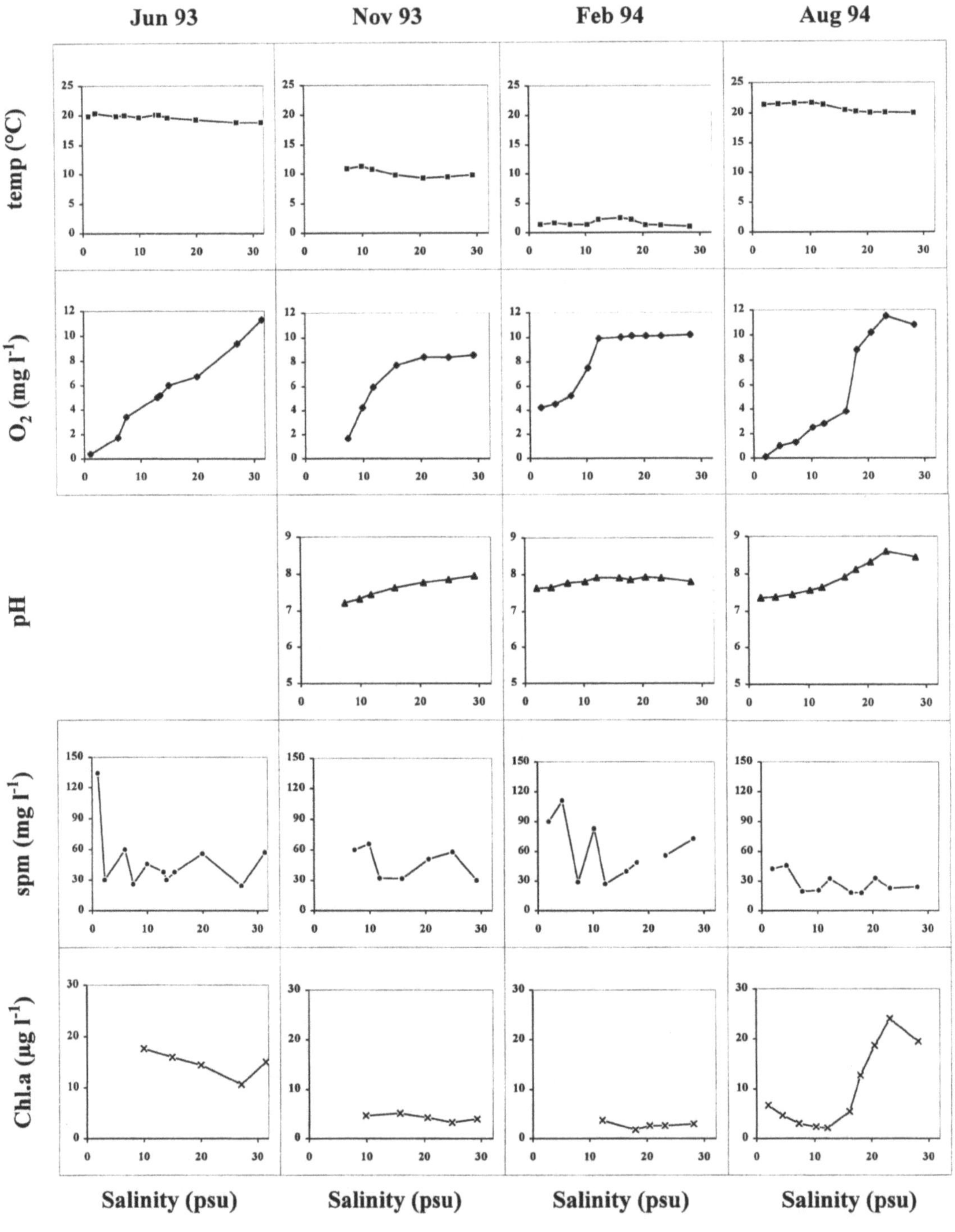

Figure 2. Profiles of temperature, dissolved oxygen, pH, suspended matter and chlorophyll *a* in surface waters of the Scheldt estuary.

Hg_R and $MMHg_D$ was observed ($r = 0.73$ and 0.82 for summer and winter respectively).

Particulate MMHg ($MMHg_P$) was also measured in February 1994, as well as in August 1994. $MMHg_P$ concentrations are higher in the low salinity range (0–

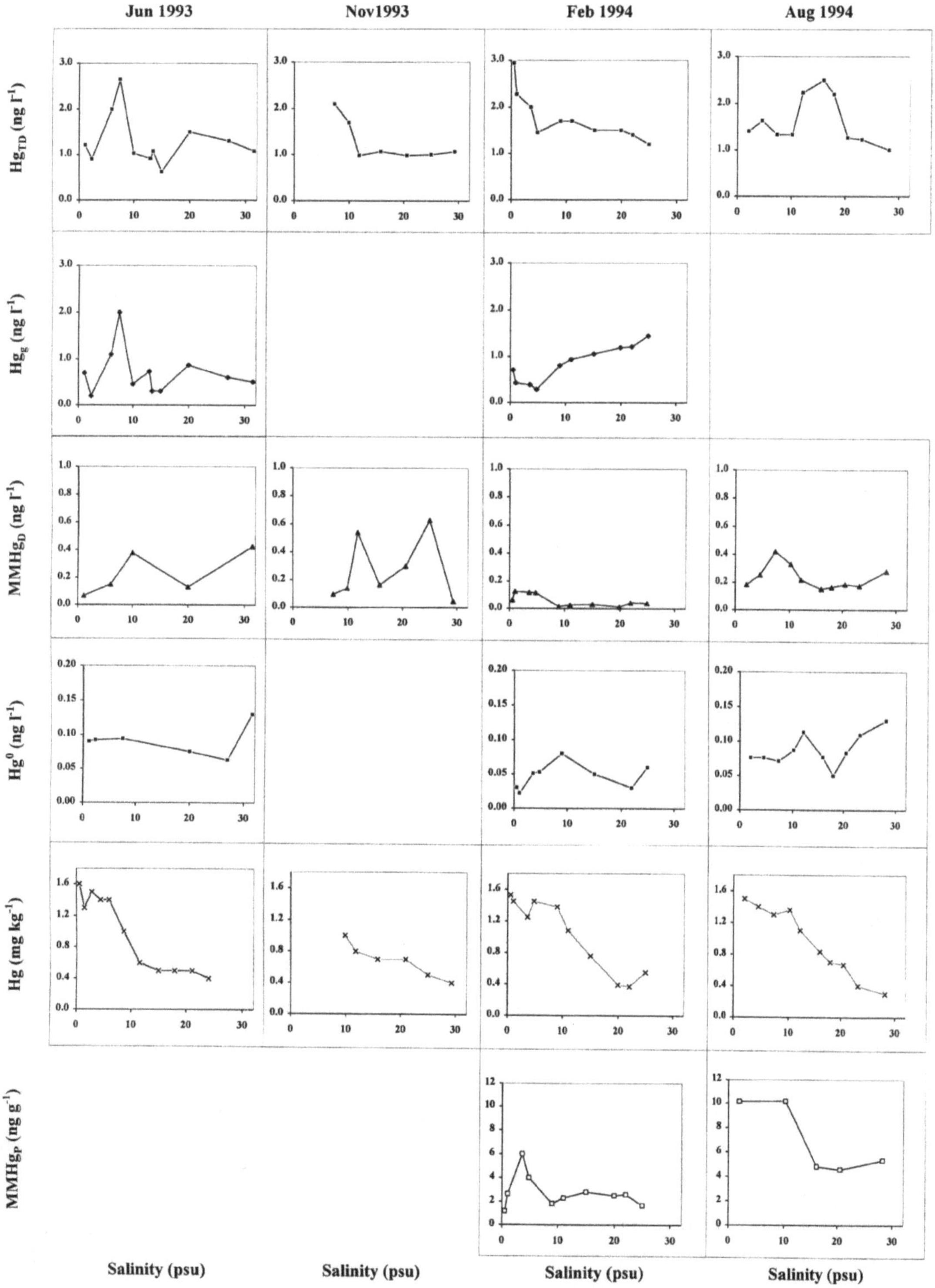

Figure 3. Profiles of dissolved and particulate Hg species in surface waters of the Scheldt estuary.

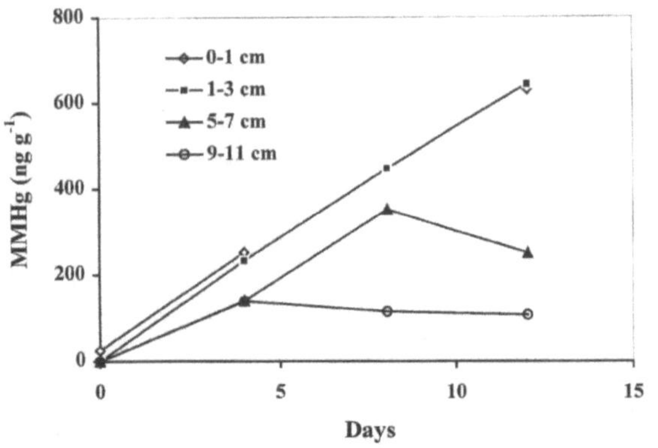

Figure 4. Anaerobic methylation of mercury in estuarine sediments (January 1990), incubation with Hg^2.

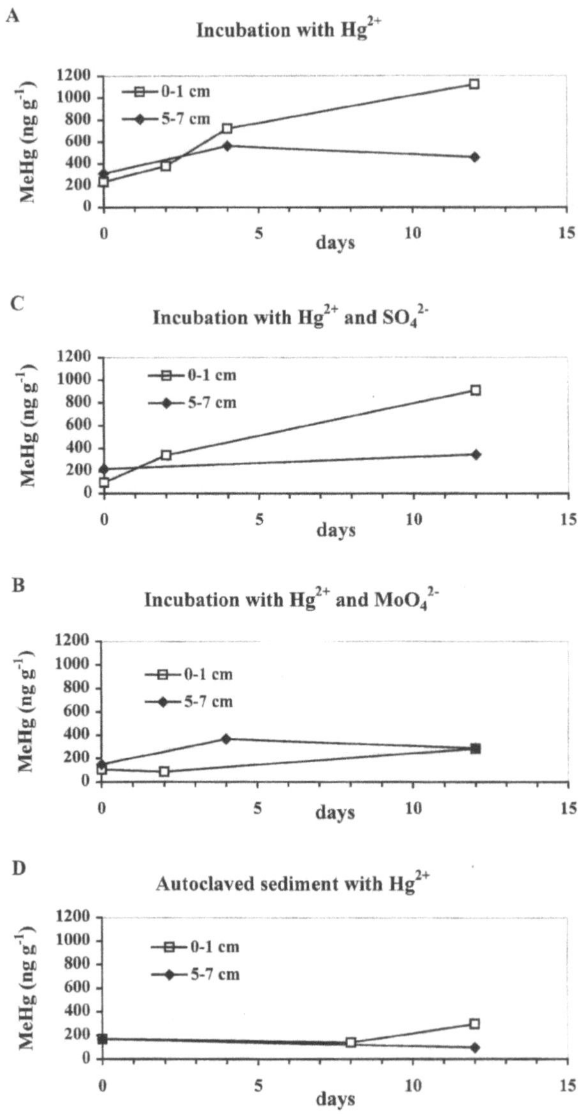

Figure 5. Anaerobic methylation of mercury in estuarine sediments (July 1990).

10 psu) and remain constant in the lower estuary. The concentrations are higher in summer than in winter (an average of 2 ng g^{-1} can be found in lower estuary during winter and 4 ng g^{-1} during summer; the upper estuary maximum is 6 ng g^{-1} in winter and 10 ng g^{-1} in summer). In contrast to total particulate mercury, a much higher fraction of MMHg is found in the dissolved phase. $MMHg_D$ accounts for 20 to 40% of the total MMHg in winter and 30 to 80% in summer. $MMHg_P$ accounts for 0.1 to 0.7% of the Hg_P.

Methylation of mercury in estuarine sediments

The results of the incubation experiments performed on sediments of the intertidal flat Groot Buitenschoor, are shown in Figures 4 and 5 and present potential methylation rates of Hg (the concentration of the Hg spike is approximately eighty times higher than the natural level). Methylation rates were found to be significantly higher in the top layer of the sediment than in the deeper layers (Figure 4). In the top layers, methylmercury concentrations are a linear function of time; saturation is not observed after fourteen days. In the deeper layers, the concentration increases during the first days and then remains constant. Methylation rates were significantly higher in summer (100 ng g^{-1} d^{-1} in the surface layer in July, Figure 5) than in winter (50 ng g^{-1} d^{-1} in the surface layer in January, Figure 4) and compare with reported methylation rates in the literature (15–137 ng g^{-1} d^{-1}; Craig, 1986). Measurements of the HgS fraction showed that about 50% of the added spike was transformed into HgS. Mercury sulphide is inert and not available for methylation under anaerobic conditions (Fagerstrom & Jernelöv, 1971; Compeau & Bartha, 1983; Berman & Bartha, 1986). Addition of Na_2MoO_4, a specific inhibitor of sulphate reducing bacteria, strongly reduced the methylation rate (Figure 5). In this case the methylation rate is comparable to the autoclaved samples. Addition of sulphate did not result in higher methylation rates of Hg although sulphate reduction rates increased from 50 mg l^{-1} d^{-1} to 300 mg l^{-1} d^{-1} (Figure 5). In autoclaved samples a constant MMHg concentration of 200 ng g^{-1} was found. This concentration is substantially higher than that found in the natural sediment (Figure 6) and is due to abiotic methylation of Hg.

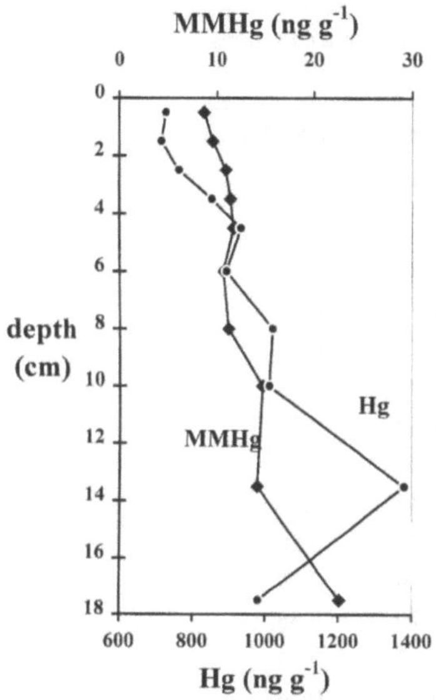

Figure 6. Total Hg and MMHg concentrations in a sediment core of the Groot Buitenschoor intertidal flat (August 1990).

Figure 6 shows the natural mercury concentrations in the sediments. The vertical sediment profiles of Hg and MMHg, obtained by sampling the muddy central part of the intertidal flat Groot Buitenschoor in August 1990, shows an increasing trend in both mercury and methylmercury with depth. Total mercury concentrations varied from 700 ng g^{-1} to 1400 ng g^{-1}, with MMHg concentrations varying from 5 to 22 ng g^{-1}. The ratio of MMHg to total Hg varied between 1 and 1.5%, which is comparable to the ratio found in the incubation experiments. In the top layer relatively higher MMHg/Total Hg ratios were found.

Seasonal variations in Hg methylation in sediments and accumulation in benthic organisms (N. diversicolor)

Total mercury concentrations in surface sediments ranged from 140 to 1900 ng g^{-1}, with 0.8 to 6 ng g^{-1} in the MMHg form accounting for 0.4 to 0.8% of the total mercury present. Both Hg and MMHg concentrations increase with increased organic matter content, increased fine grain size fraction and suboxic to anoxic conditions (Table 2). In contrast Hg accumulation in *N. diversicolor* was higher at stations with sandy sediments compared to stations with muddy sediments. MMHg concentration in *N. diversicolor* ranged from 2.2 to 20.9 ng g^{-1} accounting for an average of 18% of the total mercury (Table 2).

Seasonal variations significantly affected total Hg and MMHg concentrations in sediments and *N. diversicolor*. Total Hg concentrations in sediments are significantly higher in autumn and winter than in spring and summer, whereas MMHg concentrations are lowest in winter compared to the other seasons (Figure 7). On the other hand, total Hg concentrations in *N. diversicolor* were lowest in spring, whereas MMHg concentrations were significantly higher in spring and summer than in autumn and winter (Figure 7).

Discussion

Distribution of Hg between particulate and dissolved phases

In surface waters total Hg ($Hg_T = Hg_{TD} + Hg_P$) ranged from 20 to 200 ng l^{-1}, of which over 90% is bound to suspended matter. The concentrations of Hg in particulate matter are comparable to values recently reported (Coquery & Cossa, 1996) for the Elbe estuary (300–1600 ng g^{-1}) and the Seine (58–1980 ng g^{-1}). Hg_P concentrations decrease gradually from the upper estuary towards the mouth due to mixing of fluvial and relatively uncontaminated marine suspended matter. K_D values, based on average concentrations of dissolved and particulate Hg range from 400 000 to 700 000 (l.kg^{-1}). These values are in good agreement with the values reported by Balls (1989). K_D values in the lower estuary reflect physical adsorption/ desorption processes whereas in the upper estuary precipitation/oxidation of iron, manganese and sulphides also have to be taken in account. As only a small fraction of Hg is present in the dissolved form, any factor affecting the distribution between the dissolved and particulate phase may have an important effect on dissolved Hg concentrations.

The redox condition of the estuary is a crucial factor determining the speciation of dissolved and particulate Hg. Especially in summer months anoxic conditions prevail in the upper estuary, leading to the formation of sulphides in suspended matter and in sediments and to the scavenging of dissolved species (Zwolsman & Van Eck, 1993). Highly anoxic and contaminated sediments have been found on intertidal flats of the upper estuary (Panutrakul & Baeyens, 1991). HgS (K_s 10^{-52}) accounts for a large fraction of the total Hg in the estuarine sediments (Leermakers et al., 1993). As oxygen is slowly restored in the estuary, the reduced compounds in suspended matter and sediments may be oxidised. This oxidation process probably explains the

Table 2. Total mercury and methylmercury concentrations in sediments *N. Diversicolor* of the intertidal flat Groot Buitenschoor (mean concentrations 1992–1993)

Station	% OM	% silt & clay	Tot Hg ng g^{-1}	MMHg ng g^{-1}	% MMHg
Sediments					
A	2.3	13	152.2	1.40	0.94
2	2.5	14.9	187.3	1.73	0.92
3	3	31.3	251.7	1.74	0.73
4	2.5	3.6	175.1	1.71	0.98
6	3.1	25.9	224.0	1.62	0.76
7	4.8	35.2	558.3	2.88	0.55
8	4	28.5	389.4	2.35	0.61
9	6.3	31.2	589.9	3.21	0.56
10	6.2	43.8	503.3	2.79	0.57
12	7.8	52.4	820.2	4.70	0.58
13	8.7	54.3	945.2	4.91	0.54
20	7.6	–	688.4	3.68	0.54
21	7.7	–	532.1	3.41	0.56
Nereis					
A			163.8	13.85	8.3
2			145.2	12.17	10.0
3			123.5	15.33	12.4
4			105.1	20.02	24.1
6			97.1	18.00	21.5
7			91.7	14.32	16.0
8			59.4	20.72	50.1
9			68.4	22.22	42.7
10			109.3	13.92	16.8
12			89.6	12.16	19.2
13			93.2	12.33	16.6
20			73.4	9.35	14.6
21			51.5	12.04	26.4

large Hg_{TD} increase at 7.5 psu observed in the summer cruise of 1993 and at 10 psu in the summer of 1994. It takes place at the dissolved oxygen range of 1.7 to 3.4 mg l^{-1} (1993) and 2.5 to 2.8 mg l^{-1} (1994). The further decrease towards the mouth cannot be described by conservative mixing alone. The sharp decrease in dissolved Hg at 10 psu (1993), respectively 18 psu (1994), corresponding to a dissolved O_2 concentration of 5.5 (1993) respectively 3.8–8.8 mg l^{-1} (1994) may be the result of precipitation of iron and manganese compounds. As an example, dissolved Fe and Mn concentrations measured during the fall 1983 survey are shown in Figure 8 (Baeyens, unpublished results). At a salinity of 12 psu dissolved Mn has dropped to 25% of the concentration prevailing at 1 psu and dissolved Fe concentrations dropped below the detection limit. Hg liberated from particulate matter due to the oxidation of sulphides, may be rescavenged onto the particulate phase by precipitating Fe- and Mn- oxihydroxides.

During the winter dissolved O_2concentrations in the water are sufficiently high to maintain Hg in solution. Elevated Hg_{TD} concentrations are found in the organic rich, high turbidity waters of the upper estuary. The sharp decrease of Hg_{TD} with increasing salinity, corresponding to an O_2 concentration of 5 mg l^{-1}, may also be attributed to the of scavenging on Fe- and Mn oxihydroxides.

Concentrations of Hg_{TD} are similar to levels measured in the Elbe estuary (0.76–3.28 ng l^{-1}; Coquery & Cossa, 1996); the St. Lawrence River (0.48–2.4 ng

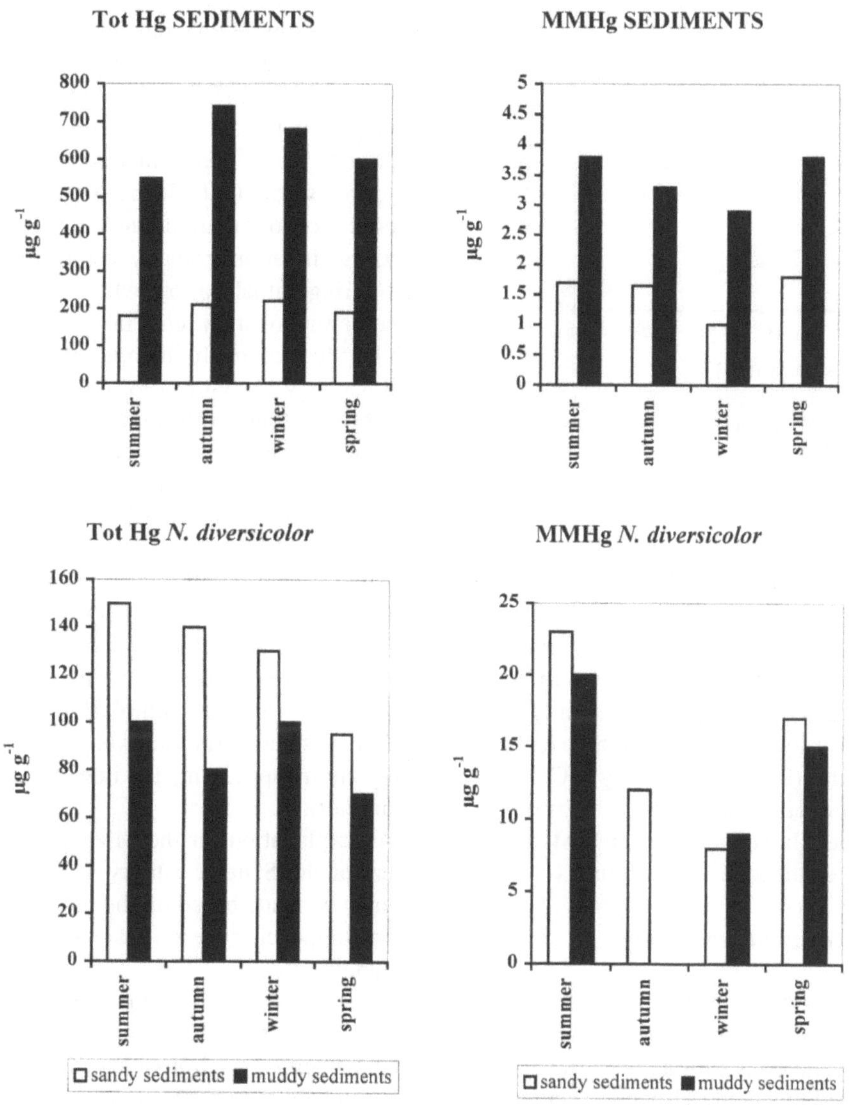

Figure 7. Seasonal variations in total Hg and MMHg concentrations in surface sediments and *N. Diversicolor* of the intertidal flat Groot Buitenschoor (1992–1993).

l^{-1}; Cossa et al., 1988); the Rhone River (0.58–3.16 ng l^{-1}; Cossa & Martin, 1991); the Pettaquamscutt estuary (0.2–1.4 ng l^{-1}; Mason et al., 1993) and the Framvaren Fjord (0.2–1.7 ng l^{-1}; Iverfeldt, 1988) and significantly higher than background levels in offshore surface waters of the North Sea (0.34 ng l^{-1}; Coquery & Cossa, 1996) or the English Channel (0.15 ng l^{-1}; Baeyens, 1998).

Speciation and complexation of Hg

In winter (the February 1994 cruise) the percentage of Hg_R relative to the total dissolved Hg amount, is lowest (10–20%) in the upstream area, where the Hg_{TD} concentrations are highest, and reaches 80 to 90% in the downstream area, where Hg_{TD} concentrations are substantially lower. The high amounts of dissolved organic matter entering the upper estuary by the tributaries of the Scheldt are responsible for the organic complexation of Hg in the upper estuary (Somville & Depauw, 1982). In the lower estuary chlorocomplexes

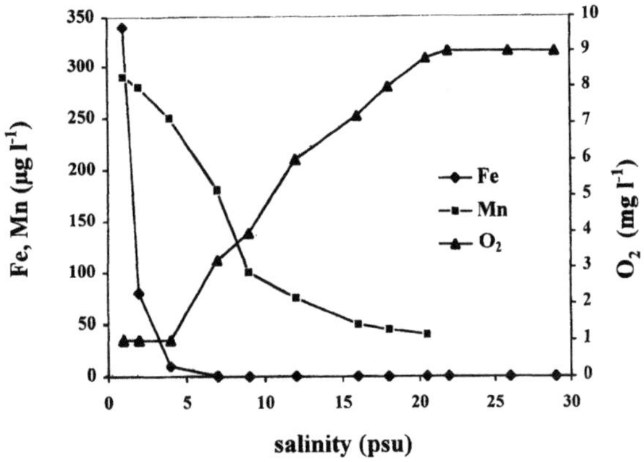

Figure 8. Dissolved Fe and Mn profiles in surface waters of the Scheldt estuary (Autumn, 1983).

of mercury dominate the speciation. During summer (the June 1993 cruise) the Hg_R fraction in the upper estuary (average 55%) was substantially higher than normally expected, which could reflect the influence of the mobilisation processes described above. In the lower estuary, on the other hand, Hg_R only accounts for 46% of the Hg_{TD}. This value is more comparable to the fraction found in the Elbe (21–40%; Coquery & Cossa, 1996). The percentage of Hg_R is also influenced by other dissolved Hg species. In summer higher $MMHg_D$, as well as higher Hg° concentrations are found.

Methylation of Hg in sediments and water

The upper estuary provides a suitable site for the methylation of Hg (low O_2 concentrations, high organic matter contents, high heterotrophic activity and high Hg concentrations in sediments and suspended matter). Methylation reactions occurring in the sediments may also be an important source of MMHg in the water column. The methylation processes in both compartments are probably not very different.

The results of the incubation experiments demonstrate the importance of sulphate reducing bacteria in the formation of MMHg and HgS in anoxic estuarine sediments as has been reported by several other authors (Compeau & Bartha, 1985; Berman et al., 1990; Craig & Moreton, 1986; Kerry et al., 1991). Sulphate reducers are also the dominant decomposers of organic matter in the sediments of Scheldt estuary (Elskens et al., 1991). Addition of Na_2MoO_4 blocked sulphate reduction as well as mercury methylation. High sulphate reduction rates and high sulphate concentrations do not necessarily favour the production of MMHg. A high sulphide production provokes the formation of insoluble HgS and depletion of Hg^2. Furthermore, MMHg reacts abiotically with S^{-2} in water and sediments to form HgS and volatile dimethylmercury (DMHg) which may escape to the atmosphere (Craig & Bartlett, 1978). Methylmercury formation was also found in a sterile (autoclaved) sediment, but in much lower proportions than in a non-sterile sediment. This abiotic methylation probably results from rupture of the cells during autoclavation, whereby enzyme involved in the biomethylation react in a purely chemical way with Hg^2 to form methylmercury (Compeau & Bartha, 1983; Craig & Moreton, 1985). Hg can also be methylated by reactions with humic and fulvic complexes (Craig & Moreton, 1985; Nagase et al., 1982 and 1984) as well as by transalkylation reactions with alkyllead and alkyltin compounds (Bellama et al., 1988; Howell & O'Connor, 1986).

Large seasonal differences were observed in the sulphate reduction rates (SRR) in sediments of the Groot Buitenschoor (Panutrakul, 1993). SRR ranged from 11.5 to 16.8 mmol $SO_4{}^{-2}$ m^{-2} d^{-1} in December and from 80.9 to 680.7 mmol SO_4^{-2} m^{-2} d^{-1} in July. Similar seasonal variations were observed in mercury methylation rates (50 ng d^{-1} in winter and 100 ng d^{-1} in summer).

An explanation for the MMHg profiles in surface waters of the Scheldt estuary in summer and winter can now be made based on the physico-chemical conditions influencing the sulphate reduction rate and Hg availability.

The summer and autumn profiles of $MMHg_D$ show two peaks; the first at a salinity between 7 and 12 psu and the second at a salinity between 25 and 30 psu. The high sulphate reduction rate may be a major process contributing to the first $MMHg_D$ maximum in the upper estuary. The lack of high $MMHg_D$ concentrations in the lowest salinity range can be explained by the presence of dissolved sulphide (0.2–0.9 μm) in the anoxic zone (Zwolsman & Van Eck, 1993) resulting in a scavenging of dissolved Hg^2 species from the water phase, significantly reducing the amount of Hg available for methylation. Both Hg_R and Hg_{TD} are low in the upper estuary during summer. MMHg produced in the summer period in the upper estuary will predominantly remain bound to particulate matter. Concentrations of $MMHg_P$ are elevated in the upper estuary in the summer (10 ng g^{-1}) and decrease to 6 ng g^{-1} in the lower estuary. $MMHg_D$ and $MMHg_P$ concentrations for the February 1994 and August 1994 surveys, both expressed in ng l^{-1} are shown in Figure 9. In

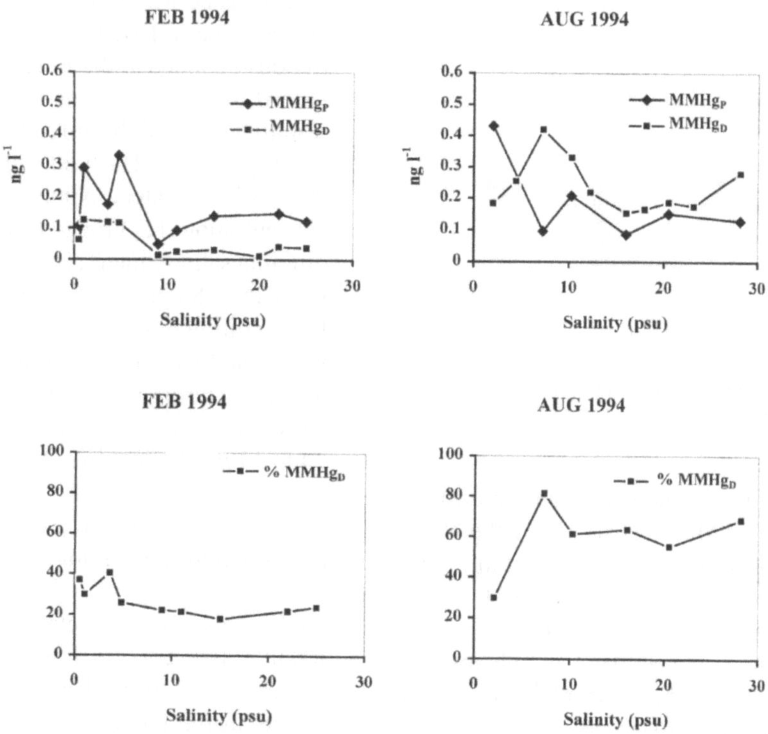

Figure 9. Particulate and dissolved MMHg in surface waters of the Scheldt estuary (February and August 1994).

the summer, the concentration of $MMHg_P$ drops in the salinity range 1–8 psu, associated with an increase in $MMHg_D$. The percentage of $MMHg_D$, relative to the total MMHg, increases sharply from 30 to 80%. In the winter, the upper estuary never becomes anoxic, though a significant undersaturation is still observed (4.2 mg O_2 l^{-1}). Both $MMHg_D$ and $MMHg_P$ increase in the suboxic upper estuarine waters. $MMHg_P$ concentrations (expressed as ng g^{-1}) range from 1–6 ng g^{-1} and are lower than in summer but, due to the higher concentrations of suspended particulate matter in the winter (average 60 mg l^{-1} compared to 25 mg l^{-1} in the summer), the particulate MMHg fraction is more important in the winter ($MMHg_D$ ranges from 20 to 40%). In the lower estuary, both $MMHg_D$ and $MMHg_P$ remain constant in winter, whereas in June 1993, November 1993 and August 1994 an increase in $MMHg_D$ was also observed between 25 and 30 psu. Although the origin of this increase is not clear, it may be related to plankton blooms and associated bacterial populations. Significantly higher amounts of Hg°, and lower amounts of Hg_R were also found in that zone. *Phaeocystis sp.*, either in its colonial or in its single cell-stage, are the algal species dominating the plankton of the North Sea in spring. Bacterial growth on the mucilage is inhibited in rapidly growing cultures by acrylic acid produced in the mucilage (Davidson and Marchant, 1987), but once the growth of *Phaeocystis sp.* and production of acrylic acid slow, bacterial numbers increase leading to the solubilisation of the mucilage and the release of metals, such as for example Mn^2 (Davidson & Marchant, 1987). Reducing conditions are created and at the same time bacterial population densities increase: such conditions are favourable for Hg methylation.

$MMHg_P$ accounts for 0.1 to 0.7% of the Hg_P in winter and 0.6 to 1.8% in summer. This is comparable to the fraction found in the surface sediments of the Scheldt (0.5 to 1.5%), as well as the fraction found in the incubation experiments, and are somewhat lower than the average of 3% found for surface waters of the North Sea (Coquery & Cossa, 1996).

Demethylation and volatilisation of Hg

Hg° concentrations range from 0.02 to 0.13 ng l^{-1} and are comparable to concentrations found in the surface waters of the North Sea (average 0.052 ng l^{-1};

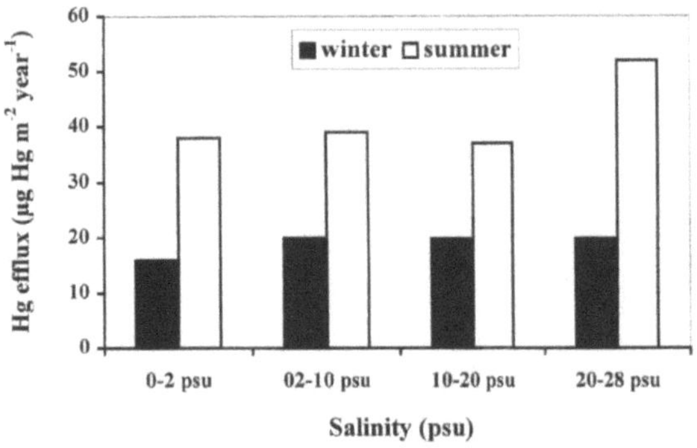

Figure 10. Gaseous mercury evasion fluxes from the Scheldt estuarine waters (1993–1994).

Coquery & Cossa, 1996), the Pettaquamscutt estuary (< 0.005–0.08 ng l^{-1}; Mason et al., 1993), the Northern Atlantic (0.014–0.17 ng l^{-1}; Mason et al., 1995) and the Equatorial Pacific (0.006–0.045 ng l^{-1}; Kim & Fitzgerald, 1986). In the lower estuary, a positive correlation was found between Hg° and phytoplankton pigments, suggesting that phytoplankton may be involved in the formation of Hg° (Mason et al., 1993). In the upper estuary the correlation between Hg° and Chl*a* is absent and Hg° may result from decomposition of methylmercury (Mason et al., 1995), the reduction of Hg by micro-organisms (Nakamura et al., 1990) or degradation of Hg humic complexes (Iverfeldt, 1984).

Surface waters were consistently greater than saturation levels, resulting in an evasion flux from the water. A gas transfer model for the evasion flux of mercury, which takes into account wind speed and wave spectra was proposed by Baeyens et al. (1991). This transfer flux of Hg° across the air-sea interface can in a general way be described by Fick's first law:

$$F = -\rho\, V\, (C_Z - C_S),$$

with:

- F: Hg flux
- C_Z: Hg concentration in the fluid with the largest resistance to the flux
- C_S: Hg° concentration at the interface between both fluids
- ρ: density
- V: transfer velocity.

Transfer velocities (V) range from 0.42 m d^{-1} for calm weather conditions to 10.9 m d^{-1} at high wind speeds and wave breaking conditions (Baeyens et al., 1991). The concentration of Hg° in the surface microlayer (C_S), in equilibrium with the atmosphere, can be calculated using the Henry coefficient of 0.32 for Hg (Sanamasa, 1975). Total gaseous mercury concentrations obtained at the Belgian coast (Decadt , 1986) ranged from 1.9 to 4.2 ng m^{-3} (average 3.0 ng m^{-3}), resulting in a equilibrium concentration at the interface of 10 pg l^{-1} (C_S). The fluid with the largest resistance to the flux is the water column. As average dissolved Hg° concentrations (C_Z) range from 55 to 90 pg l^{-1} in the different salinity intervals, the Hg sursaturation ranges from 45 to 80 pg l^{-1}. In Figure 10 the evasion fluxes for the different salinity zones have been calculated in winter and summer using an average wind speed of 8 m s^{-1}. During winter Hg evasion fluxes range from 17 to 24 μg Hg° m^{-2} yr^{-1} and during summer from 24 to 33 μg Hg° m^{-2} yr^{-1}.

Accumulation of methylmercury in benthic organisms in relation to methylation of mercury in sediments

High levels of heavy metals such as Cd and Pb (Panutrakul & Baeyens, 1991; Srinetr, 1992) and Hg (Leermakers et al., 1993) have been recorded in the sediments in the intertidal mudflat Groot Buitenschoor, and benthic organisms have significantly decreased in species diversity and biomass in the last thirty years (Van Impe, 1985; Ysebaert et al., 1993). Of the remaining species, *N. diversicolor* and *Corphium volutator* account for up to 95% of the benthic population.

Accumulation of heavy metals in benthic organisms is a complex phenomenon, especially in contaminated estuaries, where the remaining organisms are resistant forms which have adapted themselves to the contaminants. Accumulation of Hg was higher in *N. diversicolor* from sandy sediments, less contaminated by Hg, than in *N. diversicolor* from organic rich sediments with higher Hg concentrations. A number of factors may be responsible for the observed inverse relationship between mercury concentrations found in sediments and in the *N. diversicolor.* Sediment bound mercury complexed to particulate organic matter may significantly reduce the bioavailability of the metal (Langston, 1986; Bryan & Langston, 1992). Porewaters may also be an important source of mercury to the polychaeta (Srinetr, 1992). Metal uptake from porewaters may occur either through membrane diffusion or via ingestion. Porewater concentrations of several heavy metals (Cd, Pb,...) have been shown to be significantly higher in the sandy stations of the Groot Buitenschoor than the muddy stations (Panutrakul, 1993). Suboxic conditions may stimulate the liberation of Hg and MMHg bound to oxy-hydroxides of Fe and Mn, whereas anoxic conditions will result in the scavenging of mercury from solution. *N. diversicolor* living in the more contaminated sediments may also be regulating Hg more efficiently, by reducing the permeability and applying an efficient secretory system, or by active demethylation (Langston, 1986; Rainbow, 1990; Miron et al., 1994). Metal binding proteins have been demonstrated in various polychaete species (Eriksen et al., 1988) and *N. diversicolor* has been shown to endure high levels of heavy metals such as Hg (Dennaï et al., 1986).

The higher total mercury concentrations in sediments found in autumn and in winter may be explained by the scavenging of Hg by sulphides produced by sulphate reducing bacteria. As Panutrakul (1993) clearly showed, sulphate reducing rates are highest in summer, but Fe and Mn oxihydroxides present in the sediments will reoxidise the produced S^{-2} so that no metal accumulation occurs. Only when these oxihydroxides are exhausted, or nearly exhausted, S^{-2} prevails and reacts with metals such as Hg, to stock them in the solid phase. The higher methylmercury concentrations in spring and summer can be explained by the influence of the temperature on the sulphate reducing rates and hence the methylation rate of mercury. Increased oxygen concentrations in spring may also oxidise and solubilise reduced mercury compounds, making these compounds available as substrates for methylation. Although there was no direct relationship between methylmercury concentrations in the sediments and methylmercury concentrations in *N. diversicolor* on yearly basis, increased methylmercury concentrations in spring and summer also resulted in increased mercury concentrations in *N. diversicolor*. This may indicate that either the freshly formed methylmercury is more available to the organisms, or that the metabolic activity and feeding behaviour of the *N. diversicolor* have important effects on the accumulation of methylmercury. Similar observations have been reported for the nymphs of the burrowing mayfly *Hexagenia rigida* (Odin et al., 1994).

Conclusions

Hg is predominantly bound to particulate matter in the surface waters of the Scheldt estuary (> 90%). Hg_P concentrations decrease progressively from the upper to the lower estuary and are not affected by seasonal changes. Dissolved mercury species on the other hand have large seasonal variations, governed essentially by the redox conditions in the estuary. Large seasonal variations are found in $MMHg_D$ concentrations with much higher concentrations in summer and autumn compared to winter. $MMHg_D$ can account for up to 60% of the Hg_{TD} in the estuarine waters in summer. $MMHg_P$ accounts for 0.1 to 0.7% of the Hg_P in winter and 0.6 to 1.8% in summer. This is comparable to the fraction found in the surface sediments of the Scheldt (0.5 to 1.5%). Supersaturation of Hg° was found throughout the whole estuary, with higher concentrations in summer than in winter. In the lower estuary a correlation between Hg° and Chl*a* was found.

Distribution of total Hg and MMHg in surface sediments and polychaete *N. diversicolor* of the intertidal flat Groot Buitenschoor is related to sediment composition, organic matter content and other factors, such as sulphate reduction rates and redox conditions. Hg and MMHg accumulation in *N. diversicolor* are not proportional to sediment concentrations due to the effect of sediment composition on the bioavailability of the Hg species and due to possible regulation of MMHg accumulation. Seasonal variations affected Hg and MMHg concentrations in sediments and *N. diversicolor*. Higher total Hg concentrations were found in the sediments in autumn and winter compared to spring and summer, whereas MMHg concentrations were higher in spring and summer, in accordance to variations in the sulphate reduction rate. In *N. diversicolor* total Hg concentrations were lowest in spring, whereas MMHg concentrations were higher in spring and summer. MMHg accumulation seems to be dependent on MMHg production and bioavailability in the sediments as well as on the metabolic activity of the *N. diversicolor*.

Acknowledgements

The authors wish to thank the Ministry of Environment Public Health and the Management Unit of the Mathematical Model North Sea (MUMM) for providing ship time on the *R.V. Belgica* and for the acquisition of physico-chemical monitoring data and the Rijksinstituut voor Kust en Zee (RIKZ, the Netherlands) for ship time on the *R.V. Argus* as well as for the acquisition of physico-chemical monitoring data. The authors would also like to thank the officers and the crew of the *R.V. Belgica* and the *R.V. Argus* for their efforts in our regard; the Belgian Science Policy Office (DWTC) for financing the EUROTRAC-ASE project; the Institute of Scientific Research for Industry and Agriculture (IWONL) for providing a research scholarship for Carine Meuleman and the Algemeen Bestuur voor Ontwikkelingssamenwerking (ABOS) for the research scolarship of Bamba Muhaya.

References

Andersen, J. & T. Niilonen (eds), 1995. North Sea Progress Report, 4th International Conference on the Protection of the North Sea, Danish EPA, Copenhagen, Denmark, 247 pp.

Baeyens, W., 1992. Speciation of mercury in different compartments of the environment. TRAC, Trends Anal. Chem. 11: 245–254.

Baeyens, W. & M. Leermakers, 1996. Particulate, dissolved and methylmercury budgets for the Scheldt estuary (Belgium and the Netherlands). In W. Baeyens, R. Ebinghaus & O. Vasiliev (eds), Global and Regional Mercury Cycles: Sources, fluxes and mass balances, Kluwer Academic Publishers: 285–301.

Baeyens, W., 1998. Evolution of trace metal concentrations in the Scheldt estuary (1978–1995). A comparison with other estuarine and ocean levels. Hydrobiologia 366: 157–167.

Baeyens, W., G. T. M. Van Eck, C. Lambert, R. Wollast & L. Goeyens, 1998. General description of the Scheldt estuary. Hydrobiologia 366: 1–14.

Balls, P. W., 1988. The control of trace metal concentrations in coastal seawater through partitioning onto suspended matter. Neth. J. Sea Res. 22: 213–218.

Bellama, M., K. L. Jewett, W. F. Manders & J. D. Nier, 1988. A comparison of the rates of methylation of Hg(II) species in aquatic media by various organitin and organosilicon moities. Sci. Total Envir. 73: 39–51.

Berman, M., T. Chase & R. Bartha, 1990. Carbon flow in mercury biomethylation by desulfovibrio desulfuricans. Appl. Envir. Microbiol. 56: 298–300.

Bloom, N. & E. Crecelius, 1983. Determination of Hg in seawater at sub nanogram per liter levels. Mar. Chem. 14: 49–59.

Bloom, N. & W. F. Fitzgerald, 1988. Determination of volatile Hg species at picogram levels by low temperature gas chromatography with CVAFS detection. Anal. Chim. Acta. 208: 151–161.

Bloom, N., 1989. Determination of picogram levels of methylmercury by aqueous phase ethylation, followed by cryogenic gas chromatography with cold vapour atomic fluorescence detection. Can. J. Fish. Aquat. Sci. 46: 1131–1140.

Bryan, G. W. & W. J. Langston, 1992. Bioavailability, accumulation and effects of heavy metals in sediments with special reference to United Kingdom estuaries: a review. Envir. Pollut. 76: 89–131.

Coquery, M. & D. Cossa, 1996. Mercury speciation in surface waters of the North Sea. Neth. J. Sea Res., in press.

Compeau, G. & R. Bartha, 1985. Sulfate reducing bacteria: principal methylators of mercury in anoxic estuarine sediments. Appl. Envir. Microbiol. 50: 498–502.

Cossa, D., C. Gobeil & P. Coureau, 1988. Dissolved mercury behaviour in the Saint Lawrence estuary. Estuar. coast. Shelf Sci. 26: 227–230.

Cossa, D. & J. M. Martin. 1991. Mercury in the Rhone delta and adjacent marine areas. Mar. Chem. 36: 291–302.

Craig, P. J. & P. D. Bartlett, 1978. The role of H2S in the environmental transport of Hg. Craig Nature 275: 635–638.

Craig, P. J., 1986. Organomercury compounds in the Environment: Principals and Reactions. Longman Group Ltd. Essex, 362 pp.

Craig, P. J. & P. A. Moreton, 1986. Total mercury, methylmercury and sulfide levels in British estuarine sediments III. Wat. Res. 20: 1111–1118.

Davidson, A. T. & H. J. Marchant, 1987. Binding of manganese by Antartic Phaeocystis pouchetii and the role of bacteria in its release. Mar. Biol. 95: 481–487.

Decadt, G., 1985. Contribution to the study of the biogeochemical cycle of mercury in the Southern Bight of the North Sea. PhD Dissertation. Vrije Universiteit Brussel, 172 pp.

Decadt, G., W. Baeyens, D. Bradley & L. Goeyens, 1985. Determination of methylmercury in biological samples by semiautomated headspace analysis. Anal. Chem. 57: 2788–2791.

Dennaï, N., N. Dhanaut-Coutois, J. M. Bouquegneau & A. Nejmeddine, 1986. Effects du cadmium et du mercure sur un ver marin (*Nereis diversicolor*). Mécanismes de détoxification. C. R. Acad. Sci., Ser. 302: 489–494.

Devolder, M., Ph. D'Hondt & G. Verreet, 1991. Stofdossier 1: Kwik. Ministerie van Volksgezondheid en Leefmilieu. Beheerseenheid Mathematisch Model Noordzee, 101 pp.

Ericksen, K. D. H., H. L. Daae & R. A. Andersen, 1988. Evidence of presence of heavy metal binding proteins in polychaete species. Comp. Biochem. Physiol. A. 91: 377–384.

Gill, G. A. & W. F. Fitzgerald, 1987. Picomolar Hg measurements in seawater and other materials using stannous chloride reduction and two stage amalgamation with gas phase detection. Mar. Chem. 20: 227–243.

Gilmour, C., E. A. Henry & R. Mitchell, 1992. Sulfate stimulation of mercury methylation in freshwater sediments. Envir. Sci. Technol. 26: 2281–2287.

Horvat, M., N. Bloom & L. Liang, 1993. Comparison of distillation with other current isolation methods for the determination of methylmercury compounds in low level environmental samples. Part 1. Sediments. Anal. Chim. Acta. 281: 135–152.

Howell, G. N. & M. J. O'Connor, 1986. Methylmercury generation in seawater by transmethylation of organo-lead and -tin. Aust. J. Chem. 39: 1167–1175.

Iverfeldt, A. & O. Lindqvist, 1984. The transfer of mercury at the air-sea interface. In W. Brusart & G. H. Jirka (eds), Gas Transfer at Water Surfaces, D. Reidel Publishing Company: 533–538.

Iverfeldt, A., 1988. Mercury in the Norwegian fjord Framvaren. Mar. Chem. 23: 441–456.

Kerry, A., P. M. Welbourn, B. Prucha & G. Mierle, 1991. Mercury methylation by sulfate reducing bacteria from sediments of an acid stressed lake. Wat. Air Soil Pollut. 56: 565–575.

Kim, J. P. & W. F. Fitzgerald, 1986. Sea-air partitioning of mercury in the Equatorial Pacific Ocean. Science 231: 1131–1133.

Langston, W. J., 1986. Metals in sediments and benthic organisms in the Mersey estuary. Estuar. coast. Shelf Sci. 23: 239–261.

Leermakers, M., P. Lansens & W. Baeyens, 1990. Storage and stability of inorganic and methylmercury solutions. Fresenius' J. Anal. Chem. 336: 656–662.

Leermakers, M., M. Elskens, S. Panutrakul, F. Monteny & W. Baeyens, 1993. Geochemistry of mercury in an intertidal flat of the Scheldt estuary. Neth. J. Aquat. Ecol. 27: 267–277.

Mason, R. P., W. F. Fitzgerald, J. Jr. Hurley, P. L. Donaghay & J. M. Sieburth, 1993. Mercury biogeochemical cycling in a stratified estuary. Limnol. Oceanogr. 38: 1227–1241.

Mason, R. P., K. R. Rolfhus & W. F. Fitzgerald, 1995. Methylated and elemental mercury cycling in surface and deep ocean waters of the Northern Atlantic. Wat. Air Soil Pollut. 80: 665–677.

Miron, G., V. Brock & E. Kristensen, 1994. Effects of mercury on the ventilation behaviour of the polychaete *Nereis virens* (Sar.). J. Exp. Mar. Biol. Ecol. 184: 67–81.

Muhaya, B., M. Leermakers & W. Baeyens, 1997. Total mercury and methylmercury in sediments and polychaete *Nereis Diversicolor* at the Groot Buitenschoor (Scheldt estuary, Belgium). Wat. Air Soil Pollut. 94: 109–123.

Nagase, H., Y. Ose, T. Sato & T. Ishikawa, 1984. Mercury methylation by compounds in humic matter. Sci. Total Envir. 73: 29–38.

Nakamura, K., M. Sakamoto, H. Uchiyama & O. Yagi, 1990. Organomercurial-volatilizing bacteria in the Hg polluted sediment of Minamata Bay, Japan. Appl. envir. Microbiol. 56: 304–305.

Odin, M., A. Feurtet- Mazel, F. Ribeyre & A. Boudou, 1994. Actions and interactions of temperature, pH and photoperiod on mercury bioaccumulation by nymphs of the burrowing mayfly Hexagenia Rigida, from the sediment contamination source. Envir. Toxicol. Chem. 13: 1291–1302.

Panutrakul, S. & W. Baeyens, 1991. Behaviour of heavy metals in a mud flat of the Scheldt estaury, Belgium. Mar. Pollut. Bull. 22: 128–134.

Panutrakul, S., 1993. Impact of the sulfur cycle on the mobilization of heavy metals in intertidal flat sediments. PhD Dissertation, Vrije Universiteit Brussel, 222 pp.

Rainbow, P. S., 1990. Heavy metal levels in marine invertebrates. In R. W. Furness & P. S. Rainbow (eds), Heavy metals in the marine environment. CRC Press, Inc. Boca Raton: 67–79.

Revis, N. W., T. R. Osborne, G. Holswoth & C. Hadden, 1989. Distribution of Hg species in soil from a Hg- contaminated site. Wat. Air Soil Pollut. 45: 105–113.

Sanemasa, I., 1975. The solubility of elemental mercury vapor in water. Bull. Chem. Soc. Jpn. 48: 1795–1798.

Somville, M. & N. De Pauw, 1982. Influence of temperature and river discharge on water quality of the Western Scheldt estuary. Wat. Res. 16: 1349–1356.

Srinetr, V., 1992. The spatial diatribution of sediment bound metals and their relationships to bioavailability. A case study on the Groot Buitenschoor, Scheldt estuary. M. S. Dissertation, Vrije Universiteit Brussel, 155 pp.

Van Impe, J., 1985. Estuarine pollution as a probable cause of increase of estuarine birds. Mar. Pollut. Bull. 16: 271–276.

Ysebaert, T., P. Meire, D. Maes & J. Buys, 1993. The benthic macrofauna along the estuarine gradient of the Scheldt estuary. Neth. J. Aquat. Ecol. 27: 327–341.

Zwolsman, J. J. G. & G. T. M. Van Eck, 1993. Dissolved and particulate trace metal geochemistry in the Scheldt estuary, S. W. Netherlands (water column and sediments). Neth. J. Aquat. Ecol. 27: 287–300.

Hydrobiologia **366:** 81–90, 1998.
W. F. J. Baeyens (ed.), Trace Metals in the Westerschelde Estuary.

Effect of organic complexation on the behaviour of dissolved Cd, Cu and Zn in the Scheldt estuary

Willy Baeyens, Leo Goeyens, Frank Monteny & Marc Elskens
Department of Analytical Chemistry, Free University of Brussels, Pleinlaan 2, B-1050 Brussels, Belgium

Abstract

Samples from the Scheldt estuary have been assayed for dissolved Cd, Cu and Zn using differential pulse anodic stripping voltammetry, either as total (after UV irradiation) or labile concentrations. Under these experimental conditions, labile concentrations ranged between 51–65% of total Cu, 16–66% of total Zn and 53–91% of total Cd. The metal–organic interactions were assessed by evaluating (a) the distribution coefficient K_d for the distribution of the metals between the liquid phase (complexation) and their binding to particulate matter, and (b) the competitive effect exerted by inorganic complexing ligands using a multi–element interaction model. The proportion of organically bound metals (strong and labile) was estimated, in this speciation scheme, to range from 86 to 99% for Cu, from 90 to 96% for Zn, and from 10 to 35% for Cd. From the dissolved organic carbon distribution in the Scheldt ($\leq$ 10 mg C l^{-1}) and taking into account competition from major cations Ca and Mg, free ligand concentrations available for heavy metal complexation were estimated to be $\leq$ 0.15 mg C l^{-1}. With these values, conditional stability constants for the chelation of Cu, Zn and Cd were calculated assuming either a single-step or a two-step complexation in the dissolved phase. Given the assumptions made in these models, stability constants in the range of $10^{7.8}$–$10^{10.6}$ for Cu, $10^{7.0}$–$10^{9.1}$ for Zn and $10^{6.9}$–$10^{8.9}$ for Cd were obtained. The relevance of these data to previous *in vitro* and *in situ* studies is discussed taking into consideration current concepts of metal binding affinity for organic ligands.

Introduction

The distribution of Cd, Cu and Zn between the dissolved and particulate matter phase has been discussed in a related paper (Baeyens et al., 1998). The behaviour of these trace metals in the Scheldt estuary is dominated by remobilisation processes from suspended matter (Wollast, 1976; Zwolsman & van Eck, 1991). During the summer period, both the dissolved Cu and Cd concentrations are found to decrease concomitantly with the dissolved oxygen content of the water in the area of maximum turbidity, suggesting that redox processes control their behaviour. Two redox processes have been identified: (a) metal sulphide formation in the water column or close to the water/sediment interface (Zwolsman & van Eck, 1991; Panutrakul & Baeyens, 1991), the sulphide being probably introduced in the water column through porewater diffusion, and (b) coprecipitation by dissolved iron and manganese compounds released from the sediments. The reduction of iron and manganese occurs in the Scheldt especially in the organic rich sediments found in the upper and mid-estuarine region (Elskens et al., 1991).

In the case of dissolved Zn, a linear decrease from the riverine end member towards the river mouth has been observed (Monteny et al., 1993). It was further suggested that scavenging of the dissolved Zn compounds was due to (a) the increasing amount of fine particles in the area of maximum turbidity offering an enhanced adsorption surface, and (b) freshly discharged or precipitated particulate organic matter containing large amounts of functional groups with high adsorption affinity (Baeyens et al., 1998).

Extensive mobilisation of dissolved metals was observed once the particles left the maximum turbidity area at 20 psu for copper, 15 psu for cadmium and zinc (Monteny et al., 1993). The mobilisation processes involved include degradation, oxidation and desorption. Exo-enzymes produced by microbial organisms can degrade particulate organic carbon and simultaneously release scavenged metals. Oxidation of authigenic metal monosulphides is a second pathway that can explain the increase of dissolved metals, in the mid to downstream estuary especially for Cd and Cu. Resuspension of fine-grained sediments can be a source of metal sulphides to the water column. Progressive shifting of the oxic borderline, hence of metal mobilisation due to oxidation further upstream, was observed from fall to winter (Monteny et al., 1993). Finally, the enhanced competition between alkaline and earth alkaline ions with the adsorbed metals and the gradual decrease of adsorption surfaces after the zone of high turbidity contribute both to the desorption flux from the particulate phase (Mantoura et al., 1978). However, to keep metal ions into solution, complexation by organic ligands might also be a leading factor (Houghton, 1979; Valenta et al., 1986, van den Berg et al., 1987). It is now well established that speciation measurements are required to investigate into the toxicity of metals for aquatic organisms, and to understand trace metal transport in rivers and estuaries (Florence, 1989).

In this report, dissolved metal-organic interactions are discussed with regard to the competitive effect exerted by complexing ligands. A limited speciation of the dissolved metal burden was carried out on the samples. It was assumed that the fraction which was determined by electrochemical analysis (DPASV) at pH = 1 consisted of free and labile bound metal, including inorganic colloids, while the fraction determined at pH = 1 after UV-destruction represented metal bound in inert organic complexes and to organic colloids, according to the usual ASV labile/non-labile discrimination (Florence, 1989). The speciation of Cd, Cu and Zn was subsequently assessed by determining ligand concentrations and conditional stability constants for these metals, using a multi-element interaction model. The competition between complexation of metals in the dissolved phase and their binding to particulate matter has also been considered.

Material and methods

Data collection and treatment

Data presented in this paper have been obtained during five cruises on the Scheldt estuary from spring 1982 and spring 1983 (12/05/1982; 25/05/1982; 24/01/1983; 08/03/1983 and 18/05/1983). Determination of trace metals was performed by differential pulse anodic stripping voltammetry (DPASV). The ASV labile metal fraction, measured after acidification of the sample at pH 1, included the free metal ion and metal dissociated from complexes and colloids in the diffusion layer, whereas total metal concentrations were determined after acidification and UV-irradiation. The difference between the two fractions was assumed to represent metal bound in inert organic complexes and to organic colloids (Florence, 1989). Full details of the analytical procedures are set out in publications by Baeyens et al. (1998).

Longitudinal profiles of total dissolved metals *versus* salinity were obtained for each cruise using polynomial regressions. Average metal concentrations at seven salinity intervals (1, 3.5, 7.5, 12.5, 17.5, 25 and 32.5 psu) were subsequently fitted and used in speciation models to determine organo-metal interactions (see Equation 3–4).

Inorganic metal complexation was calculated with thermodynamic equilibrium data of Woods & Garrels (1987). In this model, the activity coefficients of all species have been calculated for consistency by means of the Davies equation (Ramette, 1981). Although this approach may be not satisfactory at high ionic strengths, it is a reasonable approximation for a preliminary survey of speciation. Unless otherwise stated, the overall relationship between the conditional stability constant β, used in the model, and the thermodynamic equilibrium constant at infinite dilution in pure water K^{∞} is as follows:

$$\beta = K^{\infty} \cdot \frac{\gamma_{metal} \cdot \gamma_{ligand}}{\gamma_{complex}} \qquad (1)$$

For hydroxide complexes, it was more convenient to use mixed stability constants defined by:

$$K' = \frac{[Complex]}{[Metal].a_{OH}} \text{ and } K' = K^{\infty} \cdot \frac{\gamma_{metal}}{\gamma_{complex}} \qquad (2)$$

The resulting set of non-linear equations provided by the mass law and the mole balance equations (see Equation 5) was solved by use of the Newton-Raphson iteration procedure (Morel & Morgan, 1972). Free metal and free organic ligand concentrations were subsequently used to assess values for the conditional stability constants of the various organo-metal complexes (see Equations 6–7).

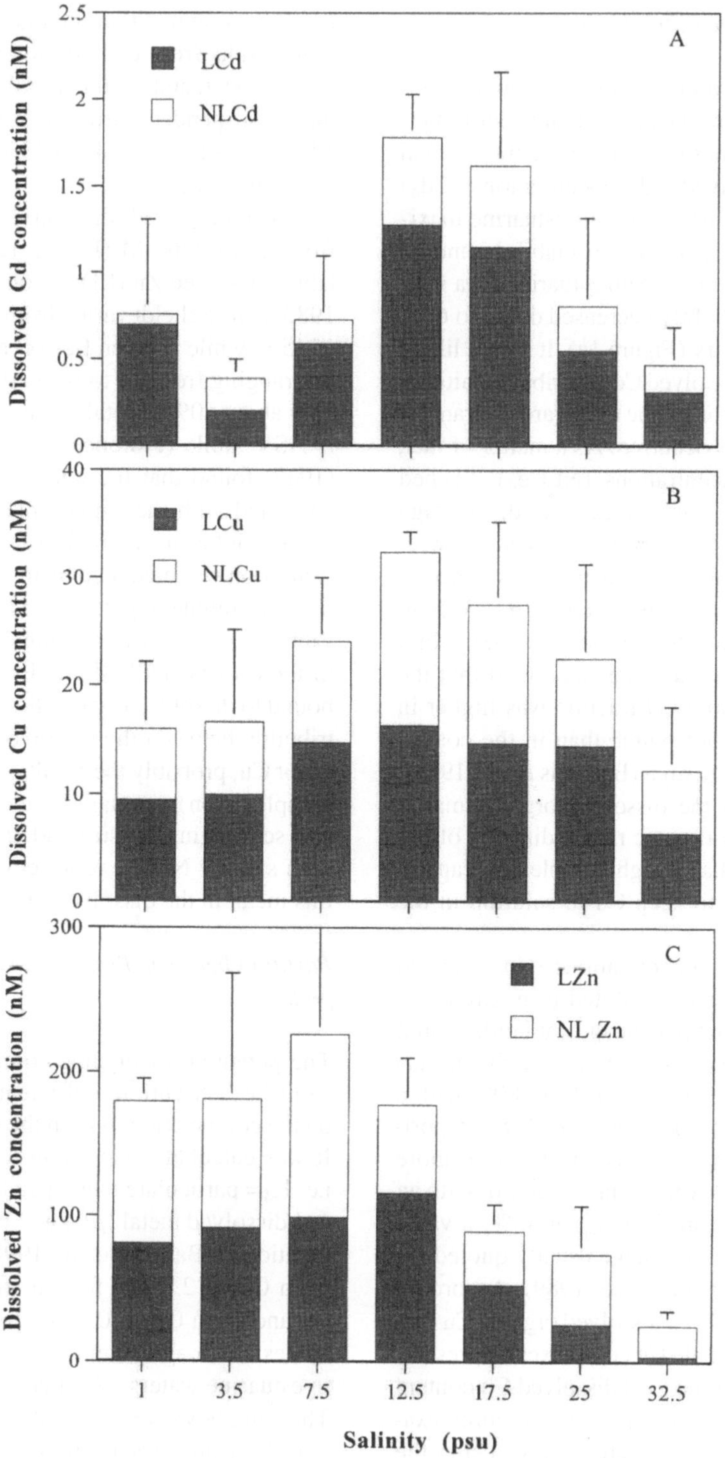

Figure 1. Dependance of total, labile and non-labile metal concentrations on salinity. LMe and NLMe represent the labile and non-labile metal fraction, respectively.

Results and discussion

Labile and non-labile metal fractions

Average longitudinal profiles of dissolved metal concentrations *vs.* salinity are shown in Figure 1 with standard deviation error bars representing the interannual variability. Total dissolved Cd concentrations (Cd_d) ranged from 0.38 to 1.8 nM with mid-estuarine maxima at 12.5–17.5 psu. The percentage of labile bound Cd (LCd_d) was higher in the upstream estuarine area with values up to 90% and slightly decreased down to 60% in the plume of the estuary (Figure 1a). It seems likely that the production of dissolved Cd contributed initially to the labile bound fraction, while afterwards, a transfer to the non-labile species occurred. As a matter of fact, the non-labile Cd concentrations ($NLCd_d$) reached a maximum, when Cd_d already decreased. Average $NLCd_d$ amounted to 28%. This is in agreement with previous anodic stripping voltammetry (ASV) speciation analyses reporting ASV non-labile Cd $< 30\%$ in both seawater and freshwaters (Florence, 1989). In a study of the Belgian coastal zone, we found that the average organically bound Cd fraction was higher in the vicinity of the Scheldt plume than in the coastal waters and in the offshore area (Baeyens et al., 1982). It is thus possible that the dissolved organic matter produced in the muddy organic rich sediments of the coastal-estuarine zone has a high complexing capacity and contributes also to keep Cd in solution in the downstream estuary.

Total dissolved Cu concentrations (Cu_d) ranged from 12.3 to 32.4 nM and exhibited a maximum at 12.5 psu (Figure 1b). The percentages of labile bound dissolved Cu (LCu_d), as observed in the Scheldt during May 1982 to 1983, varied from 14 to 80% with a mean of 57%, while van den Berg et al. (1987) reported values between 5 and 58%. Cu appeared to be more strongly bound to complexing ligands than Cd, with an average non-labile fraction ($NLCu_d$) of 43%, a value falling within the 40 to 60% range usually quoted for coastal surface seawater (Florence, 1989). According to van den Berg et al. (1987), dissolved organic Cu (the sum of labile and non-labile Cu complexes) represented more than 99.99% of the total dissolved Cu content in their samples, suggesting that organic complexation alone almost controls the behaviour of Cu in the Scheldt estuary. The total Cu-complexing ligand concentrations were generally higher than the dissolved Cu concentrations, but assuming a two-site model, van den Berg and co-workers (1987) estimated that the concentrations of the strong Cu-complexing ligands should be similar to that of the dissolved Cu. Hence, it can be assumed that the dissolved to particulate Cu ratio results from competition between complexation by organic ligands in the dissolved phase and scavenging by suspended particles. It was also observed that $NLCu_d$ was highest (about 50%) when Cu_d reached its maximum value.

Total dissolved Zn concentrations (Zn_d) ranged from 25 to 226 nM (Figure 1c). The percentages of labile dissolved Zn (LD_{Zn}) found during May 1982 to 1983 in the Scheldt varied from 26 to 68% with a mean of 45%, while van den Berg et al. (1987) provided values ranging from 34 to 69%. It was also reported that only about 50% of total Zn in seawater and river water is ASV labile (Florence, 1989). van den Berg et al. (1987) found that the concentration of dissolved Zn co-varied with the Zn-complexing ligand throughout the Scheldt estuary, and it was argued that the Zn concentration was mediated by interactions with dissolved organic complexing ligands. On average the non-labile Zn fraction ($NLZn_d$) amounted to 55% indicating that, in the estuary, half of the dissolved Zn was strongly bound to dissolved organic ligands. Therefore, the distribution between the dissolved and particulate Zn is, as for Cu, probably the result of a competition between complexation by organic ligands in the dissolved phase and scavenging by suspended particles. However, at high salinity NLD_{Zn} represented the main fraction of this metal in the dissolved phase.

Partition between dissolved and particulate matter phase

The potential for organic complexation was assessed from the distribution coefficient (K_d), since K_d depends on the composition of both the liquid and solid phases. It was calculated according to Valenta et al. (1986): i.e. K_d = particulate metal [μg kg^{-1} of suspended matter]/dissolved metal [μg kg^{-1} of water]. As previously mentioned (Baeyens et al., 1998), the K_d-values varied from 0.2 to 2.7 10^5 for Cu, from 0.1 to 6.1 10^5 for Cd and from 0.3 to 0.7 10^5 for Zn (Figure 2). These values are in agreement with those reported by others in estuarine waters (Valenta et al., 1986; Balls, 1989). They are, however, substantially lower, except for Cd, than those observed in areas with low suspended matter (ocean) with 1–4 10^6 for Cu, 0.3–1 10^5 for Cd and 0.4–1 10^6 for Zn. Since estuarine water is generally richer in dissolved organic matter than oceans, the lower K_d values observed here were assumed to be the result

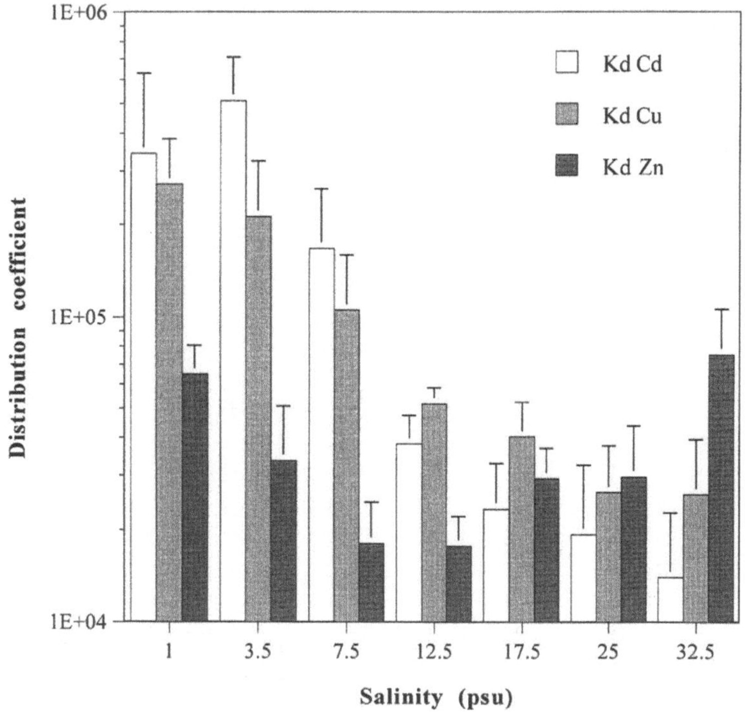

Figure 2. Dependance of the distribution coefficient K_d for Cd, Cu and Zn on salinity. K_d = the ratio of particulate metal (μg kg^{-1}) of suspended matter) over dissolved metal (μg kg^{-1} of water).

of organic complexation (Valenta et al., 1986). Hence the fraction of dissolved organo-metal complexes was assessed according to:

$$K_d = \frac{Me_p}{Me_d} = \frac{Me_p}{MeL_{in} + MeL_{or}} \qquad (3)$$

where K_d is the actual distribution coefficient measured in the Scheldt (Figure 2), Me_p the particulate metal concentration (Baeyens et al., 1998), Me_d the dissolved metal concentration (Figure 1), MeL_{in} the inorganic metal species and MeL_{or} the organic metal complexes. Let K'_d be the distribution coefficient observed in the absence of dissolved organic species = Me_p/MeL_{in} with values close to those reported in the surface water of the oceans, i.e. 2.10^6 for Cu and $0.7.10^6$ for Zn (Valenta et al., 1986; Balls, 1989). Then from (3):

$$MeL_{or} = Me_p \cdot (1/K_d - 1/K'_d) \qquad (4)$$

For both Cu (12–31 nM) and Zn (23–215 nM), MeL_{or}-values are significantly higher than the corresponding $NLMe_d$ ones (Table 1). This might be expected since speciation schemes based on physicochemical separations and ASV measurements have demonstrated the electrochemical lability of several Cu and Zn organic complexes (Batley & Florence, 1976). Moreover, it is known that with DPASV, the labile metal fraction increases dramatically as pH is lowered (Duinker & Kramer, 1977; van den Berg & Dharmvanij, 1984). Therefore, at natural pH values, a greater potential for organic complexation than the one resulting from our DPASV measurement at pH 1 might be expected. Accordingly, it is suggested to subdivide the organic complexing ligands of Cu and Zn into two groups, mono and bidentate, corresponding to weak (labile) and strong (non-labile) complexing sites (Figure 3). The same reasoning can not hold for Cd. The stability constants for organic complexation of Cd are substantially smaller than those of analogous Cu and Zn complexes (Mantoura et al., 1978; Valenta et al., 1986). This suggests that only $NLCd_d$ has a chance to act as a competitive ligand in keeping Cd into solution.

Inorganic and organic metal speciation

Considering dissolved metal speciation in estuarine waters, it should be borne in mind that the extent of complexation markedly fluctuates with the nature and the concentration of the inorganic and organic compounds, as well as with the values of the stability con-

Table 1. Speciation of Cu, Zn and Cd (Equation 4–5) and conditional stability constants (M^{-1}) from complexes formed with humic (Equation 6) and amino acids (Equation 7). Data were calculated as described in the text

Salinity (psu)	Experimental Me_d (nM)	$NLMe_d$ (nM)	Equation (4) MeL_{or} (nM)	Equation (5) Me_f (nM)	MeL_{or}/Me_f	Equation (6) L_f (μg C l^{-1})	Equation (7) L_f (μg C l^{-1})
Copper						$\beta_{Hum} = 10^{10.6}$	$\beta_1 = \beta_2 = 10^{7.8}$
1	16.0 ± 6.0	6.6 ± 2.6	14.3 ± 1.0	0.06 ± 0.03	251 ± 149	21 ± 12	25 ± 9
3.5	16.7 ± 8.3	6.7 ± 4.8	14.0 ± 1.3	0.33 ± 0.16	42 ± 21	5 ± 3	15 ± 4
7.5	24.1 ± 11.8	9.3 ± 5.3	21.6 ± 2.0	0.42 ± 0.34	52 ± 42	7 ± 6	19 ± 10
12.5	32.4 ± 1.7	16.0 ± 5.2	31.3 ± 0.1	0.16 ± 0.01	194 ± 18	32 ± 3	45 ± 2
17.5	27.5 ± 7.6	12.7 ± 3.5	27.1 ± 0.5	0.04 ± 0.01	721 ± 52	121 ± 9	91 ± 3
25	22.5 ± 8.7	7.9 ± 4.4	22.2 ± 0.1	0.02 ± 0.01	891 ± 421	146 ± 69	98 ± 25
32.5	12.3 ± 5.7	5.8 ± 3.4	11.9 ± 0.4	0.04 ± 0.03	272 ± 186	43 ± 29	49 ± 20
Zinc						$\beta_{Hum} = 10^{9.1}$	$\beta_1 = \beta_2 = 10^{7.0}$
1	179 ± 15	98 ± 9	167 ± 8	7.9 ± 5.3	21.2 ± 14.2	50 ± 33	38 ± 16
3.5	180 ± 87	89 ± 28	172 ± 3	6.4 ± 2.3	26.9 ± 9.5	94 ± 33	67 ± 14
7.5	226 ± 73	127 ± 16	215 ± 5	8.5 ± 3.7	25.4 ± 11.2	103 ± 45	76 ± 20
12.5	177 ± 32	61 ± 34	170 ± 2	5.2 ± 1.4	32.5 ± 8.8	149 ± 40	99 ± 14
17.5	90 ± 17	37 ± 11	86 ± 1	2.7 ± 0.6	31.6 ± 7.5	149 ± 36	101 ± 13
25	74 ± 32	48 ± 16	69 ± 3	2.8 ± 1.5	24.7 ± 13.1	114 ± 61	85 ± 26
32.5	25 ± 9	21 ± 2	23 ± 1	1.5 ± 0.5	15.1 ± 5.1	67 ± 23	62 ± 12
Cadmium						$\beta_{Hum} = 10^{8.9}$	$\beta_1 = \beta_2 = 10^{6.9}$
1	0.77 ± 0.53	0.06 ± 0.02	–	0.31 ± 0.01	0.2 ± 0.1	1 ± 0.3	2 ± 1
3.5	0.38 ± 0.11	0.19 ± 0.06	–	0.05 ± 0.02	3.4 ± 1.5	22 ± 10	29 ± 8
7.5	0.73 ± 0.36	0.14 ± 0.10	–	0.09 ± 0.02	1.5 ± 1.1	11 ± 9	18 ± 11
12.5	1.78 ± 0.24	0.50 ± 0.16	–	0.13 ± 0.02	3.9 ± 1.4	33 ± 12	41 ± 9
17.5	1.62 ± 0.53	0.47 ± 0.15	–	0.08 ± 0.01	6.2 ± 2.1	54 ± 19	56 ± 12
25	0.81 ± 0.50	0.26 ± 0.12	–	0.02 ± 0.01	10.5 ± 5.3	90 ± 45	74 ± 23
32.5	0.46 ± 0.22	0.16 ± 0.05	–	0.01 ± 0.00	18.4 ± 6.2	150 ± 51	99 ± 19

stants. To provide a better insight into the speciation of the dissolved metal burden, the potential for inorganic complexation was first examined with the following relation:

$$Me_d = Me_f \cdot \left(1 + \sum_i K_i \cdot L_i^{\alpha}\right) + MeL_{or} \qquad (5)$$

where Me_f = the free metal ion concentration, L_i the concentration of free inorganic complexing ligands ,K_i the stability constant for the ith complex, valid at the ionic strength of the sample, α the stoichiometric coefficient of the ith reactant and MeL_{or} the organic metal complexes given by Equation (4) for Cu and Zn, and set equal to $NLCd_d$ for Cd.

As discussed later, an estimation of Me_f was required to asses the stability constants involved in the organic complexation. Calcium, magnesium, sulphate and chloride were considered to behave conservatively during mixing of river water and sea water (Mantoura et al., 1978 and references cited), while the concentrations of carbonate, bicarbonate and hydroxide were obtained from longitudinal profiles measured in the Scheldt estuary by Frankignoulle et al. (1996). The stability constants of the inorganic metal complexes were taken from Woods & Garrels (1987).

The results of the computation are given in Table 1 and Figure 4. According to previous estimates (Mantoura et al., 1978; Zuehlke & Kester, 1983; Florence, 1989), the major inorganic Cu complexes are the hydroxide and carbonate species. They vary between 1 and 10% of Cu_d throughout the estuary (not illustrated), while the free Cu ion still exceeds 2% at a salinity of 3.5 psu (Figure 4a). On the other hand, Cu exists to a significant extent (> 90%) as organic complex even at a salinity of 30 psu (Table 1). Zn and Cd exhibit a different behaviour. The major inorganic Zn species in

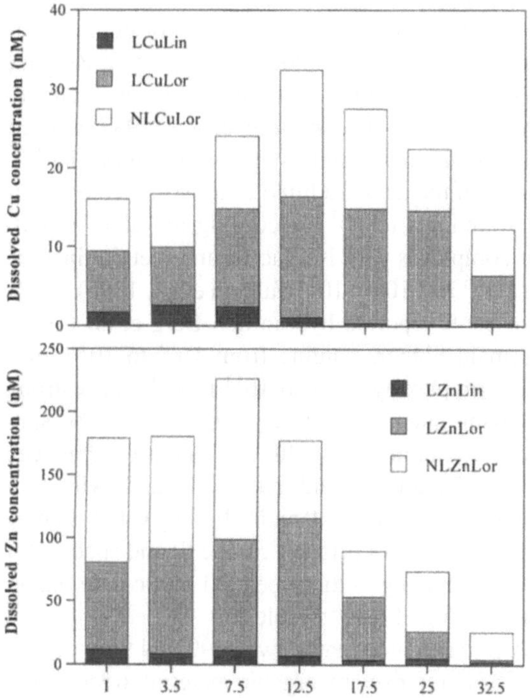

Figure 3. Speciation of Cu and Zn according to results of Equation (3–4). $LMeL_{in}$, $LMeL_{or}$ and $NLMeL_{or}$ represent the labile (inorganic and organic) and the non-labile organic metal fractions.

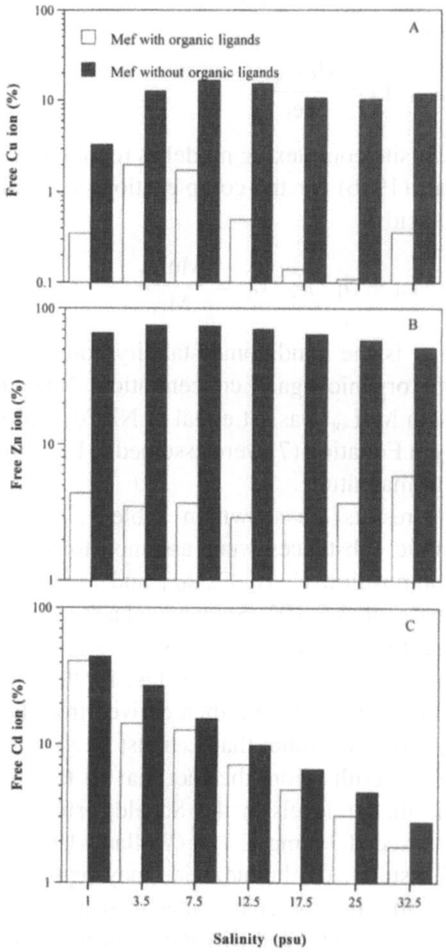

Figure 4. Predicted fraction of free Cu, Cd and Zn ions *versus* salinity in the presence of dissolved organic ligands and in their absence, i.e. MeL_{or} was set equal to zero in Equation (5).

the estuary is the free ion with values between 3 to 6% (Figure 4b). Only in the downstream area do the sulphate, mono and dichloride Zn species exceed the 3% limit (not illustrated). As for Cu, the organic complexes are predominant with values > 90% (Table 1). In the upstream area, the free Cd-ion was the major species with 40%, and this percentage decreased gradually to the mouth where only 2% of the free ion was left (Figure 4c). The monochloride complex is important in the whole estuary. At 1 psu it represents 26% of Cd_d, increasing to 60% in the middle estuarine area where Cd_d is highest, and decreases slightly at the mouth to 24%. The dichloride complex shows an reverse trend: in the upstream estuary where the chloride concentration is very low, it represents only 2% of Cd_d, whilst it increases up to 31% at the mouth of the Scheldt (not illustrated). As previously mentioned, the organically bound fraction, equal to $NLCd_d$, represents on average less than 30% of Cd_d (Table 1).

A sensitivity analysis of the model (Equation 5) shows that even small fluctuations in the concentration of MeL_{or} results in large variability of Cu^{2+} (C.V: 10–72%) and Zn^{2+} (CV: 23–67%). Obviously, the uncertainties on the inorganic metal fraction increase dramatically as the concentration of MeL_{or} approaches that of Me_d. Yet, when MeL_{or} was set equal to zero in Equation (5), a drastic increase was observed in the free Cu and Zn ions, presumably the most toxic forms for aquatic organisms, while the distribution of Cd^{2+} remained almost unaffected (Figure 4). This suggests that chloride acts as a significant competitor for organic ligands with Cd.

Conditional stability constants and ligand concentrations for organic complexation

Data from Table 1 allow calculations of the product between the free organic ligand concentration and the conditional stability constant assuming either a single site complexing model, as proposed by Mantoura et al. (1978) for the complexation of metals with humic materials;

$$\beta_{Hum} \cdot L_f = \frac{MeL_{or}}{Me_f} \tag{6}$$

or a two-site complexing model as reported by Valenta et al. (1986) for the complexation of metals with amino-acids;

$$\beta_1 \cdot L_f + \beta_1 \cdot \beta_2 \cdot L_f^2 = \frac{MeL_{or}}{Me_f} \tag{7}$$

where β is the conditional stability constant and L_f the free organic ligand concentration. Note that for cadmium MeL_{or} was set equal to $NLCd_d$ and that β_1 and β_2 in Equation (7) were assumed to be of the same order of magnitude.

The results are shown in Table 1. In Equation (6) humic substances were assumed to be of average composition $C_{10}H_{11}O_6N_{0.3}$ and average molecular mass of 3.5 10^3 (Schwarzenbach et al., 1993) yielding the following relationship: 1 mM of humic material = 1817 mg of C l^{-1}. Values for the conditional stability constant were then derived from Equation (6) with the constraint that L_f must be $\leq$ 0.15 mg C l^{-1}. This results from the fact that (i) the dissolved organic matter levels in the Scheldt estuary usually do not exceed 10 mg C l^{-1} (Wollast, 1997; unpublished results), (ii) humic acids may represent up to 30% of the total organic matter in seawater (Stumm & Morgan, 1981), and (iii) the competition from Ca and Mg is sufficiently high to utilise > 95% of the complexing capacity of the humic ligands (Mantoura et al., 1978). A same procedure was applied to assess interactions between heavy metals and amino acids. For the two-site model, however, an average composition of $C_{10}H_{18}O_5N_3$ and an average molecular mass of 134 were considered yielding the following relationship: 1 mM of amino acids = 62 mg of C l^{-1}. Furthermore, since amino acids rarely exceed 20% of the dissolved organic matter levels in seawater (Stumm & Morgan, 1981), the constraint applied to Equation (7) was 0.1 mg C l^{-1}.

Throughout the various calculations, the ratio of the activity coefficients $\gamma_{ligand}/\gamma_{complex}$ has been assumed to be unity, therefore, $\beta = K^{\infty} \cdot \gamma_{Me}$ where K^{∞} is the thermodynamic equilibrium constant at infinite dilution and γ_{Me} was assessed with the Davies equation as aforementioned. The longitudinal profiles of organic ligands obtained for the various metals with the corresponding minimum values of β at infinite dilution is given in Table 1.

With copper, β_{Hum} amounts to $10^{10.6}$ which is in close agreement with values reported for humic interactions $10^{8.9}$–$10^{11.4}$ (Mantoura et al., 1978; Zuehlke & Kester, 1983), but is significantly lower than values reported by van den Berg et al. (1987) for the Scheldt, assuming a single-step complexation: $10^{11.8}$–10^{14}. When the data are fitted to a two-site model, β_1 and β_2 amounts to $10^{7.8}$ according to values reported for complexes with isoleucine and methionine, i.e. $10^{8.1}$–$10^{8.6}$ and $10^{6.7}$–10^7 (Valenta et al., 1986), while the constants reported by van den Berg et al. (1987) still remain higher, ranging from 10^{12} to 10^{13} for a weaker complexing site and 10^{13} to 10^{15} for a strong complexing site. From these results, it can be stated that free humic acids with concentrations between 0.01 to 0.15 mg C l^{-1} and free amino acids with concentrations between 0.019 and 0.100 mg C l^{-1}, as may occur in the Scheldt estuary, can be ligand species for copper, competing with suspended particulate matter for the binding of Cu^{2+} (Table 1).

It was calculated that between 90 and 96% of zinc should be complexed by organic material to reproduce the pattern of K_d *versus* salinity. Therefore, given the assumptions made in the models (Equation 6–7), Zn^{2+} required β_{Hum} of $10^{9.1}$ and β_1, β_2 of $10^{7.0}$ (Table 1). These values are two to three orders of magnitude higher than the corresponding stability constant of Zn complexes with humic compounds, i.e. up to $10^{5.9}$ (Mantoura et al., 1978) and amino-acids, i.e. up to 10^5 (Stumm & Morgan, 1981), but fall within the range reported by van den Berg et al. (1987) in the Scheldt, i.e. $10^{8.6}$–$10^{10.6}$. Since the organic Zn fraction (MeL_{or}) reported in Table 1 is in agreement with the one reported by others in estuarine waters (van den Berg & Dharmvanij, 1984; van den Berg et al., 1987), it seems that models considering only humic and/or amino acid interactions, excluding the formation of mixed ligand complexes and polynuclear complexes, can underestimate the potential of Zn for organic complexation. The position of the metal ions within the Iving-Williams series does not alone determine metal binding affinity. Other factors, including hard-soft acid-base and ligand-field stabilisation energy effects can be important determinants as well. Fulvic acids, for instance, exhibit stronger interactions with Zn than with Cu and Cd (Piotrowicz et al., 1983). This was also demonstrated in studies investigating the metal binding specificity of Zn-finger peptides (Lippard & Berg, 1994).

Tabulated stability constants of the Cd complexes with humic and amino acids are about 10^5 and 10^4, respectively (Mantoura et al., 1978; Valenta et al., 1986) and are thus substantially smaller than the values reported in Table 1: $\beta_{Hum} = 10^{8.9}$ and $\beta_1, \beta_2 = 10^{6.9}$. Investigations of the required ligand concentrations for complexation of Cd^{2+} by amino acids in seawater have shown that to achieve a complexation of about 20%, the organic ligand concentrations should reach the mM level (Valenta et al., 1984; Sugawara et al., 1984). This is much higher than the ligand concentrations provided in Table 1 and those reported in the Scheldt estuary for the complexation of Zn and Cu (van den Berg et al., 1987). Yet, a significant fraction of the inert Cd can be inorganic, probably adsorbed on colloidal particles of Fe_2O_3, Al_2O_3 and MnO_2 coated with humic acid (Florence, 1989). Hence, only 28% of the total dissolved Cd concentration should be associated with colloidal particles coated by humates to fit the $NLCd_d$ fraction reported in Table 1. Alternatively, as pointed out for Zn, higher values for stability constants might be expected with respect to the hard-soft acid-base effect. One example are the complexes with metallothioneins where Cd^{2+} exhibits higher affinity for cysteinyl residues than Zn^{2+} and Cu^{2+}.

To conclude, we have related our model calculations to electrochemical measurements of metals by grouping the metal species into two categories: electrochemical labile and non-labile. For Cd, the ratio of particulate metal over labile dissolved metal provide K_d-values close to those reported in oceans, while for Cu and Zn this ratio was much lower. This discrepancy can be explained by a high affinity of Cu and Zn for complexation with components of natural dissolved organic matter. In this context, the speciation model provides estimates for conditional stability constants and ligand concentrations in the Scheldt. It supports the data of van den Berg et al. (1987) suggesting that in the Scheldt, interactions of Cu and Zn with dissolved organic complexing ligands determine almost completely the biogeochemical pathway (toxicity and transport) of these metals. On the contrary, for Cd, competition from chloride ion is the leading factor since not only are the stability constants of its chlorocomplexes very high, but also because the concentration of the *n*th complex is dependent on the *n*th power of the chloride concentration.

Acknowledgement

The authors are indebted to G. Gillain for the analyses of the samples. They thank the MUMM for providing shiptime and for the acquisition of physico-chemical monitoring data and the crew of the R.V. Mechelen for technical assistance during sampling.

References

Baeyens, W., M. Bogaert, G. Decadt, H. Dedeurwaerder, F. Dehairs, M. Dejonghe, G. Gillain, L. Goeyens & S. Wartel, 1982. Distribution, transport and fate of Bi, Cd, Cu, Hg, Pb, Sb, and Zn in the Belgian coastal marine environment. In A. Distèche & I. Elskens (eds), Distribution, transport and fate of heavy metals in the Belgian coastal marine environment, Projet Mer, Rapport final, 2. Services du Premier Ministre, Programmation de la Politique Scientifique: 7–138.

Baeyens, W., G. Gillain, G. Decadt & I. Elskens, 1987. Trace metals in the eastern part of the North Sea. I: Analyses and short-term distributions. Oceanol. Acta. 10: 169–179.

Baeyens, W., M. Elskens, G. Gillain & L. Goeyens, 1998. Biogeochemical behaviour of Cd, Cu, Pb and Zn in the Scheldt estuary during the period 1981–1983. Hydrobiologia 366: 15–44.

Balls, P. W., 1989. The partition of trace metals between dissolved and particulate phases in European coastal waters: a compilation of field data and comparison with laboratory studies. Neth. J. Sea Res. 23: 7–14.

Batley, G. E. & T. M. Florence, 1976. Determination of the chemical forms of dissolved cadmium, lead and copper in seawater. Mar. Chem. 4: 347–353.

Duinker, J. C. & C. J. Kramer, 1977. An experimental study on the speciation of dissolved zinc cadmium, lead and copper in river Rhine and North Sea water, by differential pulsed anodic stripping voltammetry. Mar. Chem. 5: 207–228.

Elskens, M., M. Leermakers, S. Panutrakul, F. Monteny & W. Baeyens, 1991. Microbial activity in sandy and muddy estuarine sediments. Geo-Marine Lett. 11: 194–198.

Florence, T. M., 1989. Electrochemical techniques for trace element speciation in waters. In G. E. Batley (ed.), Trace element speciation: Analytical Methods and Problems. CRC Press, Inc. Boca Raton, Florida: 77–116.

Frankignoulle, M., I. Bourge & R. Wollast, 1996. Atmospheric Co2 fluxes in a highly polluted estuary (The Scheldt). Limnol. Oceanogr. 41: 365–369.

Houghton, R. P., 1979. Metal complexes in organic chemistry. Cambridge University Press, Cambridge, 308 pp.

Leermakers, M., M. Elskens, S. Panutrakul, F. Monteny & W. Baeyens, 1993. Geochemistry of mercury in an intertidal flat of the Scheldt estuary. Neth. J. Aquat. Ecol. 27: 267–277.

Lippard, S. J. & J. M. Berg, 1994. Principles of bioinorganic chemistry. University Science Books, Mill Valley, California, 409 pp.

Mantoura, R. F. C., A. Dickson & J. P. Riley, 1978. The complexation of metals with humic materials in natural waters. Estuar. coast. mar. Sci. 6: 387–408.

Monteny, F., M. Elskens & W. Baeyens, 1993. The behaviour of copper and zinc in the Scheldt estuary. Neth. J. Aquat. Ecol. 27: 279–286.

Morel, F. & J. Morgan, 1972. A numerical method for computing equilibria in aqueous chemical systems. Envir. Sci. Technol. 6: 58–67.

Panutrakul, S. & W. Baeyens, 1991. Behaviour of heavy metals in a mud flat sediment of the Scheldt estuary, Belgium. Mar. Poll. Bull. 22: 128–134.

Piotrowicz, S. R., G. R. Harvey, M. Springer-Young, R. A. Courant & D. A. Boran, 1983. Studies of cadmium, copper, and zinc interactions with marine fulvic and humic materials in seawater using anoding stripping voltammetry. In C. S. Wong, E. Boyle, K. W. Bruland, J. D. Burton & E. D. Goldberg (eds), Trace Metals in Sea Water, Plenum Press, New York and London: 699–718.

Ramette, R. W., 1981. Chemical equilibrium and analysis. Addison-Wesley Publishing Compagny, Massachusetts, 765 pp.

Schwarzenbach, R. P., P. M. Gschwend & D. M. Imboden, 1993. Environmental Organic Chemistry. John Wiley & Sons, INC. New York/Chichester/Brisbane/Toronto/Singapore, 681 pp.

Stumm, W. & J. J. Morgan, 1981. Aquatic Chemistry. John Wiley & Sons, INC. New York/Chichester/Brisbane/Toronto/Singapore, 780 pp.

Sugawara, M., P. Valenta, H. W. Nurnberg & T. Kambara, 1984. Voltammetric study of the speciation of cadmium (II) with L-aspartic acid in sea water. J. Electroanal. chem. 180: 343–354.

Valenta, P., E. K. Duursma, A. G. A. Merks, H. Rutzel & H. W. Nurnberg, 1986. Distribution of Cd, Pb, and Cu between the dissolved and particulate phases in the Eastern and Western Scheldt estuary. Sci. Total Envir. 53: 41–76.

Valenta, P., M. L. S. Simoes-Goncalves & M. Sugawara, 1984. Voltammetric studies on the speciation of cadmium and zinc by amino acids in sea water. In C. J. M. Krammer & J. C. Duinker (eds), Complexation of Trace Metals in Natural Waters, Martinus Nijhoff/W. Junk Publ., The Hague, Boston, Lancaster: 357–366.

van den Berg, C. M. G. & S. Dharmvanij, 1984. Organic complexation of zinc in estuarine interstitial and surface water samples. Limnol. Oceanogr. 29: 1025–1036.

van den Berg, C. M. G., A. G. A. Merks & E. K. Duursma, 1987. Organic complexation and its control of the dissolved concentrations of copper and zinc in the Scheldt estuary. Estuar. coast. mar. Sci. 24: 785–797.

Wollast, R., 1976. Transport and accumulation de polluants dans l'estuaire de l'Escaut. In J. C. J. Nihoul & R. Wollast (ed.), L'estuaire de l'Escaut, Projet Mer, rapport final, 10. Services du Premier Ministre, Programmation de la Politique Scientifique: 191–128.

Woods, T. L. & R. M. Garrels, 1987. Thermodynamic values at low temperature for natural inorganic materials. An Uncritical Summary. Oxford University Press, Oxford, New York, 242 pp.

Zuehlke, R. W. & D. R. Kester, 1983. Copper speciation in marine waters. In C. S. Wong, E. Boyle, K. W. Bruland, J. D. Burton & E. D. Goldberg (eds). Trace Metals in Sea Water. Plenum Press, New York and London: 773–788.

Zwolsman, J. J. G. & G. T. M. van Eck, 1991. The behaviour of dissolved Cd, Cu and Zn in the Scheldt estuary. Estuarine Water Quality Management, Springer-Verlag, Berlin: 413–420.

Hydrobiologia **366**: 91–108, 1998.
W. F. J. Baeyens (ed.), Trace Metals in the Westerschelde Estuary.

The impact of the Scheldt input on the trace metal distribution in the Belgian coastal area (results of 1981–1983 and 1995–1996)

Willy Baeyens, Marc Elskens, Ramses Van Ryssen & Martine Leermakers
Department of Analytical Chemistry, Free University of Brussels, Pleinlaan 2, B-1050 Brussels, Belgium

Key words: Trace metals, North Sea, coastal waters, fluxes, residual flows, river inputs

Abstract

Offshore fluxes of Cu, Zn, Cd, Pb and Hg were calculated based on residual flow patterns and salinity gradients along the Belgian coast. The residual flow lines along the Belgian coast are more or less parallel to the coast except in the area where the north-easterly flowing watermass coming from the Channel encounters the south-westerly-oriented Scheldt outflow, forming a residual hydrodynamical front. From the steady-state salinity pattern, diffusion coefficients perpendicular to the residual flow were deduced; they ranged from 21 to 108 $m^2 s^{-1}$.

Offshore fluxes of dissolved and particulate trace metals based on diffusive and mixing processes are calculated. The steady state profiles of dissolved metals show a dilution effect in the coastal waters, reaching an almost constant concentration in the marine watermass in the 1981–1983 dataset. The ratios of the Scheldt input of trace metals to the total dissolved offshore flux vary from 38 to 55% (1981–1983), depending on the kind of metal, and from 55 to 91% (1995–1996). The ratio of the Scheldt input to the dissolved metal flow parallel to the coast, is in both periods (1981–1983 and 1995–1996), smaller than 1%. The steady-state concentration profiles of particular metals versus salinity are fairly constant in the coastal-estuarine and marine watermasses, but decrease very abruptly from the first to the second watermass. Assuming a conservative behaviour of the particular metals, offshore fluxes and the resulting concentration increases agree fairly well with the observed values. The ratios of the Scheldt input to the particulate trace metal offshore flux vary between 30 to 46% (1981–1983) and 13 to 37% (1995–1996). The contribution of the Scheldt estuary to the flows parallel to the coast ranges from 1.6 to 2.9% (1981–1983) and from 0.6 to 1.6% (1995).

Introduction

Unlike the open ocean, coastal systems such as the North Sea have attracted considerable attention where trace metal studies are concerned. The explanation appears to be related to several factors. Coastal waters are relatively accessible, the type of research vessel required for most work is relatively unsophisticated, trace metal levels are sufficiently enhanced and moreover there has been a growing concern over the last 15 years regarding the effects of industrial discharges on trace metal levels in nearshore waters and estuaries.

Most of the trace metal studies in coastal areas fall into the category of 'snap shot' studies. Such short-term metal distribution studies are insufficient to derive metal flows through a coastal system. The time scales of hydrodynamical and chemical data need to be matched, e.g. when one uses, for example, a stationary winter flow, the metal distribution over the same period also needs to be known.

The assessment of the various metal flows through the Belgian coastal area can thus only be achieved if information exists about: (1) the residual steady-state currents; (2) the diffusion coefficients related to these residual currents; and (3) the steady-state metal distributions. In this paper we will discuss these three items for the distribution and fluxes of Cu, Zn, Cd and Pb through the Belgian coastal zone based on eleven surveys in the period 1981–1983 and five surveys in the period 1995–1996. In addition, in the 1995–1996 surveys Hg was also included.

Methods and materials

A detailed description of the sampling and analytical methods for trace metal analysis used during the 1981–1983 and the 1995–1996 surveys can be found in related articles in this volume (Baeyens et al., 1998a and Baeyens et al., 1998b for Cu, Zn, Cd and Pb in 1981–1983 and 1995–1996 respectively and Baeyens et al., 1998c for Hg). Briefly, during the 1981–1983 surveys, seawater was collected by continuous pumping and on-line filtration and analysed by differential pulse anodic stripping voltametry (DPASV). For the analysis of the total dissolved metal fraction, filtered seawater acidified to pH1 was exposed to UV light overnight before analysis. Detection limits were 0.1 $\mu g\ l^{-1}$ for Cu, 0.01 $\mu g\ l^{-1}$ for Zn, 0.005 $\mu g\ l^{-1}$ for Cd and 0.03 $\mu g\ l^{-1}$ for Pb. The coefficient of variation is estimated at 12% for a concentration of 0.5 $\mu g\ l^{-1}$ Cu, 7% for 1 $\mu g\ l^{-1}$ Zn, 14% for 0.05 $\mu g\ l^{-1}$ Cd and 13% for 0.1 $\mu g\ l^{-1}$ Pb. Particulate matter was digested with a HNO_3, H_2O_2, HF (4:4:1) mixture in a PTFE bomb at 160 °C. After cooling the solution was evaporated to dryness and dissolved in HCl and diluted for DPAVS analysis.

During the 1995–1996 surveys, surface water samples were collected by hand from a rubber boat and filtered on board of the ship in a clean lab container. For the dissolved metal fraction (Cu, Zn, Cd, Pb), seawater is preconcentrated by solvent extraction. Seawater is buffered to pH 4.5, complexed with ammonium pyrolidine dithiocarbamate (APDC) and diethylammonium diethyl dithiocarbamate (DDDC) and extracted into freon. After back extraction into HNO_3 and dilution with Milli-Q water samples are analysed by inductively coupled plasma mass spectrometry (ICPMS) or graphite furnace atomic absorption spectrometry (GFAAS). Blank extractions and certified reference samples (NASS-2) are run in every sample batch. Typical detection limits are 0.02 $\mu g\ l^{-1}$ for Cu, 0.03 $\mu g\ l^{-1}$ for Zn, 0.001 $\mu g\ l^{-1}$ for Cd and 0.002 $\mu g\ l^{-1}$ for Pb. The coefficient of variation is estimated at 14% for a concentration of 0.4 $\mu g\ l^{-1}$ Cu, 14% for 0.5 $\mu g\ l^{-1}$ Zn, 10% for 0.03 $\mu g\ l^{-1}$ Cd and 13% for 0.04 $\mu g\ l^{-1}$ Pb. Particulate matter was digested with a HNO_3/HCl (4:1) acid mixture at 60 °C during 12h and analysed by ICPMS or GFAAS. Blank filters and refence samples (MESS-1, BCSS-1) are run in every sample batch.

Hg was analysed by cold vapour atomic fluorescence (CVAFS) using a gold amalgamation preconcentration step. Total dissolved Hg was analysed by BrCl oxidation and reduction with $NH_2OH{\cdot}HCl$ prior to reduction with $SnCl_2$. The detection limit is 50 pg l^{-1} and the coefficient of variation is 15% at a concentration of 0.2 ng Hg l^{-1}. Particulate matter was digested with a HNO_3/HCl (4:1) acid mixture at 60 °C during 12h and analysed by CVAFS.

Residual flows and salinity patterns

The residual flows are defined as the mean flows over a sufficiently long period of time, in order to cancel, to a large extend, the influence of transitory wind currents and tidal oscillations (Nihoul, 1975). The circulation pattern in the Belgian coastal area (Figure 1) has been inferred from the residual flow lines calculated by Djenidi & Ronday (1985) with a coarse grid mesh ($10' \times 10'$) in combination with a smaller scale (500 m × 500 m), covering the eastern area of the Belgian coast. The different flow lines separate a flow rate of 10 000 $m^3\ s^{-1}$ or 315 $km^3\ y^{-1}$, except in the coastal zone which is subdivided in a streamtube of 85 $km^3\ y^{-1}$ closest to the coast, and a second one of 230 $km^3\ y^{-1}$, for a better assessment of the exchange of material (mixing and diffusion) from inshore to offshore water masses. Exchange through diffusion is calculated along lines A to D (Figure 2), perpendicular to the flow lines. The flow lines are more or less parallel to the coast, except in the zone around stations 13 and 23 (Figures 3 and 4 show the monitoring network in the period 1981–1983 and 1995–1996 respectively). In the neighbourhood of stations 13 and 23, the north-easterly flowing water mass coming from the Channel encounters the south-westerly oriented Scheldt outflow. As a result, a residual hydrodynamical front is established with, at the northern side a less saline estuarine water mass, and at the southern side a more saline marine water mass. Once the coastal-estuarine water mass has turned at the hydrodynamical front, it progressively mixes with the adjacent streamtube I of 85 $km^3\ y^{-1}$, increasing the flow rate to 89.4 $km^3\ y^{-1}$. The composition of streamtube I can be modified: in the upstream part (south of the front) by coastal inputs and the river Yser (0.4 $km^3\ y^{-1}$), at the hydrodynamical front (close to station 13) by diffusion processes with the coastal estuarine water mass, and north of this front by mixing with the coastal estuarine-water mass. Exchange between streamtubes I and II and also between further streamtubes can only occur through diffusion, since the flow lines are parallel to the current.

The residual current appears to be fairly constant over the whole area and amounts to 2.7 cm s^{-1}. The

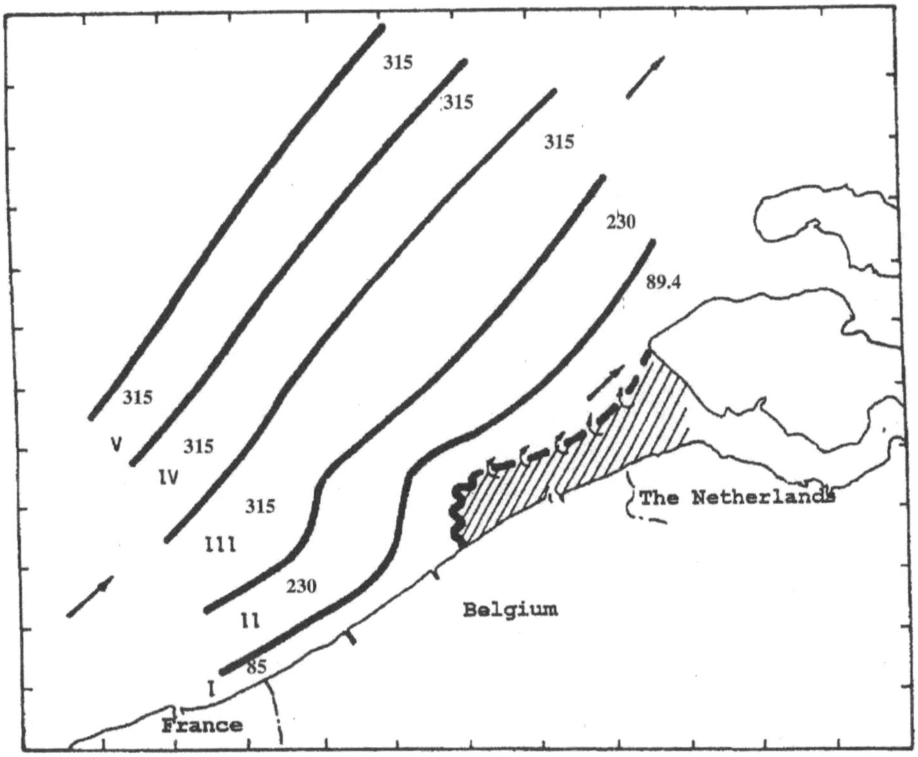

Figure 1. Residual streamflows through the Belgian coastal area (km^3 y^{-1}). The shaded area indicates the coastal-estuarine water mass and the undulating line the residual hydrodynamical front.

residence time of the water masses flowing through our coastal zone (length of 70 km) can thus be estimated to be 1 month.

The salinity pattern is strongly influenced by the residual circulation pattern (Figure 2). The isohalines run parallel to the residual flow lines except downstream of the hydrodynamical front, where streamtubes I and II (85 and 230 km^3 y^{-1} respectively) show a decrease in salinity. An investigation of the impact of the three processes capable of modifying the composition of streamtube I as mentioned above, shows that diffusion should be the dominant one. Diffusion coefficients were calculated based on the salt mass balance equation (McDowell & O'Connor, 1977). Diffusion coefficients along the lines B, C and D (Figure 2), ranged from 21 to 108 m^2 s^{-1}. The diffusion coefficient in the mouth of the Scheldt (line A) was slightly higher: 180 m^2 s^{-1}. In Table 1 the average salinity and the salinity range measured at the fixed sampling stations during the different surveys are prestented. The average salinities measured are in good agreement with the residual salinity patterns.

Distribution of dissolved and particulate trace metals

Short-term trace metal distributions in the Belgian coastal waters may be quite different from other coastal areas. Duinker & Nolting (1982) reported that although the concentration gradients of dissolved metals in the transects perpendicular to the Dutch and Belgian coasts are similar (but opposite) to the salinity gradient, the Belgian coast transect is quantitatively different as a tongue of increased metal concentration occurs. At some stations increased offshore concentrations for a number of elements have also been observed (Baeyens et al., 1987).

Possible sources are local inputs (dumping or contributions from bottom sediments), variations in the composition of the mixing end-members or atmospheric input. Wollast (1976) and Mommaerts et al. (1984) suggested that particles derived from the Scheldt containing increased levels of trace metals and organic matter may accumulate in bottom sediments off the Belgian coast, as a result of particular current con-

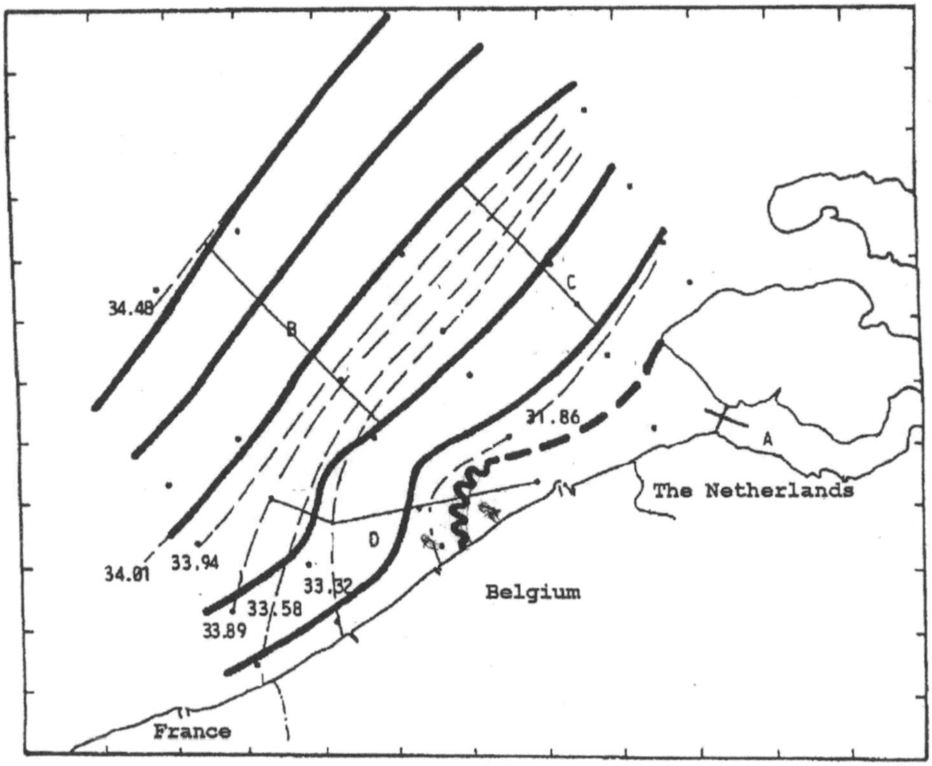

Figure 2. Salinity patterns (- - - -) and residual flow lines in the Belgian coastal area. Diffusion coefficients are calculated along the lines A to D.

ditions. Under certain conditions of wind field, and watermass supply from the English Channel and the river Scheldt, a residual gyre develops off the Belgian coast, strongly increasing the residence time of included material (e.g. Nihoul, 1980). The good correlation between trace metals and organic matter in the sediment column (Wollast, 1976) may indicate that metals are mobilised from the sediments after diagenetic transformation of organic matter. This may lead to an increase of trace elements in the watercolumn. The relative importance may, however, differ between metals and its effect may be time-dependent.

The simultaneous assessment of all of these possible sources influencing the short-term metal distribution in the sampling area is quite an ambitious, difficult and complex task. Averaging the system over a time period corresponding to that of the residual currents and the steady-state salinity distribution, firstly smoothes out or even eliminates all time-dependent effects on the metal distribution, and secondly permits the quantification of metal flows through the coastal zone. The steady-state metal distribution is obtained by averaging the results obtained during eleven cruises in the period 1981–1983 (sampling stations are shown in Figure 3). Stations sampled less than six times were withdrawn from the calculations. In 1995, surveys were carried out in March, October and December and in 1996 in May and October, in approximately the same stations as in the 1981–1983 surveys (Figure 4). Although the approximation of the steady-state conditions in the mid nineties, by the average values obtained from these five cruises is much lower than for the period 1981–1983, some preliminary observations can be made.

Mixing curves of dissolved and particulate metal concentrations plotted against a suitable index of mixing such as salinity make it possible to distinguish between conservative and non-conservative behaviour. In a previous paper (Baeyens et al., 1987) we had major problems interpreting such curves, due to e.g. fluctuations in end-member concentrations. The steady-state approach eliminates these fluctuations. For the 1981 to 1983 surveys, the coefficient of variation of the dissolved and particulate trace metal concentrations at fixed sampling stations, which is a measure of these fluctuations in time, is:

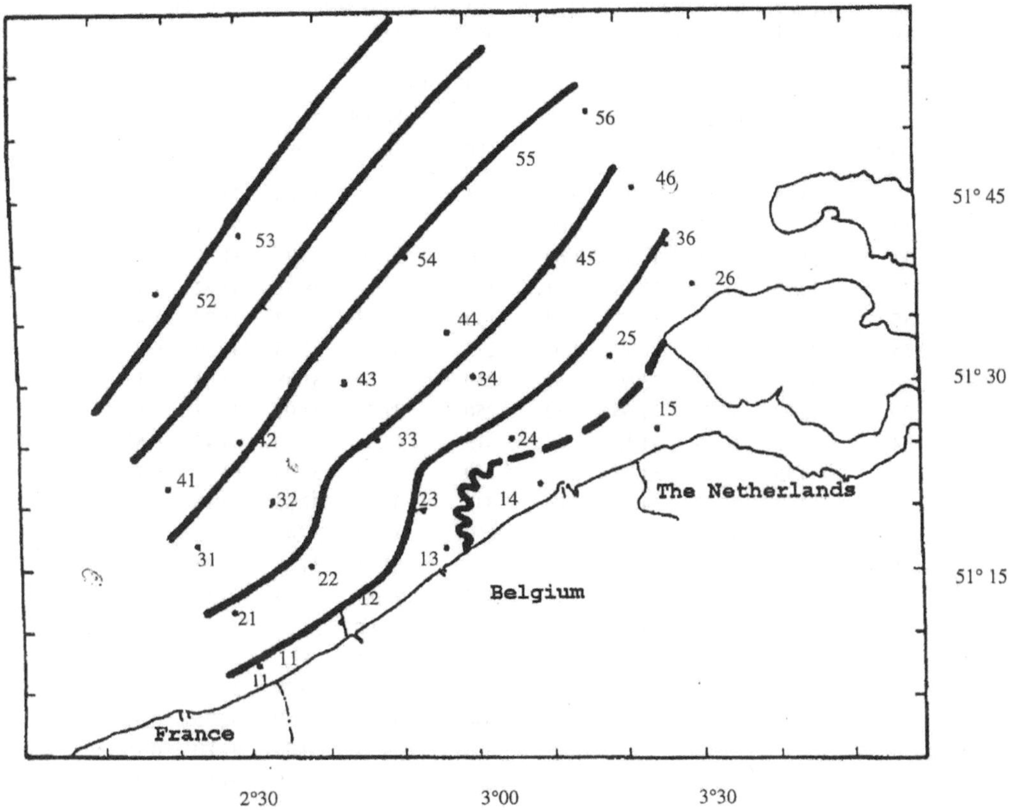

Figure 3. Location of the sampling stations of the surveys between 1981–1983.

- for dissolved metals of the order of 78% (Zn), 60% (Pb), 41% (Cu), 62% (Cd) in the coastal-estuarine stations and 83% (Zn), 65% (Pb), 54% (Cu), 43% (Cd) in the offshore stations;
- for particulate metals of the order of 66% (Zn), 106% (Pb), 56% (Cu), 80% (Cd) in the coastal-estuarine stations and 99% (Zn), 127% (Pb), 96% (Cu), 112% (Cd) in the offshore stations.

Dissolved metals

For the 1981 to 1983 surveys, concentration ranges measured in the coastal waters are 0.25–3.2 $\mu g\ l^{-1}$ for Cu, 0.25–17 $\mu g\ l^{-1}$ for Zn, 0.011–0.38 $\mu g\ l^{-1}$ for Cd and 0.04–1.5 $\mu g\ l^{-1}$ for Pb. Average concentrations and standard deviations are shown in Table 2. The profiles of dissolved metal concentrations versus salinity are all very similar (Figure 5), and can be subdivided into two parts: 1) The concentration profiles progressively decrease from coastal station 15 with lowest salinity (30.4 psu) to station 46 with a salinity of about 32.4 psu. The concentrations at station 15 are approximately the same as the average concentration in the mouth of the river Scheldt. The hydrodynamical front (31.8 psu) is also located in this salinity range. 2) In the marine watermass the concentrations of Cd and Pb remain almost constant, whereas Cu and Zn show an increase at a salinity of 33.5 psu. This increase is due to the relatively higher values observed in nearshore stations 11 and 12, where concentrations may be influenced by inputs from the area around Dunkerque and the river Yser. The concentrations at the offshore station 45 are much higher than at surrounding stations 33, 34 and 46 and this for all metals. Dumping activities and accompanying processes such as dissolution and desorption of particulate metals may explain this aberrant result. Indeed, several hundred thousand tons of waste material from TiO_2 industries were dumped in that area during the period of sampling. The trace metal concentrations measured in the Belgian coastal waters were significantly higher than those measured in Calvi, Corisca, in the same period using identical sampling and analytical techniques. The trace metal assessment in this region revealed dissolved levels of Cu ranging

Table 1. Average salinity (psu), salinity ranges and standard deviation measured during 11 surveys between 1981–1983 and five surveys between 1995–1996

1981–1983					1995–1996				
	Salinity (psu)					Salinity (psu)			
Station	av	min	max	std	Station	av	min	max	std
11	33.58	30.55	34.34	0.83	115	33.54	32.08	34.28	1.01
12	33.32	29.97	34.40	1.09	120	33.48	31.90	34.14	0.92
13	31.72	29.58	34.33	1.69	130	31.06	30.76	31.35	0.31
14	30.70	28.43	32.63	1.11	140	30.21	28.51	31.22	1.48
15	30.41	27.33	34.27	1.72	150	30.38	27.87	31.27	1.45
25	31.46	29.46	33.49	1.02	215	33.38	32.15	34.39	0.95
26	31.50	28.27	33.60	1.87	230	31.67	31.48	31.84	0.15
33	32.82	30.79	34.64	1.12	250	31.73	31.07	32.37	0.54
34	32.63	30.68	34.13	1.07	315	34.00	33.33	34.45	0.55
41	34.19	31.28	34.75	0.75	330	32.65	31.95	33.72	0.73
42	34.01	29.52	34.71	1.06	340	32.55	32.35	32.76	0.29
45	32.79	31.19	34.98	0.88	350	32.46	32.41	32.52	0.08
46	32.45	31.41	34.04	1.12	421	34.51	34.29	34.66	0.14
52	34.48	34.24	34.72	1.19	435	33.87	33.53	34.13	0.25
53	34.43	34.10	34.57	0.16	545	33.69	33.05	34.12	0.45
54	34.03	33.17	34.37	0.34	700	30.31	28.66	31.20	0.98
55	33.92	33.17	34.35	0.51	710	31.24	29.79	32.12	0.97
					780	31.78	30.59	32.56	0.81
					800	34.57	34.38	34.80	0.15
					b03	31.59	31.33	31.84	0.37
					b07	31.34	31.29	31.39	0.07
					b10	30.13	27.78	31.34	2.03

Table 2. Average concentrations and standard deviations of dissolved Cu, Zn, Cd and Pb in the Belgian coastal zone (1981–1983)

Station	Cu diss $\mu g\ l^{-1}$		Zn diss $\mu g\ l^{-1}$		Cd diss $\mu g\ l^{-1}$		Pb diss $\mu g\ l^{-1}$	
	av	stdev	av	stdev	av	stdev	av	stdev
11	0.92	0.39	2.90	1.70	0.052	0.017	0.12	0.08
12	0.90	0.34	2.80	1.20	0.038	0.019	0.12	0.12
13	0.95	0.36	2.80	1.70	0.088	0.040	0.11	0.06
14	1.30	0.48	3.30	1.90	0.106	0.044	0.19	0.11
15	1.50	0.70	4.10	1.60	0.108	0.100	0.25	0.19
25	1.00	0.31	2.80	1.10	0.070	0.036	0.19	0.13
26	1.20	0.71	2.30	0.58	0.097	0.048	0.14	0.08
33	0.72	0.23	1.90	0.70	0.041	0.017	0.09	0.03
34	0.76	0.57	2.20	1.10	0.049	0.024	0.14	0.05
41	0.53	0.15	2.50	1.90	0.040	0.028	0.10	0.18
42	0.81	0.40	2.30	1.40	0.049	0.036	0.14	0.10
45	1.00	0.31	3.30	1.10	0.074	0.037		
46	0.75	0.54	1.70	0.49	0.046	0.010	0.09	0.04
52	0.74	1.00	1.60	0.39	0.047	0.021	0.11	0.07
53	0.70	0.60	2.50	1.30	0.041	0.027	0.09	0.09
54	0.60	0.21	2.40	1.10	0.040	0.010	0.11	0.06
55	0.66	0.10	3.00	1.90	0.045	0.010	0.09	0.05

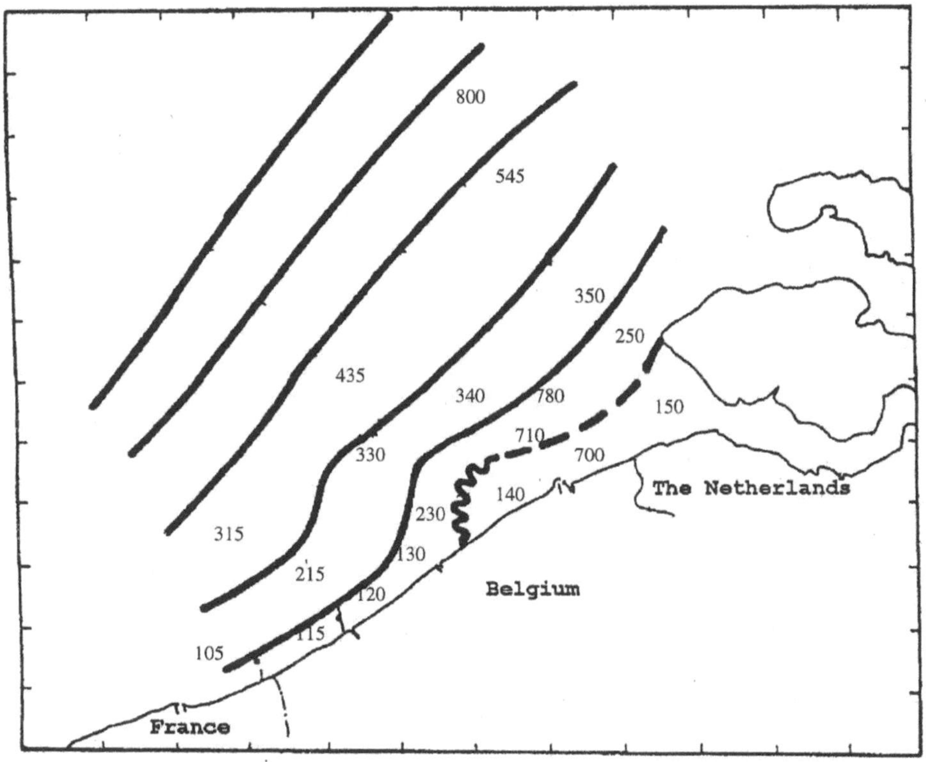

Figure 4. Location of the sampling stations of the surveys between 1995–1996.

from 0.04–0.10 $\mu g\ l^{-1}$, of Zn from 0.05–0.18 $\mu g\ l^{-1}$, of Cd from 0.005–0.02 $\mu g\ l^{-1}$ and of Pb from 0.015–0.08 $\mu g\ l^{-1}$. We can therefore be confident that the elevated results found in the Belgian coastal zone are not due to inadequate sampling and analytical techniques. Moreover, the trace metal concentrations measured in the Scheldt estuary and the Belgian coastal zone are in good agreement with those reported by other authors in the same period as is discussed in a related paper in this volume (Baeyens, 1998). A decreasing trend in trace metal concentrations in the Scheldt estuary between 1980–1995 ranging from 20–66% has been observed. For the 1995–1996 surveys concentrations range from 0.08–1.17 $\mu g\ l^{-1}$ for Cu, 0.06–2.72 $\mu g\ l^{-1}$ for Zn, 0.005–0.057 $\mu g\ l^{-1}$ for Cd, 0.016–0.152 $\mu g\ l^{-1}$ for Pb and 0.13–1.89 ng l^{-1} for Hg. Average concentrations and standard deviations are shown in Table 3. Trace metal-salinity profiles are shown in Figure 6. A progressively decreasing trend is apparent between 31.5 and 34.5 psu. Only Pb shows very little decrease in concentrations. The data are scattered around an average concentration of 0.04 $\mu g\ l^{-1}$. It should be noted that the reported concentrations refer to surface waters (30 cm depth) which may be influenced by atmospheric deposition. The dissolved Hg concentrations (Figure 7) show a progressive decrease from salinity 30.5 psu to 34.5 psu. A number of coastal stations in the vicinity of harbours (Zeebrugge, Oostende, Nieuwpoort) show enhanced concentrations.

The ratio between dissolved trace metal concentrations in the coastal estuarine watermass and the marine watermass is a factor of 2 for Cu, 1.7 for Zn, 1.8 for Pb, 2.3 for Cd in the period 1981–1983 and a factor 3.9 for Zn, 3.3 for Cu and Cd, 1.6 for Pb and 2.7 for Hg in 1995–1996. These ratios are much smaller than for particulate metals concentrations.

Particulate metals

In contrast to the dissolved trace metals, particulate metal concentrations observed in the 1981–1983 surveys (Table 4), remain almost constant in the coastal-estuarine watermass and abruptly decrease in the salinity range of 31.4–31.8 psu to the much lower values of the marine watermass (Figure 5). This sharp decrease in concentration shows a strong deviation from the ide-

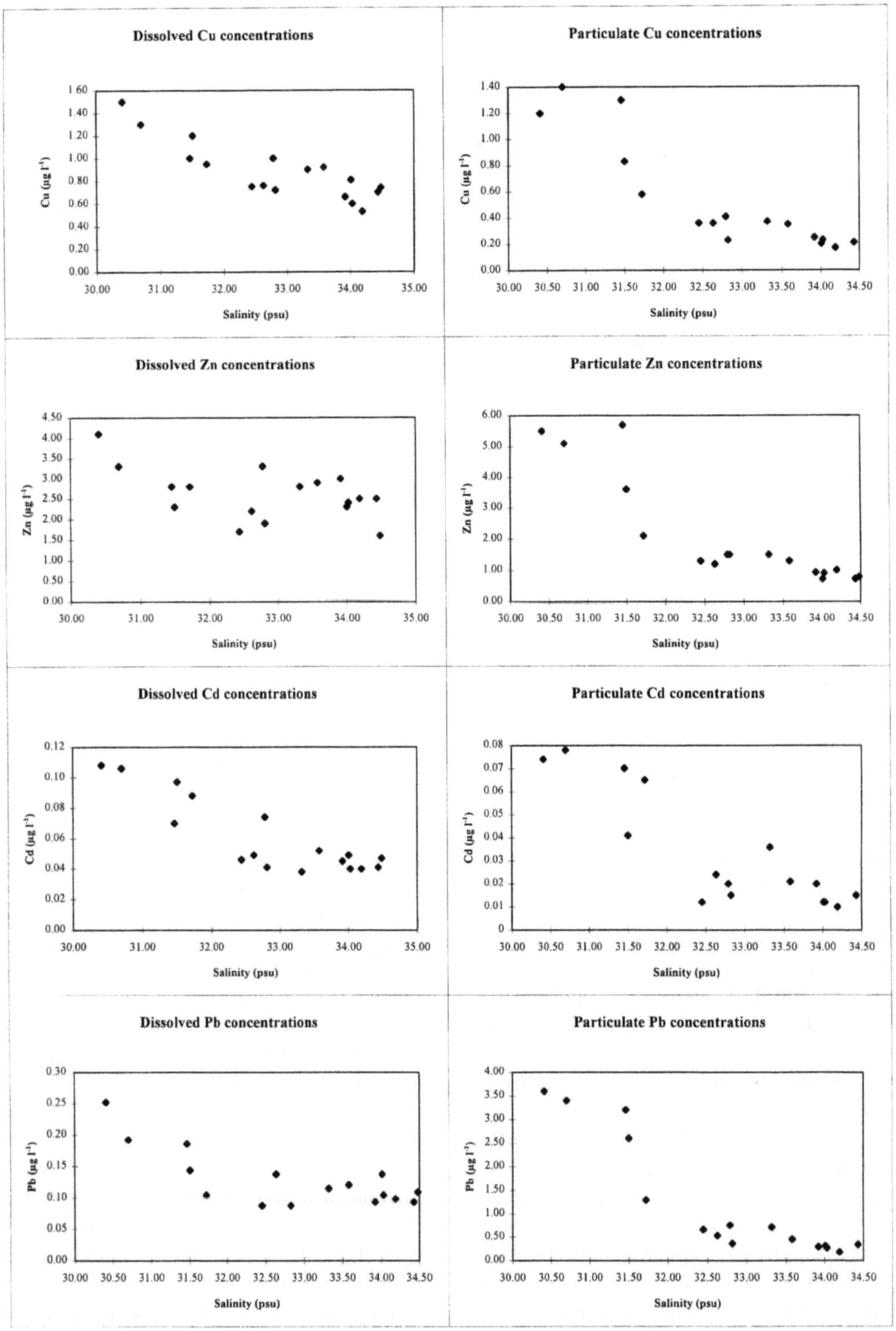

Figure 5. Stationary dissolved and particulate metal-salinity plot for the data of surveys between 1981–1983.

Table 3. Average concentrations and standard deviations of dissolved Cu, Zn, Cd, Pb and Hg in the Belgian coastal zone (1995–1996)

Station	Cu diss $\mu g\ l^{-1}$		Zn diss $\mu g\ l^{-1}$		Cd diss $\mu g\ l^{-1}$		Pb diss $\mu g\ l^{-1}$		Hg diss $ng\ l^{-1}$	
	av	stdev	av	stdev	av	stdev	av	stdev	av	stdev
115	0.39	0.23	0.55	0.13	0.013	0.004	0.037	0.015	0.72	0.16
120	0.49	0.30	0.56	0.15	0.015	0.007	0.039	0.020	0.57	0.20
130	0.70	0.28	1.30	0.57	0.031	0.022	0.048	0.028	0.66	0.24
140	0.80	0.27	1.25	0.44	0.026	0.016	0.033	0.006	0.61	0.12
150	0.74	0.29	1.38	0.88	0.028	0.015	0.040	0.009	1.12	0.69
215	0.48	0.35	0.60	0.26	0.019	0.010	0.044	0.019	0.52	0.14
230	0.71	0.31	1.34	0.56	0.030	0.016	0.048	0.017	0.82	0.53
250	0.62	0.19	1.08	0.42	0.025	0.007	0.042	0.032	0.84	0.69
315	0.33	0.17	0.50	0.25	0.016	0.003	0.048	0.024	0.58	0.21
330	0.40	0.15	0.71	0.32	0.020	0.005	0.047	0.020	0.54	0.11
340	0.73	0.24	0.60	0.00	0.025	0.005	0.036	0.010	0.69	0.48
350	0.72	0.19	1.02	0.42	0.030	0.012	0.046	0.008	0.37	0.18
421	0.25	0.09	0.39	0.19	0.013	0.001	0.060	0.054	0.36	0.27
435	0.28	0.13	0.66	0.25	0.016	0.003	0.045	0.021	0.39	0.03
545	0.30	0.11	0.53	0.28	0.018	0.005	0.033	0.019	0.40	0.07
700	0.75	0.27	1.26	0.75	0.033	0.014	0.065	0.016	0.99	0.49
710	0.76	0.16	1.32	0.73	0.033	0.013	0.055	0.023	0.79	0.36
780	0.58	0.13	1.26	0.55	0.027	0.005	0.057	0.031	0.53	0.07
800	0.21	0.10	0.31	0.16	0.012	0.005	0.029	0.019	0.24	0.02
b03	0.84	0.11	1.01	0.23	0.023	0.014	0.062	0.018	0.75	0.32
b07	0.81	0.14	1.04	0.46	0.024	0.022	0.046	0.007	0.72	0.46
b10	0.87	0.27	1.82	0.40	0.030	0.016	0.051	0.016	0.69	0.48

Table 4. Average concentrations and standard deviations of particulate Cu, Zn, Cd and Pb in the Belgian coastal zone (1981–1983)

Station	turbidity $\mu g\ l^{-1}$		Cu part $\mu g\ l^{-1}$		Zn part $\mu g\ l^{-1}$		Cd part $\mu g\ l^{-1}$		Pb part $\mu g\ l^{-1}$	
	av	stdev	av	stdev	av	stdev	av	stdev	av	stdev
11	13.9	7	0.35	0.30	1.30	1.00	0.021	0.013	0.45	0.21
12	19.9	7.6	0.37	0.22	1.50	1.50	0.036	0.031	0.71	0.32
13	26.5	28	0.58	0.39	2.10	2.10	0.065	0.066	1.3	0.57
14	51.8	57	1.40	0.88	5.10	4.00	0.078	0.096	3.4	4.6
15	44.2	22	1.20	0.53	5.50	5.80	0.074	0.046	3.6	7.6
25	57.4	34	1.30	0.92	5.70	3.00	0.070	0.037	3.2	2.1
26	36.1	19	0.83	0.79	3.60	2.90	0.041	0.020	2.6	1.7
33	11.7	5.2	0.23	0.34	1.50	2.50	0.015	0.024	0.36	0.48
34	15.8	9.7	0.36	0.59	1.20	0.96	0.024	0.051	0.53	0.38
41	6.8	1.4	0.17	0.10	1.00	2.40	0.010	0.005	0.18	0.14
42	9.1	8.4	0.20	0.28	0.71	1.30	0.012	0.018	0.31	0.32
45	14	4.6	0.41	0.41	1.50	0.49	0.020	0.010	0.75	0.46
46	17	8.4	0.36	0.27	1.30	1.40	0.012	0.010	0.66	0.54
52	9.2	7.8			0.79	1.20				
53	7.4	7.7	0.21	0.24	0.72	0.72	0.015	0.031	0.34	0.97
54	10.5	9	0.23	0.31	0.90	1.50	0.012	0.018	0.27	0.56
55	9.1	20	0.25	0.55	0.93	2.00	0.020	0.056	0.29	0.58

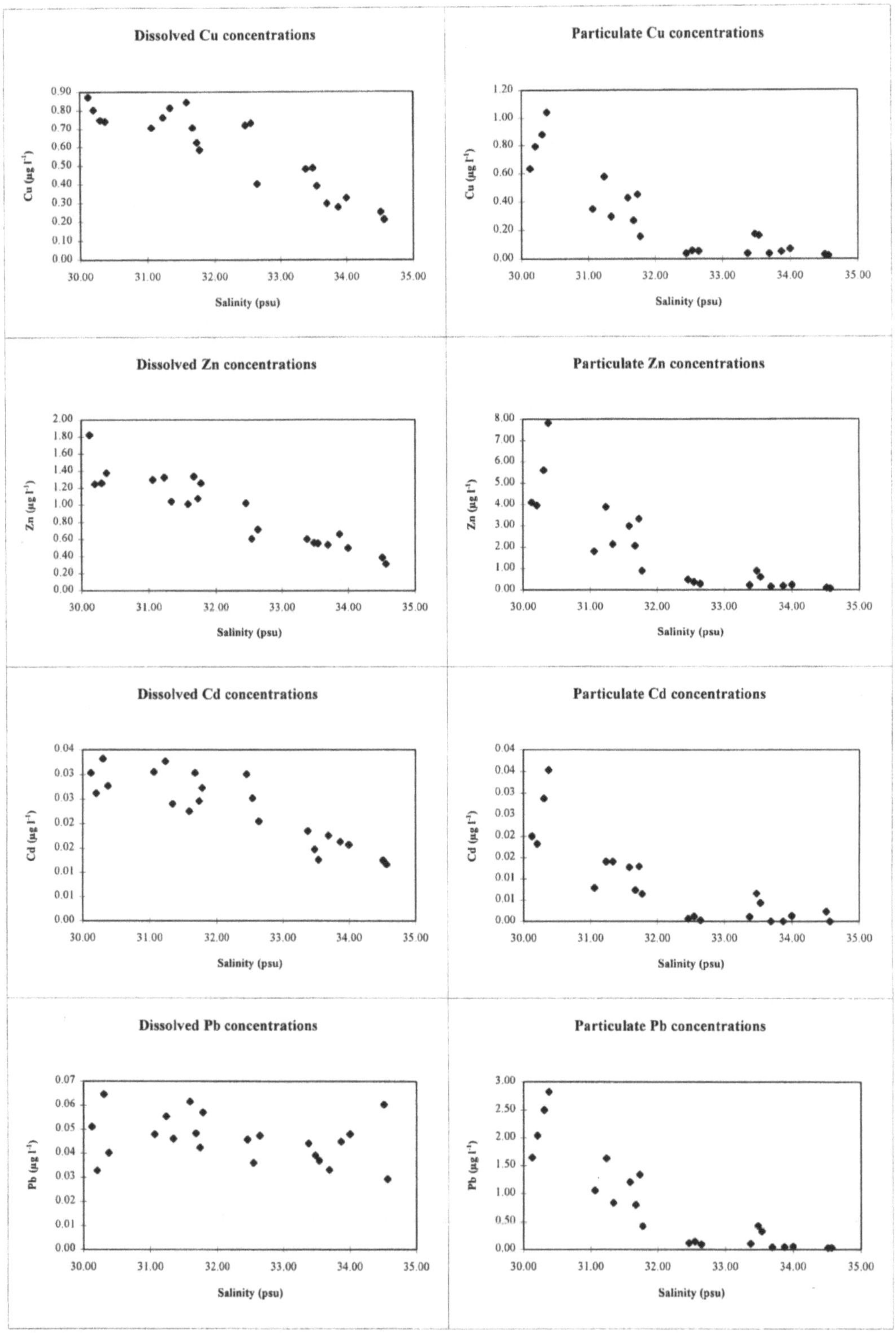

Figure 6. Stationary dissolved and particulate metal-salinity plot for the data of surveys between 1995–1996.

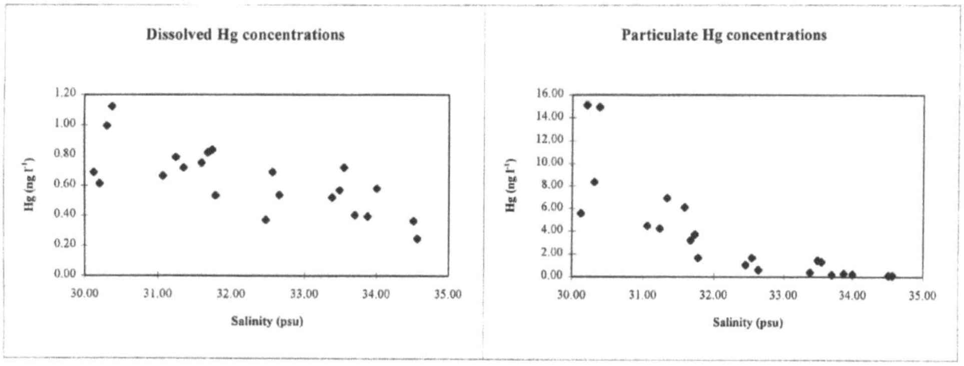

Figure 7. Stationary dissolved and particulate mercury-salinity plot for the data of surveys between 1995–1996.

al dilution line between 30.4 and 32.4 psu. The offshore transport of trace metals is clearly more limited than that of dissolved trace metals.

The particulate concentrations of Zn, Cu and Pb in the coastal-estuarine watermass stations (30.7 to 31.4 psu) are also comparable to the average concentrations in the mouth of the Scheldt (26.5 psu), with the remark that these latter values rely on a smaller number of measurements. The suspended matter content (Table 4), however, increases slightly from 40 mg l^{-1} (at the mouth of the estuary) to 57 mg l^{-1} (31.46 psu, station 25), which can be attributed to local differences in sedimentation and erosion characteristics (e.g. Adam et al., 1981). This implies a slight decrease in metal concentration in the suspended matter ($\mu g\ g^{-1}$). In the marine watermass (salinity $>$ 32.4 psu), particulate Zn, Pb and Cu follow the same pattern as the turbidity, smoothly decreasing with increasing salinity. The fluctuations around this curve are much smaller than in the case of dissolved metals. The levels in the coastal-estuarine watermass exceed the lower values in the marine watermass by factors of about 5 to 6 for Cd, Cu and Zn, comparable to the turbidity, but of 12 for Pb. Pb differs from the other metals in that it is predominantly present in the particulate phase. Its K_D ratio (particulate to dissolved metal concentration) changes more than that of the other metals from inshore to offshore waters (Baeyens et al., 1987). Pb seems to be associated with iron and manganese hydrous oxides and these kind of particles, introduced in the coastal system via the estuary, are not found in the marine water mass. Similar observations can be made for Hg in the 1995–1996 surveys. The levels of particulate Hg in the coastal-estuarine watermass exceed the values in the marine watermass by factors of 16. However, the results from 1995 do not confirm the different behaviour of Pb compared to the other three metals. Zn, Cd and Cu levels in the coastal-estuarine watermass exceed the values in the marine watermass by factors of about 9 to 11.5, while for lead this factor is only slightly higher (13).

Scheldt input

Trace metal flows from the Scheldt river into the coastal waters for the 1981–1983 data were obtained by extrapolating the steady state coastal concentrations to a salinity of 26.5 psu and using a residual riverine flow of 4.4 $km^3\ y^{-1}$. These inputs are shown in Table 6 and are in good agreement with the box model calculations of trace metal inputs from the Scheldt estuary (Baeyens et al., 1998d). For the 1995–1996 data the box model calculations were used in the Scheldt input estimates.

Offshore transport

As discussed in the section on residual currents, three processes may affect the concentrations in streamtube I: 1) coastal inputs upstream the hydrodynamical front; 2) diffusion through the front; and 3) mixing with the coastal-estuarine watermass downstream of the front. Exchange of material between streamtube I and streamtube II is only possible through diffusion. These processes shall be discussed for dissolved and particulate metal flows. Uncertainties associated with the calculations of fluxes in this manner are associated with analytical uncertainties, uncertainties associated with the approximation of steady state conditions as

Table 5. Average concentrations and standard deviations of particulate Cu, Zn, Cd, Pb and Hg in the Belgian coastal zone (1995–1996)

Station	Turbidity mg l^{-1}		Cu part $\mu g\ l^{-1}$		Zn part $\mu g\ l^{-1}$		Cd part $\mu g\ l^{-1}$		Pb part $\mu g\ l^{-1}$		Hg par ng l^{-1}	
	av	stdev	av	stdev	av	stdev	av	stdev	av	stdev	av	stdev
115	8.0	6.9	0.16	0.15	0.60	0.66	0.004	0.004	0.33	0.33	1.27	1.56
120	8.6	5.1	0.17	0.05	0.89	0.61	0.007	0.001	0.43	0.24	1.41	1.00
130	19.5	9.5	0.35	0.17	1.80	1.61	0.008	0.003	1.06	0.51	4.46	4.12
140	40.6	24.1	0.79	0.48	3.94	2.92	0.018	0.011	2.03	1.21	15.12	8.62
150	53.2	56.9	1.04	1.26	7.81	9.94	0.035	0.041	2.82	3.25	14.95	15.09
215	5.5	3.4	0.04	0.03	0.22	0.18	0.001	0.000	0.11	0.04	0.36	0.42
230	19.0	14.7	0.27	0.23	2.05	1.59	0.007	0.005	0.80	0.57	3.19	3.02
250	27.1	21.7	0.45	0.46	3.31	3.52	0.013	0.011	1.34	1.25	3.68	3.89
315	4.0	2.5	0.07	0.03	0.23	0.18	0.001	0.000	0.05	0.02	0.22	0.17
330	5.8	3.5	0.05	0.03	0.28	0.21			0.09	0.08	0.59	0.66
340	5.1	2.4	0.06	0.06	0.36	0.26	0.001	0.001	0.14	0.08	1.63	0.82
350	4.1	2.6	0.04	0.01	0.48	0.16	0.001	0.000	0.12	0.02	1.00	0.46
421	2.3	1.1	0.03	0.02	0.10	0.07	0.002		0.03	0.02	0.09	0.07
435	5.1	1.6	0.05	0.04	0.18	0.18			0.05	0.03	0.27	0.03
545	3.5	1.2	0.04	0.01	0.14	0.07			0.04	0.03	0.16	0.06
700	42.4	18.8	0.88	0.52	5.60	2.88	0.029	0.015	2.50	1.53	8.34	8.46
710	30.2	27.0	0.58	0.59	3.88	3.60	0.014	0.010	1.63	1.60	4.21	0.81
780	10.2	5.7	0.15	0.17	0.88	0.88	0.006	0.002	0.42	0.41	1.63	0.72
800	2.5	1.0	0.02	0.02	0.06	0.02			0.03	0.03	0.09	0.02
b03	27.6	5.2	0.43	0.05	2.98	0.77	0.013	0.006	1.21	0.15	6.08	3.75
b07	20.4	3.5	0.29	0.01	2.13	0.28	0.014	0.004	0.84	0.04	6.91	5.99
b10	32.6	15.1	0.63	0.43	4.09	2.37	0.020	0.007	1.65	1.05	5.57	3.60

Table 6. Trace metal flows (tons y^{-1}) through the Belgian coastal zone (1981–1983)

	Dissolved metal flows				Particulate metal flows			
	Cu	Zn	Cd	Pb	Cu	Zn	Cd	Pb
	tons y^{-1}	tons y^{-1}	tons y^{-1}	tons y^{-1}	tons y^{-1}	tons y^{-1}	tons y^{-1}	tons y^{-1}
Scheldt input	7.04	18.04	0.55	1.10	5.28	24.20	0.44	15.40
Diffusion hydrodynamic front	3.30	9.90	0.22	0.57	7.40	31.00	0.45	20.00
% Scheldt	*46.88*	*54.88*	*40.00*	*51.82*	*140.15*	*128.10*	*102.27*	*129.87*
Mixing	4.80	12.10	0.40	0.71	5.50	23.80	0.30	14.10
% Scheldt	*68.18*	*67.07*	*72.73*	*64.55*	*104.17*	*98.35*	*68.18*	*91.56*
Diffusion I to II	5.22	10.70	0.83	0.86	3.30	14.00	0.20	17.00
Flow parallel to the coast	900	3000	59	142	340	1400	21	520
English channel	530	930	38	180	120		6	190
[1]	280	760	13					
[2]	430	980	52					
Northern North Sea	690	2500	39	300	78		5.4	140
[1]	330	3400	88					
[2]	1300.00	1300.00	38.00					

[1] Jones, in Topping et al. (1980)
[2] Duinker & Nolting (1982)

well as non-conservative behaviour of the elements. The temporal variations in trace metal concentrations largely exceed analytical variance and thus the approximation of steady state conditions (by averaging the results of the different surveys) is probably the limiting factor.

Dissolved trace metals

Coastal inputs

These seem to be important for Cu and Zn for the 1981 to 1983 surveys as well as for Pb and Hg in 1995. However, no quantified data of such inputs exist.

Diffusion at the hydrodynamic front

For the calculation of the concentration gradients at the hydrodynamical front, we made use of the salinity at the stations at each side of the front and the corresponding metal concentrations inferred from the metal-salinity best fit curve. For the 1981 to 1983 surveys, the concentration gradients at the front (the length of the front is about 10 km) amount to 0.022 (Cu), 0.065 (Zn), 0.0014 (Cd), 0.0037 (Pb) in tons km^{-4}. Using the diffusion coefficient deduced for salinity at the front (1018 km^2 y^{-1} or 32 m^2 s^{-1}), diffusion fluxes of 3.3 (Cu), 9.9 (Zn), 0.22 (Cd), 0.57 (Pb) in tons y^{-1} are obtained (Table 6). These diffusion fluxes represent between 40 and 55% of the dissolved metal supply by the river Scheldt. The resulting increases in streamtube I are 0.039 (Cu), 0.12 (Zn), 0.0025 (Cd), 0.0067 (Pb), in $\mu g\ l^{-1}$.

In 1995–1996, diffusion fluxes of 1.14 (Cu), 1.27 (Zn) in tons y^{-1}, and 0.78 kg y^{-1} Hg are calculated (Table 7). The fluxes were negligible for Cd and Pb and represent from 0 to 24.3% of the dissolved metal supply by the river Scheldt. The resulting increases in streamtube I are 0.011 (Cu), 0.01 (Zn) in $\mu g\ l^{-1}$and 0.01 ng l^{-1}Hg.

Mixing

The increases in streamtube I due to mixing with the adjacent coastal-estuarine water mass (4.4 km^3 y^{-1} and metal concentrations observed at S = 31.46) yield for the 1981 to 1983 surveys: 0.057 (Cu), 0.14 (Zn), 0.0047 (Cd), 0.0084 (Pb), in $\mu g\ l^{-1}$. These mixing fluxes (4.8 (Cu), 12.1 (Zn), 0.40 (Cd) and 0.71 (Pb), in tons y^{-1}) represent between 64% and 73% of the dissolved metal supply by the river Scheldt (Table 6). The input flux from the river Scheldt can almost account (between 87% and 90%) for the total supply (diffusion and mixing) from the coastal-estuarine water mass to streamtube I. When we compare the concentrations observed at station 13 (S = 31.72), increased by the contributions from diffusion and mixing calculated above, with the concentrations at station 26 (salinity 31.5 psu), we find a good agreement (Table 8).

In 1995–1996, mixing fluxes of 2.75 (Cu), 4.73 (Zn), 0.11 (Cd) and 0.19 (Pb), in tons y^{-1} and 3.68 kg y^{-1} Hg were calculated. These fluxes represent between 48.8 and 81% of the dissolved metal supply by the river Scheldt (Table 7). The input flux from the river Scheldt can completely account (between 120–200%) for the total supply (diffusion and mixing) from the coastal-estuarine water mass to streamtube I. The comparison between observed and calculated concentrations at the outflow of streamtube I was not carried out, since sampling station 26, located at the outflow of streamtube I was not sampled during the 1995–1996 surveys.

Exchange of material between streamtube I (85 km^3 y^{-1}) and streamtube II (230 km^3 y^{-1}) is only possible through diffusion. For the 1981 to 1983 surveys, metal concentration gradients upstream of station 13 are very small and diffusion is negligible. The diffusion fluxes downstream of station 13 amount to 5.22 (Cu), 10.7 (Zn), 0.83 (Cd), 0.86 (Pb), in tons y^{-1}. They are calculated analogous to the diffusion fluxes from the coastal-estuarine water mass to streamtube I, but now over a length of 30 km and using the concentration gradients between stations 13 and 33. The resulting increases in streamtube II are then 0.02 (Cu), 0.05 (Zn), 0.004 (Cd), 0.004 (Pb), in $\mu g\ l^{-1}$, and represent a concentration increase in streamtube II of less than 3%, except for Cd (8%). The input of the river Scheldt is between 1.4 and 1.7 times larger than the diffusion flux to streamtube II, except for Cd (only 0.7 times this diffusion flux). Since the total dissolved metal input of the Scheldt estuary is required to balance the metal supply to streamtube I, an additional source (atmosphere, bottom,. . .) is necessary to support the flows to streamtube II.

In 1995–1996, the diffusion flux from streamtube I to streamtube II respectively amounts to 2.21 (Cu), 12.42 (Zn), 0.13 (Cd) and 0.16 (Pb), in tons $y^{-1}$and 4.73 kg y^{-1} Hg. The resulting increases in streamtube II are 0.009 (Cu), 0.053 (Zn), 0.0005 (Cd), and 0.0007 (Pb), in $\mu g\ l^{-1}$ and 0.02 ng l^{-1} Hg. These fluxes represent 46% (Cu), 127% (Zn), 59% (Cd), 69% (Pb) and 77% (Hg) of the dissolved metal input of the river Scheldt. Hence in 1995–1996, also an additional

Table 7. Trace metal flows (tons y^{-1} except for Hg in kg y^{-1}) through the Belgian coastal zone (1995–1996)

	Dissolved metal flows					Particulate metal flows				
	Cu tons y^{-1}	Zn tons y^{-1}	Cd tons y^{-1}	Pb tons y^{-1}	Hg kg y^{-1}	Cu tons y^{-1}	Zn tons y^{-1}	Cd tons y^{-1}	Pb tons y^{-1}	Hg kg y^{-1}
Scheldt input	4.70	9.70	0.22	0.23	6.10	1.60	12.20	0.14	8.10	40.5
Diffusion hydrodynamic front	1.14	1.27	0.00	0.00	0.78	2.87	18.34	0.11	7.11	54.5
% Scheldt	*24.3*	*13.1*	*0.00*	*0.0*	*12.7*	*179.4*	*150.3*	*75.8*	*87.8*	*134.4*
Mixing with streamtube I	2.75	4.73	0.11	0.19	3.68	1.98	14.58	0.06	5.88	16.2
% Scheldt	*58.4*	*48.8*	*49.2*	*81.0*	*60.3*	*124.0*	*119.5*	*40.7*	*72.6*	*40.0*
Diffusion I to II	2.21	12.42	0.13	0.16	4.73	7.41	48.68	0.21	22.19	60.3
Diffusion II to III	2.90	1.25	0.09	0.050	1.68	0.00	3.57	0.00	1.59	16.1
Flow Parallel to the coast	650	1040	28.00	58.00	740	250	1500	8.50	680	2400
Dover straight										
total flux (flow 2850 km^3 y^{-1}) [1]	894	1680	61.7	168.0		422	5770	12	1000	
normalised to 1260 km^3 y^{-1}	380	717	26.2	71.4		179	2452	5	425	
total flux (flow 2964 km^3 y^{-1})	820	2200	55.0	55.0						
(from data Jones, 1994) [2]						490	1310	32	3240	
normalised to 1260 km^3 y^{-1}	363	973	24	24		217	580	14	1434	
(from data Howarth et al., 1993) [2]						84	75	8	1681	
total flux (flow 2850 km^{-3} y^{-1}) [3]					1980					2300
normalised to 1260 km^{-3} y^{-1}					694					807

[1] Statham et al., 1993
[2] Prandle et al., 1996
[3] Cossa et al., 1994

Table 8. Comparison of observed and calculated metal concentrations in streamtube I

Station	Salinity	Dissolved metal concentrations					Particulate metal concentrations				
	psu	Cu $\mu g\ l^{-1}$	Zn $\mu g\ l^{-1}$	Cd $\mu g\ l^{-1}$	Pb $\mu g\ l^{-1}$	Hg ng l^{-1}	Cu $\mu g\ l^{-1}$	Zn $\mu g\ l^{-1}$	Cd $\mu g\ l^{-1}$	Pb $\mu g\ l^{-1}$	Hg ng l^{-1}
1981–1983											
13	31.72	1.00	2.50	0.08	0.15		0.54	2.05	0.03	1.29	
26	31.50	1.09	2.70	0.09	0.16		0.78	3.00	0.04	2.10	
Calc	31.50	1.10	2.76	0.09	0.16		0.65	2.70	0.04	1.70	

source (atmosphere, bottom,...) is necessary to support the total offshore flows.

Diffusion to the outer streamtubes (III, IV, V) is insignificant as a result of negligible concentration gradients for the 1981 to 1983 surveys. For 1995–1996, diffusion fluxes of 2.9 (Cu), 1.25 (Zn), 0.09 (Cd), 0.05 (Pb) tons y^{-1} and 1.68 kg y^{-1} Hg are calculated. This results in increases of 0.012 (Cu), 0.005 (Zn), 0.0004 (Cd), 0.0003 (Pb) in $\mu g\ l^{-1}$ and 0.0007 ng l^{-1} Hg.

Transport parallel to the coast

The total flow through streamtubes I to V (Figure 1) equals 1260 km^3 y^{-1}; about 25% of the flow entering the Southern Bight of the North Sea through the English Channel. The cumulative dissolved metal flow through streamtubes I to V equals 3000 (Zn), 142 (Pb), 900 (Cu), 59 (Cd) in tons y^{-1} for the 1981–1983 surveys. The contribution of the Scheldt to this flow is very small and ranges from 0.6 to 0.9%. In 1995–1996, the cumulative dissolved metal flow through streamtubes I to V equals 1036 (Zn), 57.89 (Pb), 645 (Cu), 27.72 (Cd) in tons y^{-1} and 742 kg y^{-1} Hg, with a contribution of the Scheldt between 0.4 and 0.9%.

To obtain an idea of the magnitude of these metal flows, we compared them with dissolved metal flows through the English Channel, and the northern North Sea, normalised to a flow of 1260 km^3 y^{-1}. Trace metal data for the 1981–1983 period are obtained from Jones in Topping et al. (1980), Duinker & Nolting (1982) and our measurements in the considered areas. It appears from Table 6 that the metal flows through the English Channel are always lower than those through the Belgian coastal area; the latter are of the same magnitude or higher as those through the northern North Sea. Much larger inputs than the Scheldt estuary are, however, required to explain the increase of dissolved metals in the northern direction. For the 1995–1996 period, we compared the fluxes with those derived through the Straight of Dover during the Fluxmanche Experiment (Statham et al., 1993 and Cossa et al., 1994) as well as with estimates made by Prandle et al., 1996. The total fluxes were again normalised to a flow of 1260 km^3 y^{-1}. For Cu and Zn, the flows parallel to the coast exceed the estimates through the Straight of Dover; whereas for Cd, Pb and Hg the fluxes are comparable. The dissolved metal flows calculated for the Scheldt and the Belgian coast in 1995–1996 are roughly a factor 2, and for the Scheldt input of Pb a factor 5, lower than in 1981–1983. The metal concentrations observed in the period 1981–1983 in the English Channel are comparable to the current metal concentrations in the Belgian coastal area. Dissolved trace metal concentrations measured in the most offshore stations at the Belgian coast are comparable to those measured in offshore stations in the English Channel and North Sea (Table 9).

Particulate metals

The offshore flux of the particles can be estimated, assuming they behave in a similar way as the fluid in which they are suspended.

Diffusion

The concentration gradients and the diffusion fluxes at the hydrodynamical front are calculated similarly to those of the dissolved metals. For the 1981–1983 surveys, the diffusion fluxes amount to 7.4 (Cu), 31 (Zn), 0.45 (Cd), 20 (Pb), in tons y^{-1}. These diffusion fluxes represent between 130 and 140% of the particulate metal supply by the river Scheldt except for Cd where the diffusion flux is balanced by the river input. The resulting increases in streamtube I are: 0.087 (Cu), 0.37 (Zn), 0.0053 (Cd), 0.24 (Pb), in $\mu g\ l^{-1}$.

In 1995–1996, the diffusion fluxes amount to 2.87 (Cu), 18.34 (Zn), 0.11 (Cd), 7.11 (Pb), in tons y^{-1} and 54.5 kg y^{-1} Hg. These diffusion fluxes represent between 75.8 and 179% of the particulate metal supply by the river Scheldt. The resulting increases in streamtube I are: 0.03 (Cu), 0.21 (Zn), 0.001 (Cd), 0.08 (Pb), in $\mu g\ l^{-1}$ and 0.61 ng l^{-1} Hg.

Mixing

The increases in streamtube I due to mixing with the coastal-estuarine water mass (4.4 km^3 y^{-1} and concentration observed at 31.46 psu) yield for the 1981–1983 data: 0.064 (Cu), 0.28 (Zn), 0.0035 (Cd), 0.17 (Pb), in $\mu g\ l^{-1}$. The mixing fluxes (0.30 (Cd), 5.5 (Cu), 23.8 (Zn), 14.1 (Pb), in tons $y^{-1)}$ represent between 70 and 100% of the Scheldt supply. In 1995–1996 the mixing fluxes are 1.98 (Cu), 14.58 (Zn), 0.06 (Cd), 5.88 (Pb), in tons y^{-1} and 16.2 kg y^{-1} Hg. They represent between 40 and 124% of the Scheldt supply.

When we compare for the 1981–1983 data, the concentrations observed at station 13 (31.72 psu), increased by the contribution from diffusion and mixing as calculated above, with the concentrations at station 26 (31.5 psu), we find for most metals a good, but not as excellent agreement as that for dissolved metals (see Table 6). The Scheldt input only accounts for 41 to 58% of the trace metal flux to streamtube I through mixing and diffusion.

In 1995–1996, the Scheldt input represents between 33 and 82% of the Cu, Zn Pb, Cd, and Hg supply from the coastal-estuarine water mass to streamtube I.

Exchange of material between streamtube I (85 km^3 y^{-1}) and streamtube II (230 km^3 y^{-1}) is only possible through diffusion. Upstream of station 13, metal concentration gradients are very small and diffusion is negligible. The diffusion fluxes downstream of station 13 amount to 0.20 (Cd), 14 (Zn), 17 (Pb), 3.3 (Cu), in tons y^{-1} for 1981–1983. They are calculated analogous to the diffusion fluxes of dissolved metals from

Table 9. Concentrations of dissolved Cu, Zn, Cd, Pb and Hg in the English Channel and North Sea (literature data)

location	Dissolved metal concentrations					Reference
	Cu $\mu g\ l^{-1}$	Zn $\mu g\ l^{-1}$	Cd $\mu g\ l^{-1}$	Pb $\mu g\ l^{-1}$	Hg $ng\ l^{-1}$	
Belgian coast (sal. > 34.4 psu)	0.23	0.35	0.01	0.04	0.30	our data. 1995–1996
Dover Straight	0.343	0.61	0.022	0.068		Stratham et al., 1993
					0.12–1.34	Cossa et al., 1994
Western English Channel	0.14–0.36	0.17–0.28	0.009–0.012	0.01–0.017		(OSPAR/ICES)
Central North Sea	0.178	0.241	0.019	0.023		(Burton, 1993)
					0.18–0.5	Coquery & Cossa, 1995
Northern North Sea	0.05–0.1	0.25–0.45	0.008–0.025	0.01–0.02	0.2–0.5	(OSPAR/ICES)

streamtube I to streamtube II. The resulting increases in streamtube II are then: 0.0009 (Cd), 0.06 (Zn), 0.07 (Pb), 0.01 (Cu), in $\mu g\ l^{-1}$ and represent a concentration increase in streamtube II of 4 to 5% for Cd, Cu and Zn, of 13% for Pb. These increases fall in the range of observed increases. The coastal values of Pb are increased more drastically over the marine water mass levels than Cu, Zn and Cd and result in much larger diffusion fluxes and concentration increases. When we compare the particulate metals input of the river Scheldt to the diffusion flux to streamtube II, we find the Scheldt input larger than the diffusion flux for Cd, Zn and Cu (between 1.6 and 2.2 times), but smaller for Pb (about 0.8 to 0.9 times).

In 1995–1996, the diffusion flux from streamtube I to streamtube II equals 7.41 (Cu), 48.68 (Zn), 0.21 (Cd), 22.19 (Pb), in tons y^{-1} and 60.3 kg y^{-1} Hg. The resulting increases in streamtube II are: 0.032 (Cu), 0.21 (Zn), 0.0001 (Cd), 0.09 (Pb), in $\mu g\ l^{-1}$ and 0.26 ng l^{-1} Hg. In the outer streamtubes (III, IV) concentration gradients are only significant for Zn, Pb and Hg. Diffusion fluxes from streamtube II to III are 3.57 for Zn, 1.59 tons y^{-1} Pb and 0.76 kg y^{-1} Hg, resulting in increases of 0.015 $\mu g\ l^{-1}$ Zn, 0.007 $\mu g\ l^{-1}$ Pb and 0.069 ng l^{-1} Hg.

In any case, if particulates move like the bulk of the fluid, our calculations of mixing and diffusion processes require an additional source (atmosphere, bottom,...) to balance these flows.

The deficit expressed in terms of the ratio of the Scheldt input of particulate metals to the total offshore flux of particulate metals varies between 33% for Cu, 30% for Zn, 46% for Cd and 35% for Pb for the 1981–1983 data. In 1995–1996 these figures are: 13% for Cu, 15% for Zn, 37% for Cd 23% for Pb, and 31% for Hg.

Transport parallel to the coast

The cumulative particulate metal flow through streamtubes I to V equals for the 1981–1983 data: 1400 (Zn), 520 (Pb), 340 (Cu), 21 (Cd), in tons y^{-1}, considering particulate metals as conservative substances. The contribution of the Scheldt to these flows is somewhat higher than in the case of dissolved metals and ranges from 1.6 to 2.9%.

In 1995–1996, the cumulative particulate metal flow through streamtubes I to V equals 252 (Cu), 1560 (Zn), 8.52 (Cd), 678 (Pb), in tons y^{-1} and 2416 kg y^{-1} Hg, making the same assumption as above. The contribution of the Scheldt to these flows varies now from 0.6 to 1.6%.

The particulate metal flows through the English Channel and the northern North Sea (1981–1983) have been calculated in a similar manner to those of dissolved metals. Both particulate metal flows are of the same magnitude, but are about three times lower than the particulate metal flows through the Belgian coastal zone (Table 6). Either additional sources or a non-conservative behaviour of the particulate metals (e.g. erosion-sedimentation cycles) must explain this result. Similar observations can be made for the 1995–1996 data. Only the Zn flux through the Straight of Dover (normalised to 1260 $km^{-3}\ y^{-1}$) reported by Statham et al. (1993) exceeds the flux parallel to the Belgian coast. On the other hand Prandle et al. (1996) calculated lower Zn fluxes and higher Pb fluxes than our estimates parallel to the coast (Table 7).

Conclusion

Based on steady state concentration distributions of dissolved and particulate trace metals in the Belgian coastal zone from 1981 to 1983, offshore fluxes of the trace metals could be calculated. Results obtained in 1995–1996 are also presented but the conclusions must be seen as preliminary, since five cruises can not represent steady-state conditions.

When we compare the offshore flux (diffusion and mixing) of metals from the Scheldt to streamtube I and then to streamtube II, with the theoretical offshore flux necessary to explain the observed increases in metal concentration in the latter streamtubes during their flow through the Belgian coastal area, it appears that the river input does not completely balance the increases in those streamtubes.

For the dissolved species, the deficit in the period 1981–1983 is of the same magnitude as the river input (the ratio of the Scheldt input to the offshore flux varies from 38 to 55% depending on the kind of metal). In 1995–1996, the dissolved metal fluxes through the Belgian coastal area as well as the input fluxes by the Scheldt seem to be lower, a factor 2 and for the Scheldt input of Pb even a factor 5, than in the period 1981–1983. Moreover, the deficit that appears when comparing the Scheldt input with the offshore flux of dissolved metals seems to be smaller than in 1981–1983. The ratio of the Scheldt input to the offshore flux varies from 67% to 83% for Cd, Cu and Zn, and for Pb even an inshore flux is observed.

For the particulate species, the deficit of the Scheldt input compared to the offshore flux is larger than in the case of the dissolved species: the ratio of the Scheldt input to the offshore flux varies from 30 to 46% in the period 1981–1983 and from 13–31% in 1995. The particulate metal flows in the Belgian coastal zone are not substantially different between 1981–1983 and 1995, except for Cd showing a decrease with approximately a factor 3.

Additional sources such as atmosphere, bottom, dumping, are possible explanations for the deficit of the Scheldt input compared to the offshore flux and should be considered in future. On the other hand, if we assume that in the shallow Belgian coastal area the high load of suspended matter is submitted to sedimentation-erosion cycles, then the particulate metals are flowing much slower than is the water mass (the residual advective and diffusive velocities of the particles are thus much smaller than the fluid in which they are suspended). For example, if – in the Belgian coastal area – flows of particulate metals are only one fourth of those we have calculated (their advective and diffusive velocity is thus four times smaller than the fluid in which they are suspended), then: 1) the flows of particulate metals parallel to the coast are of the same order as those observed in the English Channel and the northern North Sea; and 2) the river input of particulate metals will now show a small excess compared to the offshore flux, excess which satisfies the small deficit of the dissolved metals offshore flux.

Testing the assumption concerning the transport velocity of particulate metals requires information about the size distribution and density of the particles, their settling rates and the size distribution of the particulate metals, while testing the assumption concerning the transfer of metals from the particulate to the dissolved phase requires information about adsorption-desorption processes in the water column and mobilisation processes in the sediments.

Acknowledgements

This research was supported by the Ministry of Scientific Policy (Geconcerteerde Acties Oceanologie) for the period 1981–1983 and the Ministry of Public Health and Environment for the period in 1995. We thank the Management Unit of the Mathematical Model of the North Sea for providing ship time.

References

Adam, Y., W. Baeyens, J. P. Mommaerts & G. Pichot, 1981. Mathematical modelling of recent sedimentology in the shallow waters along the Belgian coast. In J. C. J. Nihoul (ed.), Ecohydrodynamics. Elsevier, Amsterdam: 333–349.

Baeyens, W., G. Gillain, G. Decadt & I. Elskens, 1987. Trace metals in the eastern part of the North Sea. I. Analyses and short term distributions. Oceanol. Acta 10: 169–179.

Baeyens, W., M. Elskens, G. Gillain & L. Goeyens 1998a. Biogeochemical behaviour of Cd, Cu, Pb and Zn in the Scheldt estuary during the period 1981–1983. Hydrobiologia 366: 15–44.

Baeyens, W., K. Parmentier, L. Goeyens, G. Ducastel, M. De Gieter & M. Leermakers 1998b. The biogeochemical behaviour of Cd, Cu, Pb and Zn in the Scheldt estuary: results of the 1995 surveys. Hydrobiologia 366: 45–62.

Baeyens, W., C. Meuleman & B. Muhaya & M. Leermakers 1998c. Behaviour and speciation of mercury in the Scheldt estuary (water, sediments and benthic organisms). Hydrobiologia 366: 63–79.

Baeyens, W., F. Monteny, R. Van Ryssen & M. Leermakers, 1998d. A box-model of metal flows through the Scheldt estuary (1981–1983 and 1992–1995). Hydrobiologia 366: 109–128.

Burton, J., M. Althaus, G. Millward, A. Morris, P. Statham, A. Tappin & A. Turner, 1993. Processes influencing the fate of trace metals in the North Sea. Phil. Trans. r. Soc. A. 343: 557–568.

Coquery, M. & D. Cossa, 1995. Mercury concentrations in surface waters of the North Sea, Neth. J. Sea Res. 34: 245–257.

Cossa, D., J. Sanjuan & J. Noel, 1994. Mercury transport in waters of the Strait of Dover, Mar. Pollut. Bull. 28: 385–388.

Djenidi, S. & F. Ronday, 1985. Tidal and residual circulation on the North-European continental shelf. In R. Van Grieken & R. Wollast (eds), Progress in Belgian oceanographic research. The Belgian Academy of Sciences, Committee of Oceanology, Brussels: 46–59.

Duinker, J. C. & R. F. Nolting, 1982. Dissolved copper, zinc and cadmium in the Southern Bight of the North Sea. Mar. Pollut. Bull. 13: 93–96.

McDowell D. M. & B. A. O'Connor, 1977. Hydraulic Behaviour of Estuaries. The MacMillan Press Ltd., London, 292 pp.

Mommaerts, J. P., G. Pichot, J. Ozer, Y. Adam & W. Baeyens, 1984. Nitrogen cycling and budget in Belgian coastal waters North Sea areas with and without river inputs. Rapp. Pv. Reun. Cons. Int. Explor. Mer 183: 57–69.

Nihoul J. C. J., 1975. Hydrodynamic models. In J. C. J. Nihoul (ed.), Modelling of Marine Systems, Elsevier Scientific Publishing Company, Amsterdam, 272 pp.

Nihoul, J. C. J. & F. Ronday, 1976. Hydrodynamic models of the North Sea. Mem. Soc. Sci. Lg. 10: 61–96.

Nihoul, J. C. J., 1980. Residual circulation, long waves and mesoscale eddies in the North Sea. Oceanol. Acta 3: 309–316.

OSPAR/ICES, 1997. Workshop on the overall evaluation and update of background/reference concentrations for nutrients and for contaminants in sea water, biota and sediments, Oslo and Paris Commissions, Report SIME, 97/7/2, 60 pp.

Prandle, D., G. Ballard, D. Flatt, A. Harrison, S. Jones, P. Knoght, S. Loch, J. McManus, R. Player & A. Tappin, 1996. Combining modelling and monitoring to determine fluxes of water, dissolved and particulate metals through the Dover Strait. Cont. Shelf Res. 16: 237–257.

Ronday, F., 1976. Modeles hydrodynamiques. In J. C. J. Nihoul & F. C. Ronday (eds), Projet Mer, Rapport Final, 3. Services du Premier Ministre, Programmation de la Politique Scientifique, Brussels, 270 pp.

Statham P. J., Y. Auger, J. Burton, P. Choisy, J-C. Fischer, R. James, N. Morley, B. Oudane, E. Puskaric, M. Wartel, 1993. Fluxes of Cd, Co, Cu, Fe, Mn, Ni, Pb and Zn through the Strait of Dover into the Southern North Sea. Oceanol. Acta 16: 541–552.

Topping, G., J. M. Bewers & P. G. W. Jones, 1980. A review of the past and present measurements of selected trace metals in sea water in the Oslo Commission and ICNAF (NAFO) areas. Coop. Res. Rep., 97. I.C.E.S., Copenhague, 43 pp.

Wollast, R., 1976. Proprietes physico-chimiques des sediments et des suspensions de la mer du Nord. In J. C. J. Nihoul & F. Gullentops (eds), Sedimentologie, Projet Mer, Rapport Final, 4. Services du Premier Ministre, Programmation de la Politique Scientifique, Brussels: 123–143.

Hydrobiologia **366:** 109–128, 1998.
W. F. J. Baeyens (ed.), Trace Metals in the Westerschelde Estuary.

A box-model of metal flows through the Scheldt estuary (1981–1983 and 1992–1995)

Willy Baeyens, Frank Monteny, Ramses Van Ryssen & Martine Leermakers
Department of Analytical Chemistry, Free University of Brussels, Pleinlaan 2, 1050 Brussels, Belgium

Key words: Scheldt estuary, trace metals, particulate, dissolved, transport model, mass balance

Abstract

The behaviour of Cu, Zn, Cd and Pb in the Scheldt estuary was investigated in the periods 1981 to 1983 and 1995. The Hg behaviour was studied between 1992 and 1994. A box model was used to establish the relationship between the trace metal inputs and the fate of these pollutants in the Scheldt estuary. Annual budgets were calculated for particulate and dissolved trace metals. For Hg, seasonal budgets of the various species (inorganic Hg, methylmercury, Hg °) could also be calculated. These budgets allow an evaluation of the consistency of each individual flux or process, such as the inputs of the various metals into the North Sea, the sedimentation flux of particulate metals in the area of high turbidity, the evasional fluxes and the formation rates of Hg °, etc. in relation to the other ones.

Introduction

The river Scheldt crosses France, Belgium and The Netherlands on its way to the North Sea (Figure 1). It has a length of 355 km and a drainage area of 22 000 km^2, mainly on Belgian territory. The average water flow is about 100 $m^3\ s^{-1}$ and varies from 20 $m^3\ s^{-1}$ during summer to 400 $m^3\ s^{-1}$ during winter. The Scheldt estuary is a macrotidal coastal plain estuary, with an average tidal range of 4 to 5 m and an average depth of 11 m. Under normal water flow conditions the estuary is well mixed. The residence time of the water in the estuary is of the order of one to three months, depending on the river flow rate.

From a morphological point of view the estuary can be divided into three zones. The lower estuary, called Western Scheldt, extends from the North Sea to the Belgian-Dutch border (km 0–60), the upper estuary is situated between the border and the river Rupel (km 60–100) and the fluvial river (fresh water zone) is situated between the Rupel mouth and Ghent (km 100–160), where a tidal range of two meter still exists. The upper estuary consists of one channel with a width between 0.4 and 2 km. The lower estuary is wider, up to 8 km, and consists of deep and shallow parts, with different ebb and flow channels. The estuary, which is the zone under investigation, is situated between Rupelmonde and Vlissingen, and has a length of about 100 km.

From a physico-chemical point of view, the estuarine mixing zone can further be subdivided (e.g. Baeyens & Monteny, 1988). Zone I, situated from Antwerp to the confluence of the rivers Scheldt and Rupel, has salinities ranging from 0 to 2 psu, clay bottom sediments, low sedimentation rates and oxygen depletion throughout a large part of the year. Zone II is situated between the Belgian Dutch border and the city of Antwerp and is characterised by a turbidity maximum, high sedimentation rates, fine grain bottom sediments (Wartel, 1977), oxygen undersaturation particularly in the summer and salinities varying from 2 to 10 psu. Zones III and IV correspond to the Western Scheldt, which is characterised by coarse bottom sediments, aerobic conditions in the water column and salinities varying from 10 to 30 psu. For a better representation of the various metal-fluxes in the downstream estuary, especially the dissolved metal-flows, this zone has been further subdivided in zone III and zone IV. In zone III the oxygen concentration progressively increases due to re-aeration, and re-oxidation

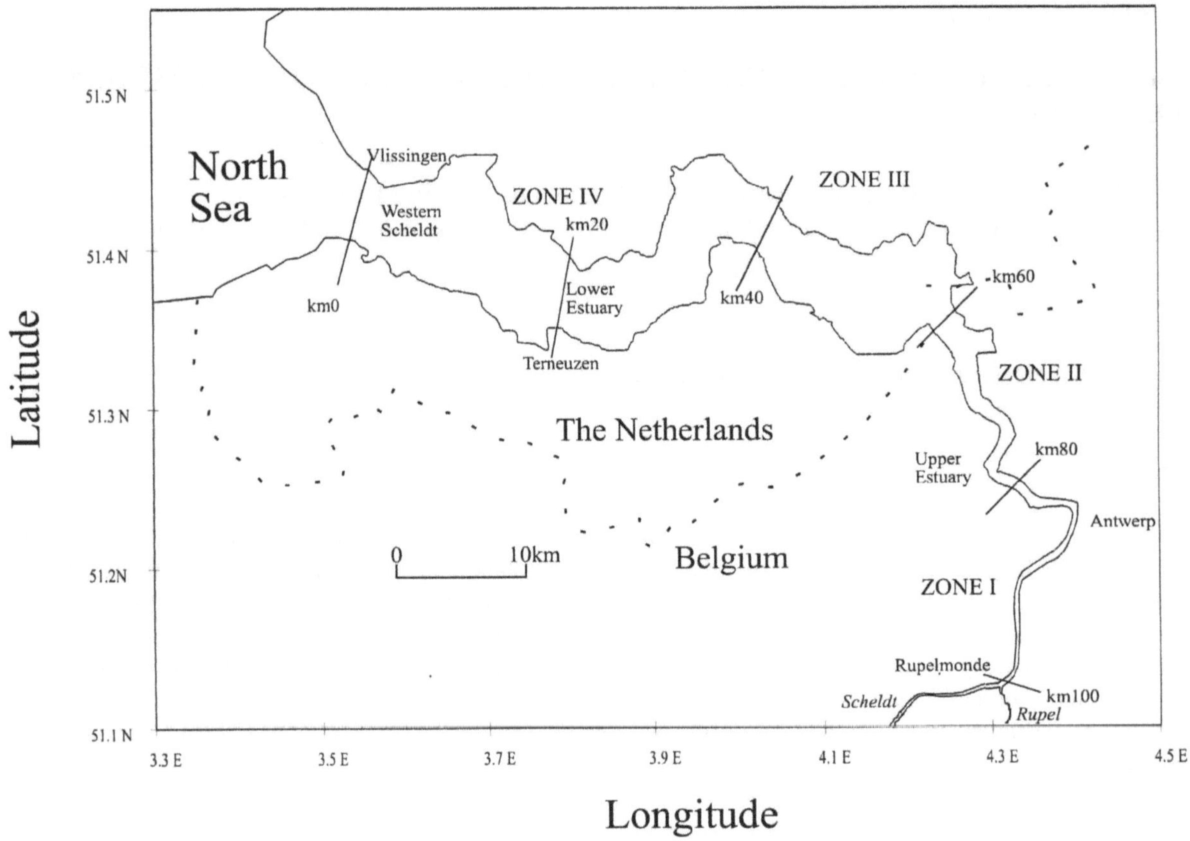

Figure 1. Map of the Scheldt estuary.

processes occur (salinity range from 10 to 20 psu). Zone IV, the mouth of the estuary, is always aerated and in the summer even oversaturated (salinity range from 20 to 27.5 psu).

Both the fluvial and the upstream river are heavily polluted as a result of large domestic, industrial and agricultural waste water discharges. A large fraction of the waste water is discharged without prior treatment, causing an intensive degradation of organic matter in the fluvial and upstream estuary. In the upper estuary a permanent oxygen undersaturation, and often a complete depletion is common in the water column (Figure 2). During summer, an anoxic region (km 60–100) from the mouth of the Rupel to the Belgian–Dutch border prevails. In winter, the entirely anoxic region is much smaller or missing, though a pronounced undersaturation of oxygen is still present.

The particulate matter flows (Figure 3) are based on average turbidity values, fresh water flow rates including side canals, discharge pipes, etc., sedimentation fluxes and bottom transport (the saline bottom current transports marine derived material into the estuary). Previous sediment mass balance estimates for the Scheldt, based on measurements from the early eighties, were made by van Eck (1991) and Baeyens et al. (1998c), but they have been updated and refined for the downstream part of the estuary, taking into account the particulate metal-mass balances for that area. The assumption was made that the sediment mass balance estimates did not change substantially over a period of ten years (early eighties to early nineties).

In zones I and II, also called boxes, two separate sedimentation fluxes are presented. This is also the case for the particulate matter flows. One sedimentation flux is related to the sedimentation of continental derived particulate matter (downstream transport), arriving from the previous box. The second sedimentation flux is due to the sedimentation of particles, transported upstream by the more saline bottom current, after encountering and mixing with the downstream transported material.

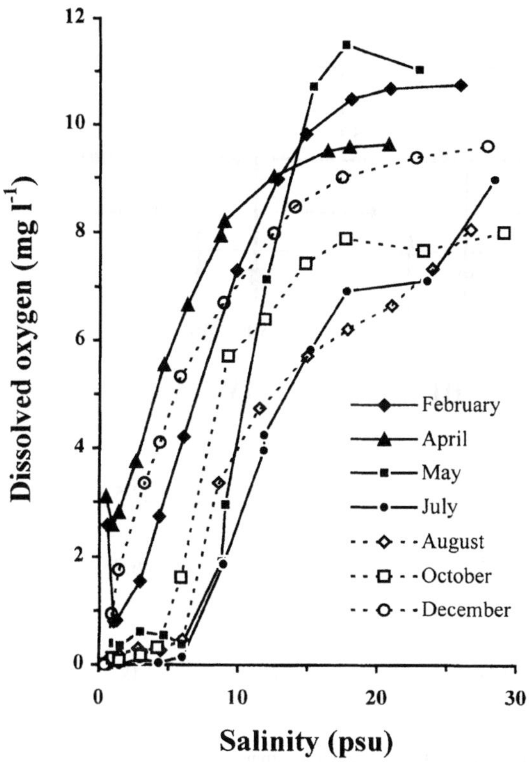

Figure 2. Dissolved oxygen concentrations in the Scheldt estuary.

Table 1. Aquatic and atmospheric emissions of metals by Belgium to the North Sea (Andersen & Niilonen, 1995)

		Water 10^3 kg yr^{-1}		Atmosphere 10^3 kg yr^{-1}	
		1985	1995	1985	1995
Hg	Point sources	0.3	0.1	13.3	7.2
	Diffuse	2.0	2.0	0.8	0.8
	Total	2.3	2.1	14.1	8.0
Cd	Point sources	14.7	2.5	11.2	3.6
	Diffuse	0.5	0.4	1.6	1.8
	Total	15.2	2.9	12.8	5.4
Cu	Point sources	42	21	132	55
	Diffuse	123	126	25	32
	Total	165	147	157	87
Pb	Point sources	58	29	320	294
	Diffuse	68	67	1387	311
	Total	126	96	1707	605
Zn	Point sources	284	127	483	414
	Diffuse	561	559	297	332
	Total	845	686	780	746

Box I also receives an important lateral input resulting from anthropogenic activities in the area around the town of Antwerp. Box II is the area of high sedimentation. About two third of the particulate metals entering this box will accumulate at the bottom.

The inputs, discharges and/or losses of heavy metals from Belgium to the aquatic system reported in the Progress Report on the Protection of the North Sea, edited by Andersen and Niilonen (1995), are much higher than the Scheldt inputs into the North Sea. Furthermore, the Scheldt estuary contributes for about 90% of the total input of Belgium to the North Sea. Table 1 shows the aquatic and atmospheric emission of metals from the Belgian territory to the North Sea. This means that they were overestimated or that most of these aquatic inputs eventually accumulated in the sediments of the Scheldt and its tributaries.

Methods and results

The sampling and analytical methods used to determine the dissolved and particulate metal species are reported in related papers: for Cd, Cu, Pb and Zn in Baeyens et al. (1998a) for Hg in Baeyens (1992) and Baeyens et al. (1998b). In addition the assessment of the atmospheric Hg concentrations as well as the calculation of the evasion fluxes are described in Baeyens et al. (1991).

Mercury

The average concentrations of the various Hg species at the corresponding salinities are shown in Table 2. These values will be used to construct three Hg species budgets (total dissolved Hg, total dissolved methylmercury (MMHg), total particulate Hg) in the Scheldt estuary.

It appears clearly from these values and also from the particulate and dissolved Hg flows (see hereafter) that the particulate phase is the major carrier of Hg. The dissolved Hg fraction, relative to the particulate one, is highest at the mouth (about 6%), and decreases to about 2% in the upstream area. The dissolved Hg budget has thus only a small impact on the total Hg budget in the upstream estuary; its importance increases somewhat towards the mouth.

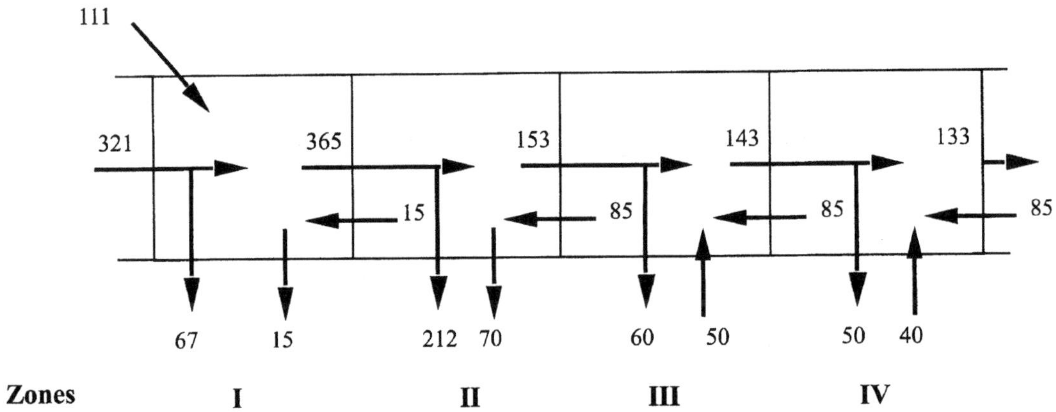

Salinity	0 psu	2 psu	10 psu	20 psu	27.5 psu
Turbidity ($mg\ l^{-1}$)	98 [102]	106 [83]	61 [57]	53.3 [50]	46.7
Water flows ($m^3\ s^{-1}$)	104 [109]	114 [119]	124 [130]	136 [142]	148

Figure 3. Particulate matter flows (10^6 kg yr^{-1}); 4 box model of the Scheldt estuary.

Cadmium, copper, lead and zinc

The average dissolved and particulate concentrations of the various metals at the corresponding salinities are shown in Table 3 for the 1981 to 1983 surveys and in Table 4 for the 1995 surveys. K_D values for the various metals and surveys are shown in Table 5.

Mercury species budgets

Mass fluxes were calculated in each of the four boxes of the Scheldt estuary for total particulate Hg, total dissolved Hg and total dissolved MMHg. The parameters used in these calculations are: (1) water flow rates for dissolved species; (2) particulate matter flows for particulate species; (3) average concentrations of the Hg species at the boundaries of the boxes and inside the boxes, which are assumed to be homogeneous; (4) an air-sea exchange model to estimate the evasion fluxes (Baeyens et al., 1991); (5) epibenthic fluxes based on Hg concentrations in the sediments (Leermakers et al., 1993) and on epibenthic fluxes of other metals in the Scheldt estuary (Baeyens et al., 1986).

Particulate mercury flows (Figure 4)

Box I: Two separate sedimentation fluxes are presented. One sedimentation flux is related to the sedimentation of continental derived particulate matter having a particulate Hg concentration of 1.46 $\mu g\ g^{-1}$ (downstream transport) which sedimentates after entering the first box without further mixing with the material transported upstream by the more saline bottom current, having a particulate Hg concentration of 1.20 $\mu g\ g^{-1}$. The second sedimentation flux is due to the sedimentation of the upstream transported particulates after mixing with the downstream transported material: the mixed material has a particulate Hg concentration of 1.44 $\mu g\ g^{-1}$. This mixed material is also the material which is transported downstream to the next box.

The first box also receives an important lateral input, estimated at 160 kg Hg yr^{-1} resulting from anthropogenic activities in the area around the town of Antwerp.

Box II: This is the area of high sedimentation. Due to this high sedimentation flux and the strong remobilisation of metals in the bottom sediments in this area (Baeyens et al., 1986; Leermakers et al., 1993), an

Table 2. Mercury concentrations in the 4 boxes of the Scheldt estuary

Species		Zone 1	Zone 2	Zone 3	Zone 4	Sea
Salinity	(psu)	0–2	2–10	10–20	20–27.5	32.5
Turbidity	(mg g^{-1})	102	83	57	50	
Sediment Hg	($\mu g\ g^{-1}$)		0.62	0.35	0.20	
Part. Hg	($\mu g\ g^{-1}$)	1.44	1.20	0.70	0.45	0.22
Part. Hg	(ng l^{-1})	147	100	40	22	22
Tot. diss. Hg						
summer	(ng l^{-1})	1.05	1.89	1.16	1.16	1.1
winter	(ng l^{-1})	2.65	1.72	1.57	1.45	1.2
Reactive Hg						
summer	(ng l^{-1})	0.45	0.78	0.55	0.55	
winter	(ng l^{-1})	0.55	0.49	1.06	1.33	
Diss. gas. Hg						
summer	(pg l^{-1})	83	85	80	108	
winter	(pg l^{-1})	41	49	48	48	
Diss. MMHg						
summer	(pg l^{-1})	183	335	181	382	
winter	(pg l^{-1})	105	83	23	39	

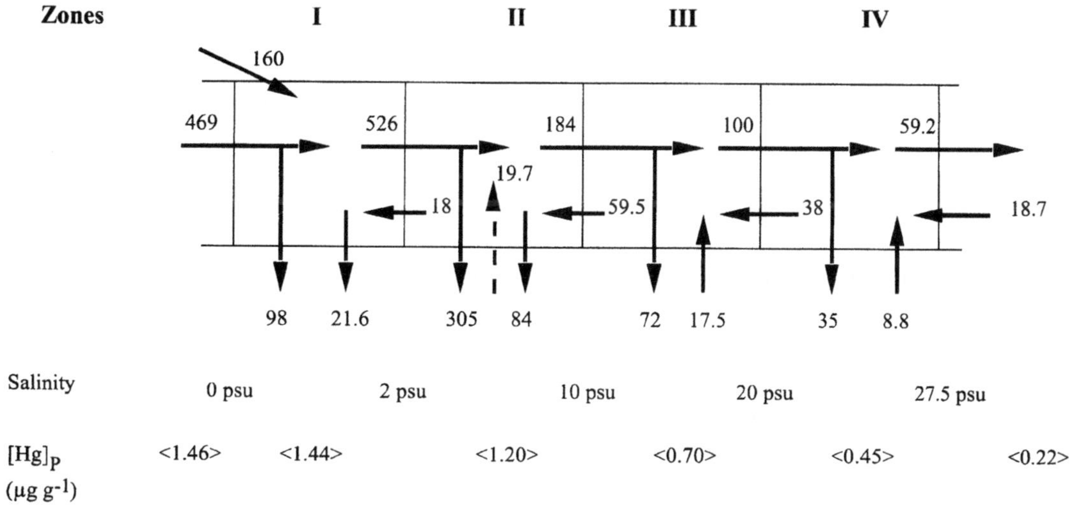

Figure 4. Particulate mercury flows (kg Hg yr^{-1}); 4 box model of the Scheldt estuary.

Table 3. Metal concentrations in the 4 boxes of the Scheldt estuary (1981–1983)

Species		Zone 1	Zone 2	Zone 3	Zone 4	Sea
Salinity	(psu)	0–2	2–10	10–20	20–27.5	32.5
Turbidity	(mg l^{-1})	102	83	57	50	
Part. Cd	(μg g^{-1})	27.9	22.8	8.2	2.8	0.70
Bottom Cd	(μg g^{-1})	5.0	5.3	0.54	0.26	
Diss. Cd	(μg l^{-1})	0.086	0.063	0.19	0.090	0.050
Part. Cu	(μg g^{-1})	278	207	95	35	26
Bottom Cu	(μg g^{-1})	29.4	33	3.5	4.6	
Diss. Cu	(μg l^{-1})	1.02	1.30	1.91	1.43	0.78
Part. Pb	(μg g^{-1})	288	254	153	101	64
Bottom Pb	(μg g^{-1})	45	55	13.2	11.9	
Diss. Pb	(μg l^{-1})	0.23	0.26	0.24	0.22	0.14
Part. Zn	(μg g^{-1})	800	397	272	157	95
Bottom Zn	(μg g^{-1})	192	216	39	39	
Diss. Zn	(μg l^{-1})	11.7	13.3	8.75	4.8	1.7

Table 4. Metal concentrations in the 4 boxes of the Scheldt estuary (1995)

Species		Zone 1	Zone 2	Zone 3	Zone 4	Sea
Salinity	(psu)	0–2	2–10	10–20	20–27.5	32.5
Turbidity	(mg l^{-1})	102	83	57	50	
Part. Cd	(μg g^{-1})	8.76	6.25	3.67	1.26	0.35
Diss. Cd	(μg l^{-1})	0.020	0.044	0.090	0.0625	0.0288
Part. Cu	(μg g^{-1})	112	80.7	47.2	21.8	15.5
Diss. Cu	(μg l^{-1})	0.68	1.17	1.37	1.04	0.37
Part. Pb	(μg g^{-1})	214	146	105	83.4	35.1
Diss. Pb	(μg g^{-1})	0.165	0.156	0.087	0.052	0.038
Part. Zn	(μg g^{-1})	610	423	275	144	82.0
Diss. Zn	(μg l^{-1})	6.24	6.04	4.48	2.16	1.06

epibenthic flux of 20 kg yr^{-1} of dissolved Hg could be estimated (in all other boxes the epibenthic diffusive flux is negligible compared to the other Hg flows). The dissolved Hg diffusing into the water column will distribute between the dissolved and particulate phases according to the distribution coefficient or the ratio of dissolved to particulate Hg flows, prevailing in box II. This yields a subsequent particulate epibenthic Hg flux of 19.7 kg Hg yr^{-1}. Regarding the other fluxes, we can hold the same reasoning as for box I.

Box III: In this box the magnitude of the flows is small-

Table 5. K_D-values in the 4 boxes of the Scheldt estuary; 1981–1983 and 1995

K_D		Zone 1	Zone 2	Zone 3	Zone 4	Sea
Salinity (psu)		0–2	2–10	10–20	20–27.5	32.5
Cd	'81–'83	324	362	43	31	14
($l\ g^{-1}$)	'95	438	142	41	20	12
Cu	'81–'83	273	159	50	24	33
($l\ g^{-1}$)	'95	165	69	34	21	42
Pb	'81–'83	1252	977	638	459	457
($l\ g^{-1}$)	'95	1300	936	1210	1600	924
Zn	'81–'83	68	30	31	33	56
($l\ g^{-1}$)	'95	98	70	61	67	77

er than in zone II; similarly, the interactions with the bottom are less. The bottom current is, however, strong enough in this box to bring a fraction of the local bottom sediments, which have a much lower Hg content than the suspended ones, into suspension where it will be mixed with inflowing material.

Box IV: In this box the processes are very similar to these occurring in box III.

Total dissolved mercury flows (Figure 5)

The behaviour of dissolved Hg in the Scheldt estuary is very sensitive to salinity and redox conditions, therefore speciation has been studied the year round as a function of these parameters (Leermakers et al., 1995; Baeyens et al., 1998b). During the summer, a large part of the upstream area becomes anoxic (Figure 2), resulting in the precipitation and co-precipitation of metal sulphides in addition to those resuspended from the reduced bottom sediments. In the downstream area, re-oxidation of these sulphides results in a temporary increase of dissolved mercury. During the winter, on the contrary, the undersaturation of oxygen does not lead to the presence of metal sulphides in the water column. Hence, total dissolved Hg budgets were calculated for summer and winter conditions.

Total dissolved Hg includes besides the inorganic mercury (II) fraction (Hg^{2+}, labile complexes with inorganic ligands, mainly chloro-complexes, and non-labile complexes with organic ligands, mainly humic acid-complexes), also gaseous and methylmercury species. The behaviour and magnitude of the dissolved gaseous fraction is of major importance, since it is directly related to the loss of Hg° and the decrease of the dissolved Hg burden in the estuary. The evasion fluxes were calculated for each of the four boxes, in summer and in winter. The basic information necessary for these calculations is shown in Table 6. An average wind speed of 29 km h^{-1} was assumed resulting in an evasion velocity of 1.44 m d^{-1}.

Winter

Box I: The amount of dissolved Hg entering and leaving this box is almost constant. The evasion flux is only about 1% of the in- and out-flow.

Box II: Almost one third of the amount of dissolved Hg entering this box disappears onto the particulate phase. The high turbidity values and sedimentation rates are the main reasons for the important scavenging of the dissolved Hg compounds. As explained above, an epibenthic flux of 20 kg yr^{-1} could be estimated, but only 0.30 kg yr^{-1} remains after equilibration with the particulate phase into solution. This flux accounts for the evasion flux.

Boxes III and IV: The dissolved Hg flows in and out these boxes are nearly balanced. The evasion fluxes which are about one fourth to one fifth of the in and out-flows, must thus be supported by desorption or mobilisation fluxes, similar in magnitude, from the particulate phase.

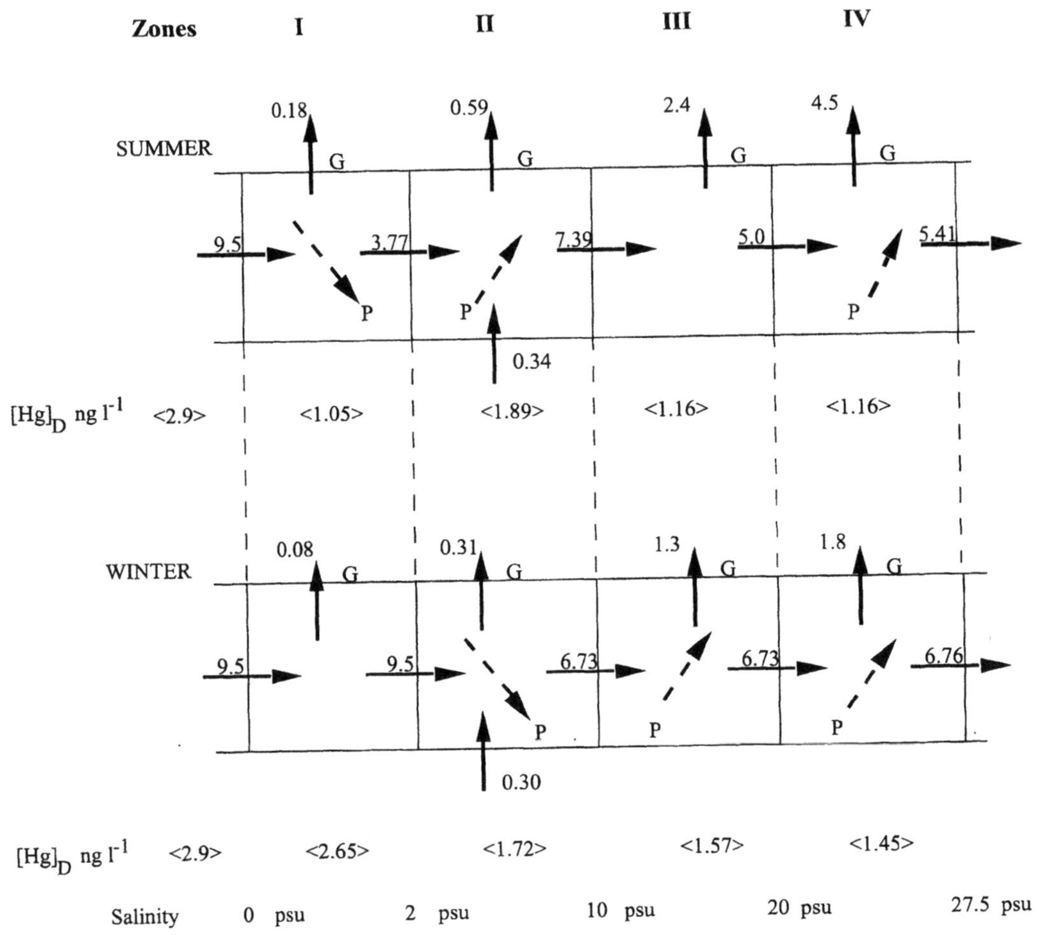

Figure 5. Dissolved Hg flows in the summer and in the winter (kg Hg yr^{-1}); 4 box model of the Scheldt estuary.

Summer

Box I: The out-flow of dissolved Hg is substantially lower than the in-flow. In this box, the water column is anoxic in summer so that insoluble, reduced Hg species such as Hg sulphides are formed. The shift from the dissolved to the particulate phase is pronounced for the dissolved species budget, but is hardly noticeable for the particulate one.

Box II: The out-flow of this box is much higher than the in-flow and again comparable to the levels observed during the winter. The dissolved Hg concentrations increase again once the oxygen levels reach 3 to 4 mg l^{-1} O_2 suggesting that oxidation and dissolution of reduced Hg compounds occurred. The process taking place here is thus the reverse of the one taking place in box I. The epibenthic flux during summer is about half the evasion flux.

Box III: The new production of dissolved Hg from box II must equilibrate with the particulate phase. That is probably the reason that in this box no dissolved Hg is transferred during the summer from the particulate phase (the K_D shows a slight increase) in contrast with the winter situation (the K_D decreases).

Box IV: In and out-flows are almost equal, so that the evasional flux should be balanced by a desorption flux from the particulate phase; that is similar to the winter situation. The high dissolved gaseous Hg concentration coincides with the presence of high plankton concentrations and a high biological activity, suggesting that Hg reduction is biologically mediated.

We also estimated the Hg ° formation rates. Based on the dissolved Hg burden contained in the depth

Table 6. Hg exchange fluxes at the air-water interface

Water column concentrations of Hg° (pg l^{-1}):

Salinity (psu)	0–2	2–10	10–20	20–27.5
Summer	83	85	80	108
Winter	41	49	48	48

Atomospheric concentrations of Hg (ng m^{-3} and pg l^{-1}):
The gaseous Hg° concentration equals 3 ng m^{-3}; this corresponds to 10 pg l^{-1} at the water surface

Fluxes in Summer

Box	Supersaturation	Transfer velocity	Flux	Surface	Total flux
Salinity (psu)	ng m^{-3}	m d^{-1}	μg Hg m^{-2} yr^{-1}	km^2	g Hg yr^{-1}
0–2	73	1.44	38	4.8	180
2–10	75	1.44	39	15	590
10–20	70	1.44	37	66	2400
20–27.5	98	1.44	52	88	4500

Fluxes in winter

BOX	Supersaturation	Transfer velocity	Flux	Surface	Total flux
salinity (psu)	ng m^{-3}	m d^{-1}	μg Hg m^{-2} yr^{-1}	km^2	g Hg yr^{-1}
0–2	31	1.44	16	4.8	80
2–10	39	1.44	20	15	310
10–20	38	1.44	20	66	1300
20–27.5	38	1.44	20	88	1800

integrated water column (between 15 and 20 μg m^{-3}), conversion rates of 0.22 to 0.35% d^{-1} during the winter (Table 7) are required to balance the estimated evasional fluxes of 16 to 20 μg m^{-2} yr^{-1}. During the summer the integrated Hg burden in the water column varies from 8 to 17 μg m^{-3}, hence conversion rates of 0.63 to 1.3% d^{-1} are required to balance the evasional fluxes of 37 to 52 μg m^{-2} yr^{-1}. These Hg ° formation rates are of the same order as those estimated by Mason et al. (1995b; 1995c) for open ocean waters (0.2 to 1% d^{-1}) but much lower than the values they found (2 to 4% d^{-1}) in the Upper Mystic Lake (Mason et al., 1995a).

Dissolved methylmercury flows (Figure 6)

The dissolved methylmercury flows are generally small compared to the total dissolved Hg flows, except during summer and fall between boxes II and III (just downstream the anoxic area) and between box IV and the sea (at the mouth where high plankton population densities were observed). In those cases the methylmercury flows can represent between 20 and 30% of the total dissolved Hg flows, which are very high fractions.

Winter
Boxes I and II: The inflow of box I is unknown, but the in and out-flows of box II are fairly similar. An epibenthic flux of 48 10^{-3} kg yr^{-1} has been estimated,

Table 7. Hg° formation rates in the 4 boxes of the Scheldt estuary

Species		Zone 1	Zone 2	Zone 3	Zone 4
Salinity	(psu)	0–2	2–10	10–20	20–27.5
Winter					
Tot. diss Hg	(ng l^{-1})	2.65	1.75	1.57	1.45
Depth	(m)	7.5	9	11	13
Integrated diss Hg	($\mu g\ m^{-2}$)	20	15.5	17.3	18.9
Evasion flux Hg°	($\mu g\ m^{-2}\ yr^{-1}$)	16	20	20	20
Hg° formation rate	% yr^{-1}	80	129	116	106
	% d^{-1}	0.22	0.35	0.32	0.29
Summer					
Tot. diss Hg	(ng l^{-1})	1.05	1.89	1.16	1.16
depth	(m)	7.5	9	11	13
Integrated diss Hg	($\mu g\ m^{-2}$)	7.88	17	12.8	15
Evasion flux Hg°	($\mu g\ m^{-2}\ yr^{-1}$)	38	39	37	52
Hg° formation rate	% yr^{-1}	482	229	289	347
	% d^{-1}	1.3	0.63	0.79	0.95

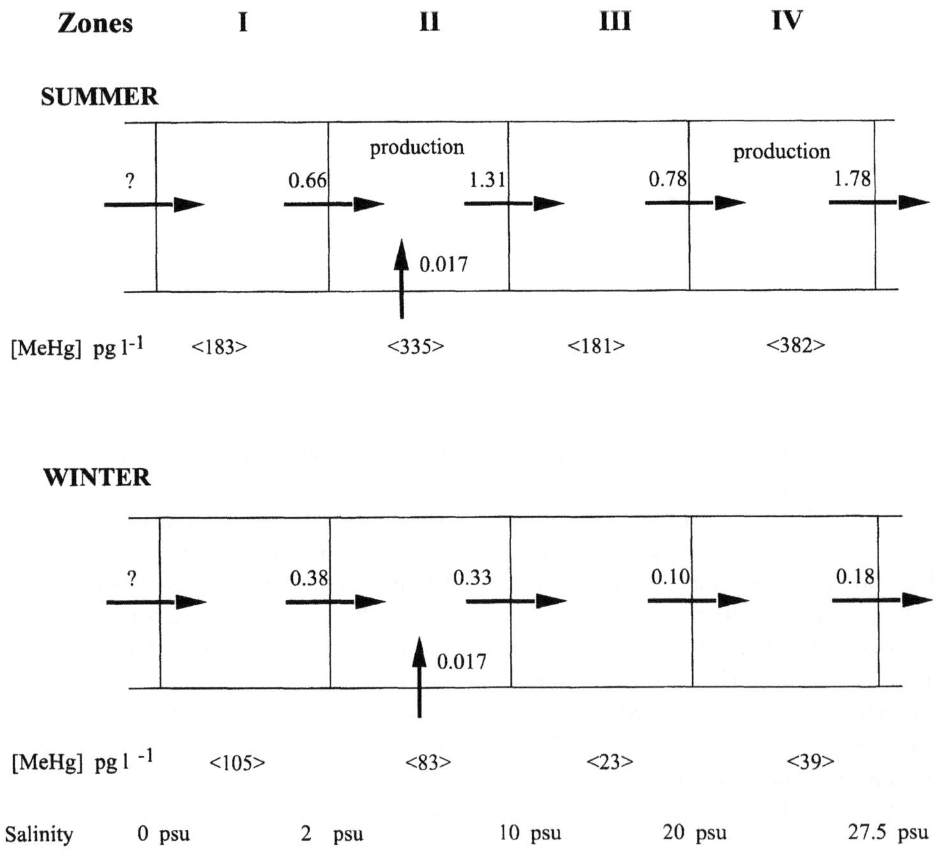

Figure 6. Dissolved methylmercury flows in summer and in winter (kg Hg yr^{-1}); 4 box model of the Scheldt estuary.

in an analogous way as for total dissolved Hg. Taking into account a particulate to dissolved methylmercury ratio (both expressed in the same units) of 1.9 (Baeyens et al., 1998a) a resulting epibenthic dissolved MMHg flux of 17 10^{-3} kg yr^{-1} was calculated.

The production of MMHg in the upper estuary is low during the winter due to the low temperature and the absence of anaerobic conditions in the water column restricting the activity of the sulphate reducing bacteria (Leermakers et al., 1993). The fraction of methylmercury in the total dissolved Hg flow is about 4 to 5%.

Boxes III and IV: In the downstream estuary the MMHg flows are still smaller than those in the upstream estuary (the fraction of MMHg in the total dissolved Hg flow is less than 2.5%).

Summer

Box II: The flows are significantly higher than during the winter (the in-flow 2 times, the out-flow 4 times). MMHg in this box is produced as a consequence of the redox gradient (the water column changes from anoxic over suboxic to oxic in the downstream direction). Sulphate reducing bacteria are very active in such conditions and have been shown to be very effective in methylating inorganic Hg (Leermakers et al., 1993). Their activity decreases, however, strongly once the system becomes more oxidised.

Box IV: A second production zone of dissolved methylmercury is observed close to and at the mouth of the estuary. The methylmercury out-flow of this box is 10 times higher in summer than in winter. The origin of this production is still under study, but the high amounts of planktonic species at the mouth during the summer, similar to the high dissolved gaseous Hg levels produced in this zone, may be involved.

Cadmium, copper, lead and zinc budgets

Mass fluxes were calculated in each of the four zones or boxes of the Scheldt estuary for total particulate and dissolved metals for the periods 1981–1983 and 1995. The parameters used in these calculations are: (1) water flow rates for dissolved species; (2) particulate matter flows for particulate species; (3) average concentrations of the metal species at the boundaries of the boxes and inside the boxes, which are assumed to be homogeneous; and (4) epibenthic fluxes reported by Baeyens et al. (1986) for the period 1981–1983. The epibenthic fluxes for 1995 were adapted taking into account the sedimentation fluxes relative to that period.

The main particulate fluxes in each of the boxes are the same as those mentioned here above when discussing the particulate matter and particulate mercury flows. For the particulate phase, box II, or the area of high sedimentation, is the most important. Most of the particulate metals accumulate at the bottom, but a fraction is remobilised in the sediments and diffuses back into the water column. Because the behaviour of Cd, Cu, Pb and Zn in the dissolved phase is much simpler than that of Hg which includes at least three species (Hg°, Hg^2 and MMHg) each characterised by particular features, we combined the dissolved and particulate metal flows into one picture. Figures 7 to 14 show these particulate and dissolved metal flows through the Scheldt estuary.

From the dissolved and particulate metal flows in each of the boxes, the mass balances were calculated (Table 8). It appears that generally the in- and outflows are fairly well-balanced, taking into account the uncertainties related to the magnitude of the particulate matter flows, especially sedimentation and resuspension, and to a lesser extent of the average particulate and dissolved metal concentrations.

The highest degrees of relative imbalance, expressed as the difference between total in- and outflows divided by the total inflow, are observed in box IV. The average relative imbalance for box IV is 20%, while for the other boxes this value is below 10%. Box IV also contains the smallest flows, which results in much stronger relative imbalances for a similar deviation.

Looking at positive (a higher input than output) and negative imbalances, one observes that boxes I (only for the particulate mass balance) and III show a slight deficit, boxes II and IV a slight excess. The net sedimentation flux (the sum of sedimentation and resuspension) in boxes II and IV is probably slightly underestimated, the opposite being true in boxes I and III. However, since a simple mass balance model of water and particles coupled to a number of metals (dissolved and particulate phase) practically yields a closed budget for all these metals in the four zones of the estuary, evidence is provided that the various components of the model are reasonably well estimated.

The Scheldt inputs into the coastal area calculated in this paper are slightly different from the values pre-

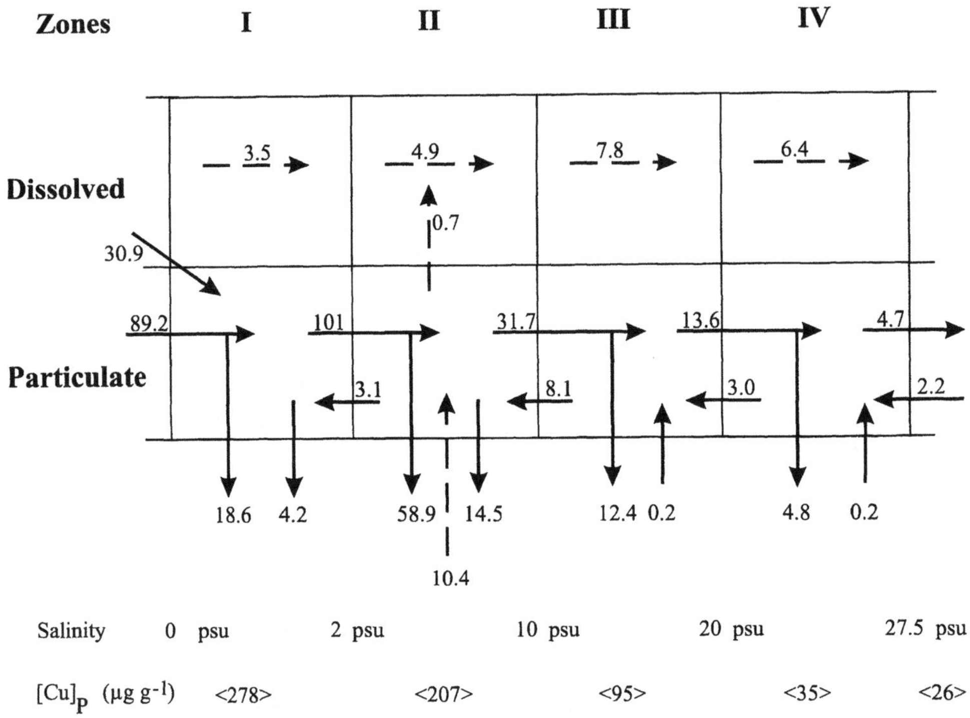

Figure 7. Dissolved and particulate Cu flows (10^3 kg yr^{-1}) in the period 1981–1983; 4 box model of the Scheldt estuary.

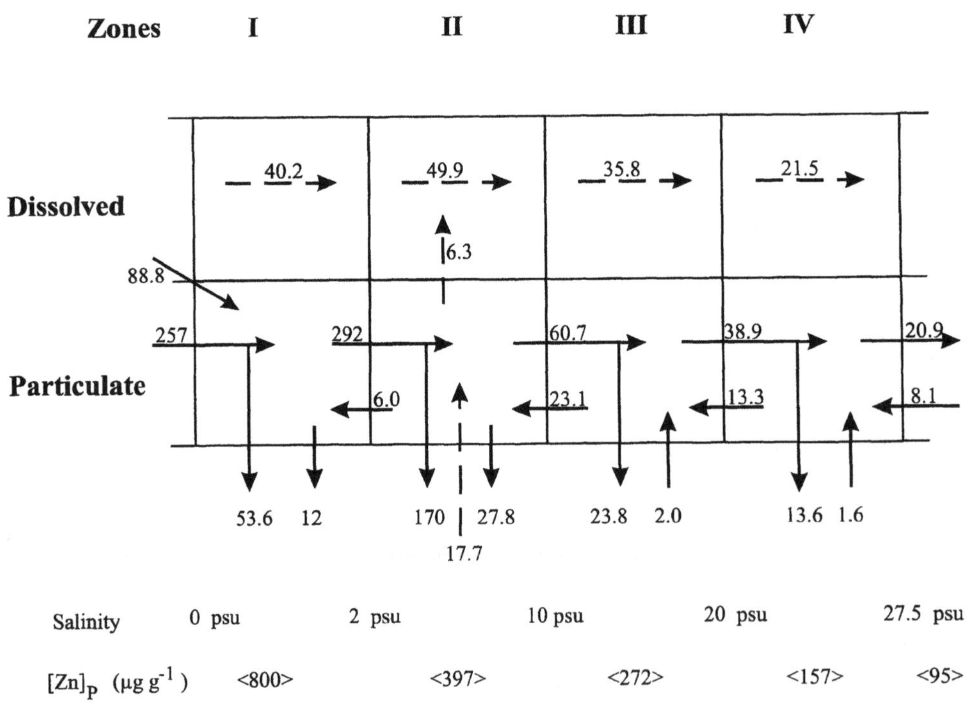

Figure 8. Dissolved and particulate Zn flows (10^3 kg yr^{-1}) in the period 1981–1983; 4 box model of the Scheldt estuary.

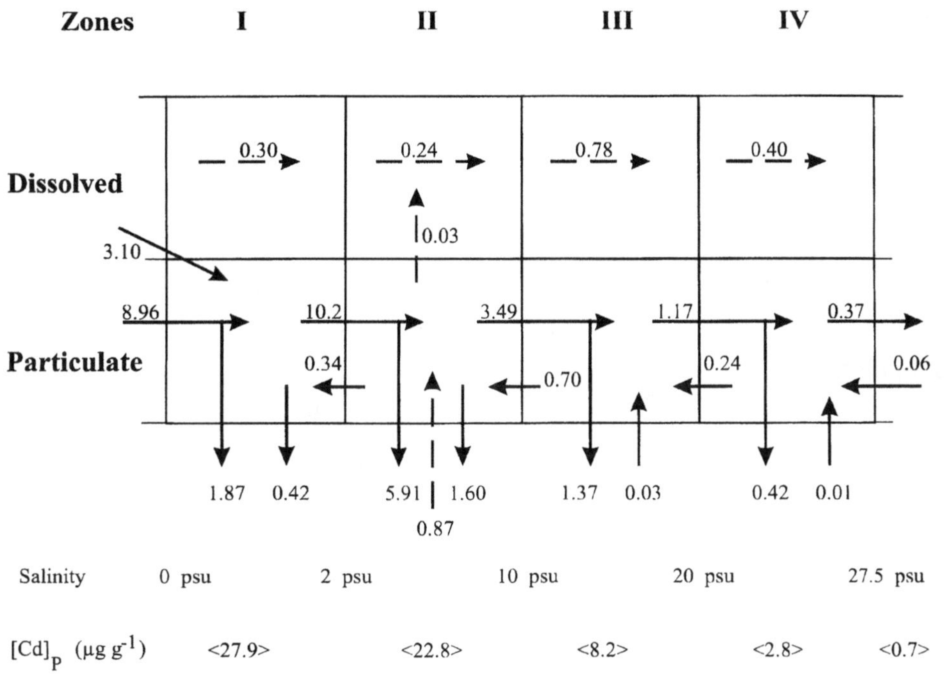

Figure 9. Dissolved and particulate Cd flows (10^3 kg yr^{-1}) in the period 1981–1983; 4 box model of the Scheldt estuary.

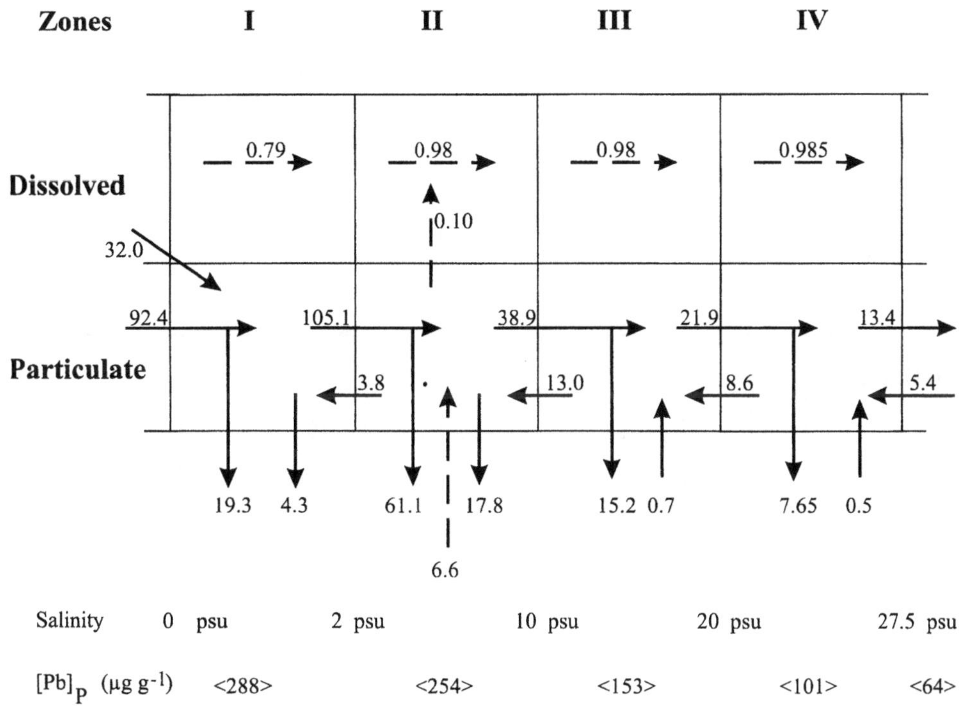

Figure 10. Dissolved and particulate Pb flows (10^3 kg yr^{-1}) in the period 1981–1983; 4 box model of the Scheldt estuary.

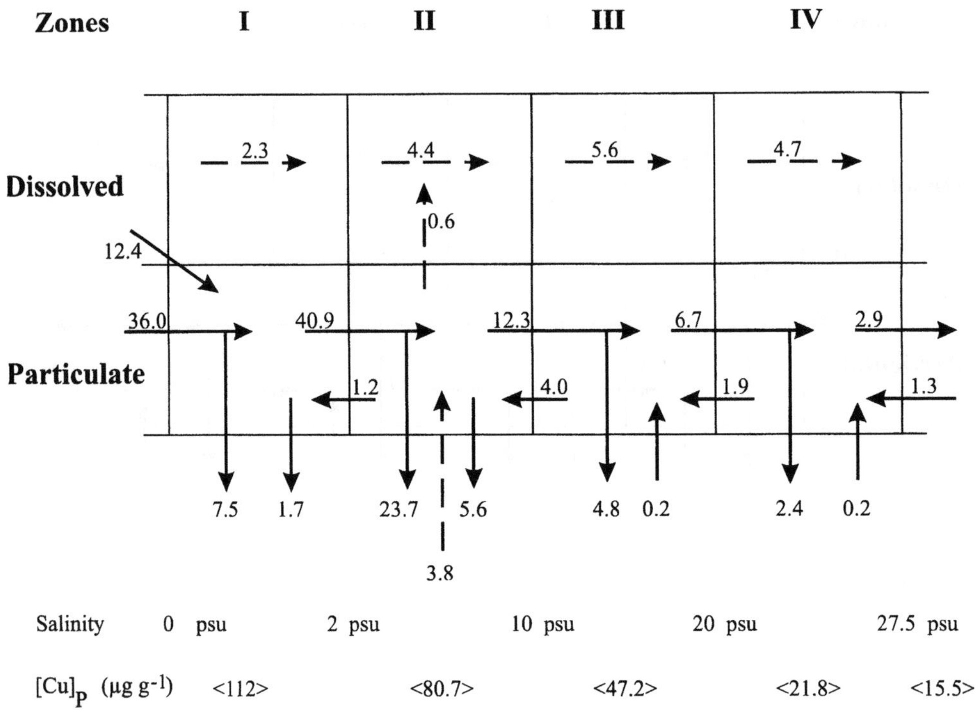

Figure 11. Dissolved and particulate Cu flows (10^3 kg yr^{-1}) in 1995; 4 box model of the Scheldt estuary.

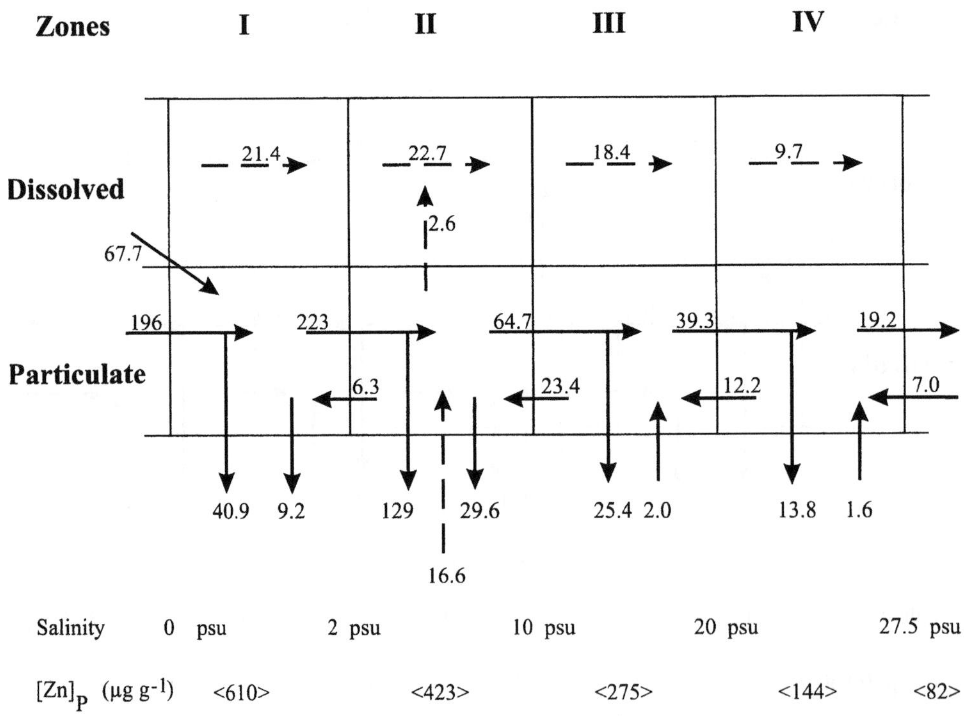

Figure 12. Dissolved and particulate Zn flows (10^3 kg yr^{-1}) in 1995; 4 box model of the Scheldt estuary.

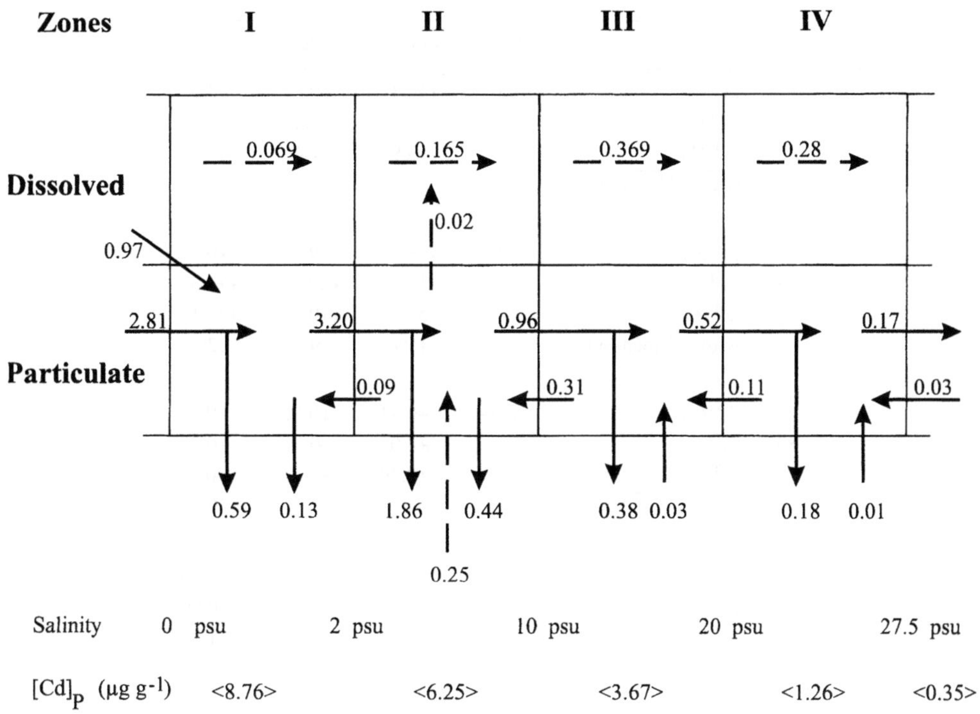

Figure 13. Dissolved and particulate Cd flows (10^3 kg yr^{-1}) in 1995; 4 box model of the Scheldt estuary.

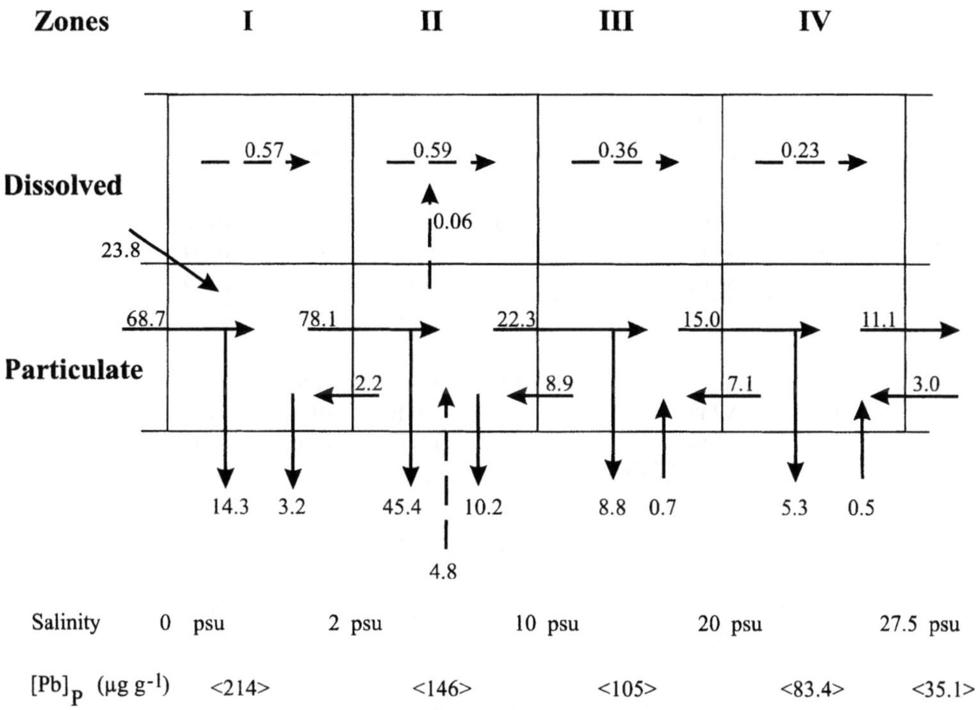

Figure 14. Dissolved and particulate Pb flows (10^3 kg yr^{-1}) in 1995; 4 box model of the Scheldt estuary.

Table 8. Mass balances of dissolved and particulate metals (tons yr^{-1}, except for Hg in kg yr^{-1}), Δ = inputs–outputs

	Species	Zone I	Zone II	Zone III	Zone IV
	Δ^P Cu	–0.6	+11.3	+0.8	+3.5
'81–'83	Δ^D Cu		–1.4	–2.2	+1.4
	Δ^T Cu		+9.9	–1.4	+4.9
	% of Δ^T/total inflow		8%	3%	21%
	Δ^P Zn	–5.8	+68.3	+9.8	+0.8
'81–'83	Δ^D Zn		–16.0	+20.4	+14.3
	Δ^T Zn		+52.3	+10.6	+15.1
	% of Δ^T/total inflow		15%	9%	22%
	Δ^P Cd	–0.09	+0.43	+0.52	+0.21
'81–'83	Δ^D Cd		+0.06	–0.51	+0.38
	Δ^T Cd		+0.49	+0.01	+0.59
	% of Δ^T/total inflow		4%	0%	29%
	Δ^P Pb	–0.5	+3.1	–1.9	–1.85
'81–'83	Δ^D Pb		–0.29	+0.1	–0.005
	Δ^T Pb		+2.81	–1.8	–1.855
	% of Δ^T/total inflow		2%	4%	6%
	Δ^P Cu	–0.5	+5.9	–1.1	+1.0
'95	Δ^D Cu		–2.7	–0.6	+0.9
	Δ^T Cu		+3.2	–1.7	+1.9
	% of Δ^T/total inflow		6%	9%	14%
	Δ^P Zn	–3.1	+33.4	–9.2	+2.7
'95	Δ^D Zn		–3.9	+6.9	+8.7
	Δ^T Zn		+29.5	–2.3	+11.4
	% of Δ^T/total inflow		10%	2%	17%
	Δ^P Cd	–0.05	+0.41	–0.11	+0.10
'95	Δ^D Cd		–0.12	–0.18	+0.09
	Δ^T Cd		+0.29	–0.29	+0.19
	% of Δ^T/total inflow		7%	19%	23%
	Δ^P Pb	–0.9	+11.7	–2.6	–5.0
'95	Δ^D Pb		–0.10	+0.3	+0.1
	Δ^T Pb		+11.6	–2.3	–4.9
	% of Δ^T/total inflow		13%	7%	26%
	Δ^P Hg	+1.4	+14.2	+8.0	–4.7
Summer	Δ^D Hg	+5.6	–3.9	0	–4.9
'92–'94	Δ^T Hg	+7.0	+10.6	+8.9	–8.0
	% of Δ^T/total inflow	1.1	1.8%	3.5%	6.9%
	Δ^P Hg	+1.4	+14.2	+8.0	–4.7
Winter	Δ^D Hg	–0.1	+2.8	–1.3	–1.8
'92–'94	Δ^T Hg	+1.3	+16.9	+6.8	–6.2
	% of Δ^T/total inflow	0.2	2.8%	2.8%	5.0%

sented in a related paper on coastal inputs (Baeyens et al., 1998d). In this paper the average concentrations of the four metals in the dissolved and particulate phases in box IV (salinity range 20 to 27.5 psu) were used to calculate the input flows, while in the other paper (see Baeyens et al., 1998d) the metal concentrations, obtained by extrapolation to a salinity of 26.5 psu of the best fitting curve through the coastal data points (concentration versus salinity), were used. For the surveys between 1981 and 1983 the dissolved metal concentrations calculated in both ways are: for Pb (0.22 and 0.252 μg l^{-1}), for Zn (4.8 and 4.1 μg l^{-1}), for Cd (0.090 and 0.125 μg l^{-1}) and for Cu (1.43 and 1.60 μg l^{-1}), while in the particulate phase they are: for Pb (4.05 and 3.5 μg l^{-1}), for Zn (7.85 and 5.5 μg l^{-1}), for Cd (0.14 and 0.10 μg l^{-1}) and for Cu (1.75 and 1.20 μg l^{-1}).

Budget uncertainties

It is clear that the metal flows presented in Figures 4 to 14, which are based on averaged values of dissolved and particulate metal concentrations, and of water and suspended sediment flows, are not exact numbers but include a varying degree of uncertainty. In Table 9 the standard deviations on the mean dissolved and particulate metal concentrations in the period 81–83 are shown. The relative errors are always lower than 50%, except for Cd (zones I and IV) in the dissolved phase, and again for Cd (zones I, III and IV) and Zn (zone I) in the particulate phase. Cd concentrations in the Scheldt are much lower than those of the three other metals, which may be an explanation for the higher standard deviations observed for this metal.

The average water flow at the river/estuary interface (Schelle, 92 km from the mouth) amounts to 104 m^3 s^{-1}, but the water flows obey a log-normal distribution. The average value minus (plus) the standard deviation equals 60 m^3 s^{-1} (180 m^3 s^{-1}), which as by hazard almost corresponds to the summer and winter averages. The average suspended sediment concentration at the river/estuary interface is calculated from 47 measurements at that location and yields a value of 101 ± 54 mg l^{-1}. Hence, the suspended sediment flow entering the estuary shows a relative error of 80%. The determination of the suspended sediment flow at the estuary/sea interface is much more complicated, since in addition of the seaward surface current, a landward bottom current exists (Baeyens et al., 1998c). D'Hondt and Jacques (1982) estimated the suspended sediment output to the sea at 300 10^3 tons y^{-1}, while Van Eck (1991) suggested a value of 136 10^3 tons y^{-1}. The latter author reported a landward flux of marine suspended matter of 79 10^3 tons y^{-1}. As explained hereabove, we estimated a seaward suspended sediment flux of 133 10^3 tons y^{-1} and a landward flux of 85 10^3 tons y^{-1}. The uncertainty on this flux is thus rather large and may be estimated at 100%.

Assuming a relative error of 50% on the average dissolved metal concentrations (period 1981–1983) and the same error on the water flow, leads to a relative error of 70% on the dissolved metal flows shown in Figures 7 to 10. The error on the particulate metal flows is larger, because the relative error on the suspended sediment flows vary from 80 to 100%. Assuming again a relative error of 50% on the average particulate metal concentrations (period 1981–1983), and an error of 90% on the suspended sediment flow results in a relative error of 100% on the particulate metal flows shown in Figures 7 to 10.

As a concrete example we calculated the minimum and maximum dissolved and particulate Cu flows (period 1981–1983) and for each box also the corresponding budget. For the calculation of the minimum (maximum) flows, the average values are decreased (increased) by the standard deviation. Results are shown in Table 10 and Figures 15 and 16.

Conclusions

From all metals under study here, the most striking decrease in aquatic metal input to the North Sea by Belgium is observed for Cd (see also Baeyens, 1997). This finding is also reported by Andersen and Niilonen (1995). Due to more stringent regulations regarding the waste water quality and the use of cadmium in particular industrial processes, the aquatic Cd input decreased in Belgium by more than a factor 5 between 1985 and 1995. These regulations had a clear effect on the dissolved and particulate Cd levels, especially in the upper and mid-estuary. Even at the mouth (zone IV and the coastal zone) a reduction of the Cd concentration by a factor 2 is still noticeable.

For Pb the aquatic input decreased by about 25% (the point sources by a factor 2), but a more drastic decrease of the atmospheric input (about a factor 3) resulted in much lower dissolved Pb values at sea and in the downstream estuary. Also the K_D values of Pb at the mouth and at sea showed the largest modifications of all metals between 1981–1983 and 1995.

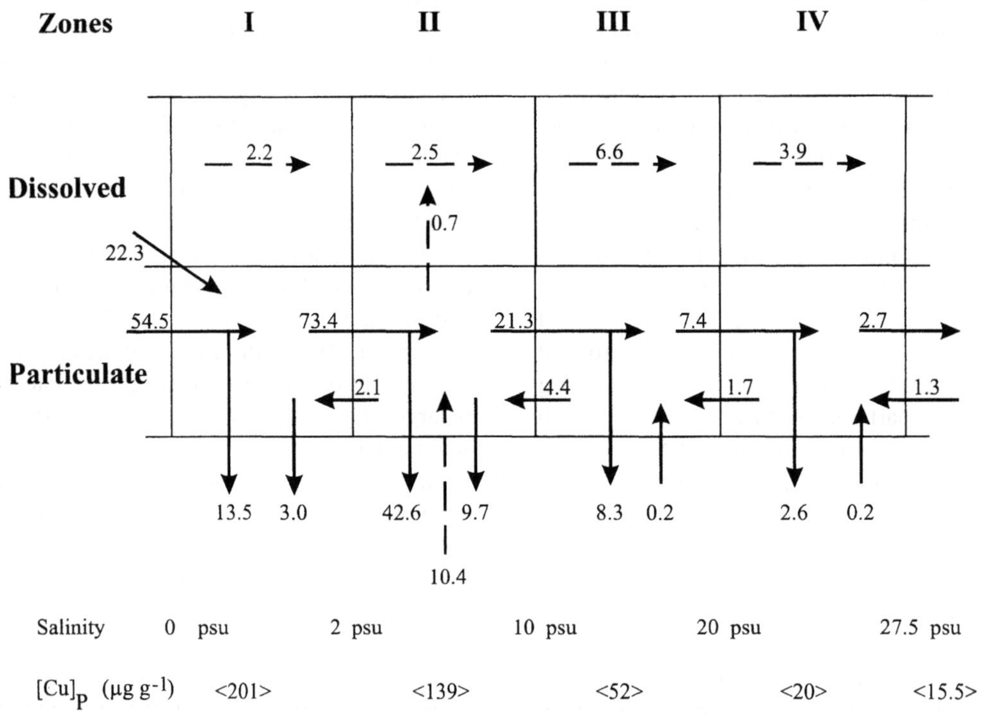

Figure 15. Minimum dissolved and particulate Cu flows (10^3 kg yr^{-1}) in the period 1981–1983; 4 box model of the Scheldt estuary.

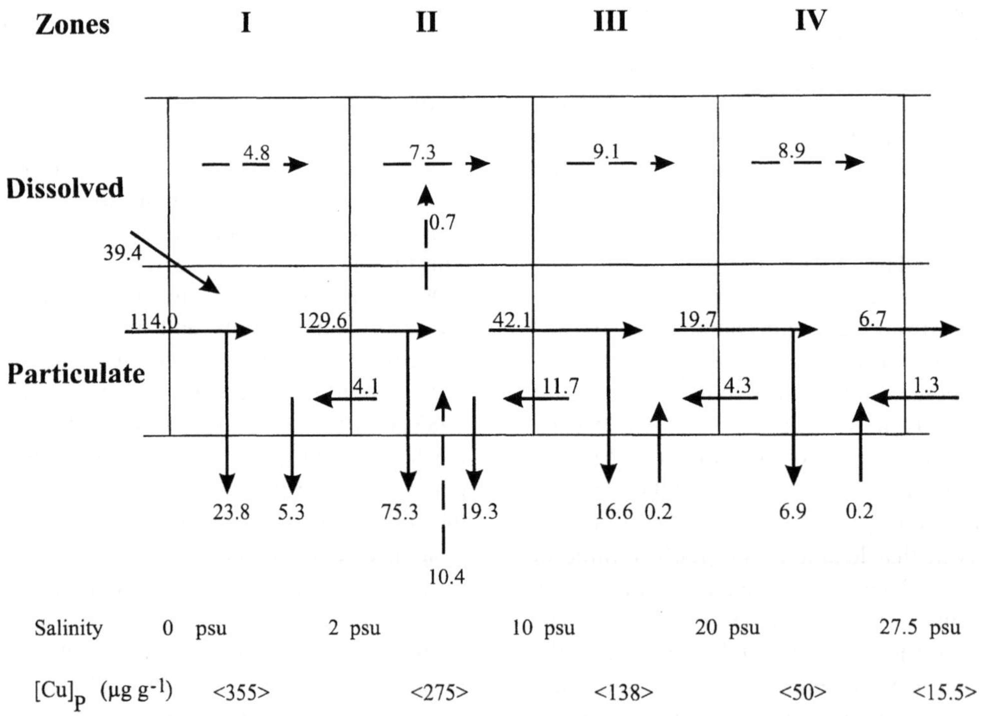

Figure 16. Maximum dissolved and particulate Cu flows (10^3 kg yr^{-1}) in the period 1981–1983; 4 box model of the Scheldt estuary.

Table 9. Average concentrations and standard deviations in the 4 zones of the Scheldt estuary (1981–1983)

	Dissolved metals ($\mu g\ l^{-1}$)							
	Cu	STD	Zn	STD	Cd	STD	Pb	STD
Zone I	1.02	0.38	11.7	1	0.086	0.067	0.23	0.05
Zone II	1.3	0.64	13.3	5.3	0.063	0.027	0.26	0.07
Zone III	1.91	0.3	8.75	1.6	0.19	0.045	0.24	0.12
Zone IV	1.43	0.55	4.8	2.1	0.09	0.056	0.22	0.07
	Particulate metals ($\mu g\ g^{-1}$)							
	Cu	STD	Zn	STD	Cd	STD	Pb	STD
Zone I	278	77	800	410	27.9	14.6	288	104
Zone II	207	68	397	120	22.8	9.5	254	58
Zone III	95	43	272	62	8.2	5.8	153	9
Zone IV	35	15	157	60	2.8	2.1	101	26

Table 10. Mass balances of average, minimum and maximum dissolved and particulate copper (tons yr^{-1}) in the period 1981–1983, Δ = inputs–outputs

	Species	Zone I	Zone II	Zone III	Zone IV
	Δ^P Cu	−1.1	11.8	0.8	3.5
Average	Δ^D Cu		−2.1	−2.2	1.4
values	Δ^T Cu		9.7	−1.4	4.9
	% of Δ^T/total inflow		8%	4%	21%
	Δ^P Cu	−11.0	12.5	3.1	1.9
Minimum	Δ^D Cu		−1.0	−3.4	2.7
values	Δ^T Cu		11.5	−0.3	4.6
	% of Δ^T/total inflow		13%	1%	30%
	Δ^P Cu	−1.2	10.9	−1.4	3.3
Maximum	Δ^D Cu		−3.2	−1.1	0.2
values	Δ^T Cu		7.7	−2.5	3.5
	% of Δ^T/total inflow		5%	5%	12%

The dissolved Zn concentrations decreased by almost a factor 2 similar to the decrease of the point sources, while the particular Zn concentrations did not change substantially. This is in contrast with Cu for which the particulate concentrations decreased stronger over the last decade than the dissolved ones. The aquatic point sources of Cu decreased by a factor of 2, but the atmospheric point sources even more.

Mercury concentrations could not be compared because the analytical methodologies changed too much over the last ten years, but it has been thoroughly studied during the past few years. The behaviour of the dissolved Hg fraction is quite different in summer than in winter. Anoxic redox conditions in the upstream estuary in summer provoke on the one hand a transfer from the dissolved to the particulate phase and on the other hand methylation of inorganic Hg. Once re-aeration occurs and the estuary becomes again oxic, the dissolved Hg level increases probably due to the oxidation of reduced, insoluble species with which it is associated or of reduced Hg species itself. At the mouth of the estuary in summer, methylation was also observed. At the same time high plankton population densities prevailed.

The production of Hg° is a very important process because it purifies the estuary via increased transfer to the atmosphere. The Hg° formation rates we calculated from our field data range from 0.22 to 0.35% d^{-1} in winter and from 0.63 to 1.3% d^{-1} in summer. These rates are in the same range as the values reported by Mason et al. (1995b; 1995c) for open ocean waters.

For all metals, the particulate flows are the major ones in the estuary except for Cu and Cd at the mouth.

High accumulation occurred in zone II, the area of maximum turbidity and high sedimentation. If we assume a linear decrease of the metal sedimentation flux between values measured in the period 1981–1983 and in 1995, following sedimentated amounts have been calculated: 49 tons of Cd, 510 tons of Cu, 670 tons of Pb and 1800 tons of Zn. Even if part of this load has been dredged or transported elsewhere in the estuary by the tidal action, it will still take many years for the estuary to recover.

The Scheldt inputs into the North Sea (particulate +dissolved), estimated from measurements in the mouth of the estuary of trace metal concentrations, residual waterflows and suspended sediments, amount to: 46.6 kg yr^{-1} for Hg, 8.9 tons yr^{-1} for Cu (81–83), 34.3 tons yr^{-1} for Zn (81–83), 0.71 tons yr^{-1} for Cd (81–83), 8.99 tons yr^{-1} for Pb (81–83), 6.3 tons yr^{-1} for Cu (95), 21.9 tons yr^{-1} for Zn (95), 0.42 tons yr^{-1} for Cd (95) and 8.33 tons yr^{-1} for Pb (95).

Acknowledgements

The authors would like to thank the Ministries of Public Health, Scientific Policy and National Defence and the Management Unit of the Mathematical Model North Sea and Scheldt Estuary for providing ship time on the M.S. Belgica. We also wish to thank the officers and the crew of the M.S. Belgica for their efforts in our regard. Martine Leermakers is indebted to the Ministry of Scientific Policy (DWTC) for financial support through the EUROTRAC Programme ASE.

References

Andersen, J. & T. Niilonen (eds), 1995. Progress Report, 4th International Conference on the Protection of the North Sea, Danish EPA, Copenhagen, Denmark, 247 pp.

Baeyens, W., G. Gillain, M. Hoenig & F. Dehairs, 1986. Mobilization of major and trace elements at the water sediment interface in the Belgian coastal area and the Scheldt estuary. In J. C. J. Nihoul (ed.), Dynamic Biological Processes at Marine Physical Interfaces, Elsevier, Amsterdam: 453–466.

Baeyens, W. & F. Monteny, 1988. Study of the degree of contamination of dredge in the Scheldt estuary (in Dutch). Synthesis report. University of Brussels, Belgium.

Baeyens, W., M. Leermakers, H. Dedeurwaerder & P. Lansens, 1991. Modelization of the mercury fluxes at the air-sea interface. Wat. Air Soil Poll. 56: 731–744.

Baeyens, W., 1992. Speciation of mercury in different compartments of the environment. Trends in Analyt. Chem. 11: 245–254.

Baeyens, W., 1998. Evolution of trace metal concentrations in the Scheldt estuary (1978–1995). A comparison with other estuarine and ocean levels. Hydrobiologia 366: 157–167.

Baeyens, W., M. Elskens, G. Gillain & L. Goeyens, 1998a. Biogeochemical behaviour of Cd, Cu, Pb and Zn in the Scheldt estuary during the period 1981–1983, Hydrobiologia 366: 15–44.

Baeyens, W., C. Meuleman, B. Muhaya & M. Leermakers, 1998b. Behaviour and speciation of mercury in the Scheldt estuary (water, sediments and benthic organisms), Hydrobiologia 366: 63–79.

Baeyens, W., B. van Eck, C. Lambert, R. Wollast & L. Goeyens, 1998c. General description of the Scheldt estuary, Hydrobiologia 366: 1–14.

Baeyens, W., M. Elskens, R. Van Ryssen & M. Leermakers, 1998d. The impact of the Scheldt input on the trace metal distribution in the Belgian coastal area (results of 1981–1983 and 1995–1996). Hydrobiologia 366: 91–108.

Leermakers, M., M. Elskens, S. Panutrakul, F. Monteny & W. Baeyens, 1993. Neth. J. Aquat. Ecol. 27: 267–277.

Leermakers, M., C. Meuleman & W. Baeyens, 1995. Wat. Air Soil Pollut. 80: 641–652.

Mason, R. P., F. M. M. Morel & H. F. Egmond,1995a. The role of microorganismsin elemental mercury formation in natural waters. In D. B. Porcella, J. W. Huckabee & B. Wheatley (eds), Mercury as a Global Pollutant. Wat. Air Soil Pollut. 80: 775–787.

Mason, R. P., J. O'Donnell & W. F. Fitzgerald, 1995b. Elemental mercury cycling within the mixed layer of the Equatorial Pacific Ocean. In C. J. Watras & J. W. Huckabee (eds), Mercury Pollution: Integration and Synthesis. Lewis Publishers, Boca Raton: 83–98.

Mason, R. P., K. R. Rolfhus & W. F. Fitzgerald, 1995c. Methylated and elemental mercury cycling in surface and deep ocean waters of the North Atlantic. In D. B. Porcella, J. W. Huckabee & B. Wheatley (eds), Mercury as a Global Pollutant. Wat. Air Soil Pollut. 80: 665–677.

Van Eck, G. T. M., 1991. The development of a water quality model for the Scheldt estuary (in Dutch), Water (Belgium), 61: 215–218.

Wartel, S., 1977. Geologie en Mijnbouw, 56: 219–233.

Hydrobiologia **366:** 129–142, 1998.
W. F. J. Baeyens (ed.), Trace Metals in the Westerschelde Estuary.

Water quality modelling of the western Scheldt estuary

Maarten R. L. Ouboter[1], Bert T. M. Van Eck[2], Jos A. G. Van Gils[1], Jean Pierre Sweerts[1] & Monique T. Villars[1]
[1]*Delft Hydraulics, P.O. Box 177, 2600 MH Delft, The Netherlands*
[2]*National Institute for Coastal and Marine Management/RIKZ, P.O. Box 8039, 4330 EA Middelburg, The Netherlands*

Key words: Western Scheldt, water quality, heavy metals, modelling, decision support system

Abstract

In 1987 the National Institute for Coastal and Marine Management (the Netherlands) initiated the SAWES project (Systems Analysis WEstern Scheldt). The main goals of the project were to acquire knowledge with respect to the pollution problems in the Scheldt estuary and to apply this knowledge to management issues of the Scheldt on an operational level. The main interest was to understand quantitatively the relation between inputs of polluting substances and effects on the ecosystem. In the framework of the project, a water quality model for the Scheldt estuary was made. In 1995, the model area was expanded to include the Belgian coastal region at the mouth of the estuary.

The model calculates the fate of discharged substances and establishes the relation between pollution inputs and the resulting water quality, including general water quality (oxygen, pH, alkalinity, major ion chemistry and nutrients) as well as pollutant concentrations. The model incorporates all chemical processes which affect these concentrations, including the precipitation/dissolution of metal sulphides which to a large extent controls the fate of trace metals. Based on calibration and verification exercises, it has been shown that the model provides a good representation of the physical and chemical processes taking place within the estuary and can therefore be used to support policy development for the estuary. For example, the model can compute how the water and sediment quality in the estuary will respond to reduced inputs of waste or to such human activities as dredging.

Introduction

The river Scheldt crosses France, Belgium and the Netherlands on its way to the North Sea. It has a length of 355 km and a drainage area of about 22 000 km^2, mostly within Belgium. The average fresh water flow at Rupelmonde is about 100 m^3/s. The estuary of the Scheldt covers a length of 95 km, between the village of Rupelmonde in Belgium and the city of Vlissingen in the Netherlands.

The upper estuary (km 0–40) consists of one channel with a width between 0.4 and 2 km. The lower estuary (km 40–95) is wider, up to 8 km, and consists of deep and shallow parts, with different ebb and flood channels. The average depth is about 11 m, while the maximum depth is 40 m. The tidal motion dominates the water movement in the estuary and significant salt intrusion may be observed over the full 95 km. Normally the estuary is well mixed, except in the upper estuary under high fresh water flow conditions. The main functions of the estuary are shipping, fishing, recreation and sand extraction. The ecological value of the area and the natural landscapes are also of significant importance to the region. However, there are two significant environmental problems in the Scheldt estuary:

(1) In order to maintain the shipping routes to Antwerpen and other smaller harbours, intensive dredging and dumping of (contaminated) sediments is necessary; and

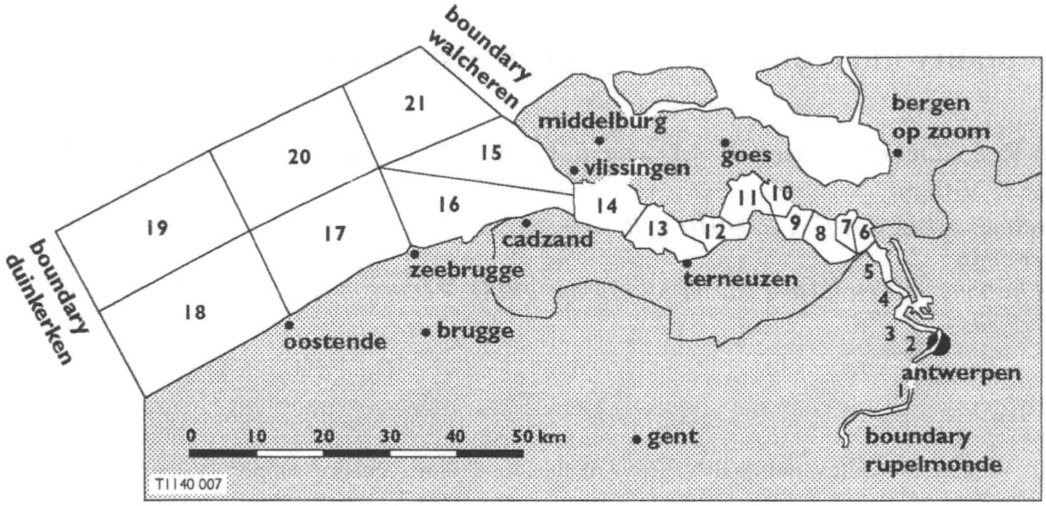

Figure 1. Physical schematisation used in the water quality model.

(2) The area is used to dispose of large amounts of domestic and industrial waste water.

In 1987 the National Institute for Coastal and Marine Management (the Netherlands) initiated the SAWES project (Systems Analysis WEstern Scheldt). The main goals of the project were to acquire knowledge with respect to the pollution problems in the estuary and to apply this knowledge on an operational level. The main interest was to understand the relation between inputs of polluting substances and effects on the ecosystem, with special attention given to heavy metals, specifically cadmium, copper and zinc. A model instrument was created to support policy development and management of the estuary by means of systems analysis. In 1995, the model was expanded to include the Belgian coastal region at the mouth of the estuary. The complete study area is shown in Figure 1.

The systems analysis approach which forms the basis for the present water quality model and its application, has the following main characteristics. The *water system* under consideration is subject to the negative impact of different *human activities*. Consequently, it suffers certain *effects* which may in turn influence the human activities. The systems analysis requires an accurate definition of the relevant impacts from human activities, as well as the changes therein, and the relevant effects. Based on those definitions the design criteria for the water quality model are established.

Description of the model instrument

The water quality model is one component of the complete model instrument, which consists of a self standing data base, the model itself, and a set of post processing, graphical and analysis tools. The entire instrument, also known as a Decision Support System (DSS) runs on an IBM compatible PC. The water and sediment quality calculations have been made using the general water quality model DELWAQ v4.02 (Delft Hydraulics, 1995a and 1995b). This is a deterministic and dynamic model and takes into account relevant polluting substances in the estuary, as well as the main physical, chemical and biological processes affecting the substances of concern.

The most important characteristics of the model are:

- based on mass balance of all modelled substances;
- transport of dissolved substances, calculated on the basis of the water flow (tidally averaged advective transport) and dispersion;
- transport of particulate substances (such as adsorbed metal species), calculated on the basis of suspended sediment transport;
- a number of non-instantaneous, slow kinetic water quality processes, particularly with respect to metal release due to sulphide oxidation, and oxygen and nitrogen processes;
- calculation of chemical equilibrium for instantaneous water quality processes dependent on macro chemistry (oxygen, pH, alkalinity, ma-

jor ion chemistry and nutrients), partitioning and speciation of metals and phosphorous;

- visualisation of substance mass balances in order to see quantitatively the importance of various different processes.

Measurements of freshwater balance, sediment balance and pollution loads into the region are part of the self standing data base of the Scheldt estuary, and serve as the input to the water quality model. The model process formulations and parameters are for the most part well accepted equations from the general literature on water quality processes. For a few processes, new formulations were written based on water quality measurements in the Scheldt estuary. The following sections describe the physical schematisation, transport of dissolved and particulate substances, the coupling with external loads, and the water quality processes per group of substances.

Physical Schematisation

The estuary and the coastal region were divided into 21 homogeneous model elements, based on the measured concentration gradients (Figure 1). The estuary is represented by 14 elements, while the coastal region is represented by the remaining 7 elements. The coastal region is schematised in two rows of elements parallel to the coast, the first of which has a width of about 15 km. The second off-shore row is wider, and corresponds to the deeper region of the coast. The division of these two rows into individual elements is based on the location of discharges into the coastal region.

Transport of dissolved and particulate substances

The transport of dissolved substances is calculated by the advection-dispersion equation, which includes terms for loads as well as chemical processes. In one spatial dimension the equation reads:

$$\frac{\partial CA}{\partial t} = -\frac{\partial CUA}{\partial x} + \frac{\partial}{\partial x}\left(DA\frac{\partial C}{\partial x}\right) + A(W+P) \quad (1)$$

with:

A = wet cross section (m^2)
C = concentration in the water phase ($g\ m^{-3}$)
D = longitudinal dispersion coefficient (m^2)
P = source and sink terms from water quality processes ($g\ m^{-3}\ s^{-1}$)
t = time (s)
U = flow velocity, averaged over the cross section ($m\ s^{-1}$)
W = pollution loads ($g\ m^{-3}\ s^{-1}$)
x = longitudinal distance (m)

The hydrodynamic coefficients in the equation are based on tidally averaged flow. The water depth in the model is constant, and is taken to be the average water depth for each segment. The exchange between model elements in the estuary is based on the mass balance of fresh water inflow into the estuary. The total fresh water discharge into the coastal region is further transported along the coast based on principles of continuity, using certain assumptions about the direction of transport, e.g. water inflow to the estuary occurs through a northern channel along the coast of Walcheren, and water outflow from the estuary occurs through a southern channel, along the coast of Zeeuws-Vlaanderen. The tidally averaged transport is based on the average measured chloride concentrations and the chloride balance per element. In the coastal zone, the advective and dispersion terms are coupled in one overall transport term. This approach is possible due to the fact that the largest tidal mixing occurs in the direction of the smallest observed concentration gradient (i.e. along the coast).

The approach for suspended sediment is similar to that for dissolved substances, with the difference that the advective term in equation (1) is replaced by a term that contains the net transport of suspended solids. The time and space dependent net transport of suspended solids is an input item to the model. It is supplied on a seasonally integrated basis, so that the effects of high discharges or storm events are included. In the coastal region, this approach could not be used, due to a lack of data. In that area the transport of suspended sediment is modelled identically to that of dissolved substances.

Additional attention has been given to the accumulation of toxic substances in the bottom sediment. This is an important process affecting the fate of these substances in the environment, specifically with the adaptation of the system to changes in the pollution loads. In the model, the bottom sediment has a thickness of 0.5 meter. The exchange between this layer and the overlying water column is determined by the presence of so-called ‘mega ripples’ which are influenced by the tidal flow over the migrating bottom ripples. The exchange between these sand dunes and the water column results in a residence time of 0.5 years for the bottom sediment at any one point. The mass of sediment in the bottom is the reason that the water quality in the model area needs approximately 15 years to come into equilibrium with the decreasing loads from upstream. Figure 2 provides an overview of the modelling of the transport of suspended solids and adsorbed pollutants.

Loads of substances

The substance loads into the estuary and coastal region are given in the SAWES data base. Since there are no measurements available, assumptions had to be made concerning the concentrations at the boundaries of the model area:

- the inflow of metals at the southern boundary of the modelled coastal region was estimated based on the available measurements of bottom concentrations;
- for the open ocean boundaries, the concentrations of sulphate, sodium, potassium, calcium and magnesium were based on measured chloride concentrations and a fixed ratio of each of the ions in sea water;
- for the upstream boundary in the estuary by Rupelmonde, as well as for the open sea boundary, the algae concentrations are based on measured chlorophyll concentrations and the assumption that half the algae are diatoms and the other half are non-diatoms;
- the CO_2 concentration at the upstream boundary in the estuary (Rupelmonde) is estimated based on the measured pH and alkalinity, and the relation between pH, alkalinity and CO_2 concentration.

Water quality processes for oxygen and nitrogen

The water quality model includes a number of water quality processes that contribute to the oxygen balance and to the nitrogen cycle. All these processes including the exchange of oxygen and carbon dioxide with the atmosphere, the mineralization of carbonaceous organic waste (BOD) and organic nitrogen, nitrification and denitrification are incorporated in the source/sink term of the model equation. The formulations used are conventional and have been derived from textbooks on water quality modelling and aquatic chemistry, such as Crank (1973), DiToro et al. (1987) and Stumm & Morgan (1981). However, due to the special character of the Scheldt estuary with a very low oxygen concentration during summer, an extra factor was added to account for the change in the reaction rates with a decreasing oxygen concentration. A complete definition of the model formulations is given by Van Gils et al. (1993).

The presence of algae affects the oxygen balance, the nutrient cycles as well as the adsorption of trace metals through the pH. Therefore algae were also included in the water and sediment quality model. The model distinguishes between diatoms and all other non-diatom species. In the present study the mortality was increased in the area with the highest salinity gradient, to account for the effect of salt stress on algae. Van Gils et al. (1993) provide a detailed description of the algae model.

The exchange of oxygen and CO_2 between the water and atmosphere is calculated based on their over or undersaturation in water relative to the atmosphere. Decay reactions are based on first order reactions including a temperature function as well as an oxygen concentration function. The pH in the system is calculated by the thermodynamic equilibrium model CHARON as a function of the alkalinity and the CO_2 concentration. This same model is also used to calculate metal speciation (Smits et al., 1985). Figure 3 shows a schematic overview of the general water quality model.

Water quality processes for heavy metals

The behaviour of trace metals in aquatic systems is influenced strongly by adsorption to organic and inorganic particles. The dissolved fraction of the metals are transported by water, while the adsorbed fraction is transported by sediment. The distribution of a trace metal into adsorbed or dissolved fractions depends on the trace metal itself, the adsorption characteristics of the particles and the local environmental conditions (e.g. temperature, pH, redox state, salinity and ionic strength). The distribution can be described by a mineral equilibrium and/or adsorption to a particulate surface. This phenomenon was included in the water and sediment model as an equilibrium computed by a chemical sub-model CHARON.

In the Belgian part of the estuary, the metals Cd, Cu and Zn often exist as precipitated sulphides (Zwolsman et al., 1990, Zwolsman & Van Eck, 1992). The field data suggest that these sulphides are present where the oxygen concentration is low. Because these sulphide minerals have a low solubility, a large fraction of the metals are in particulate form. In a seaward direction, as the oxygen concentration increases, the sulphides become oxidised. However, the field data

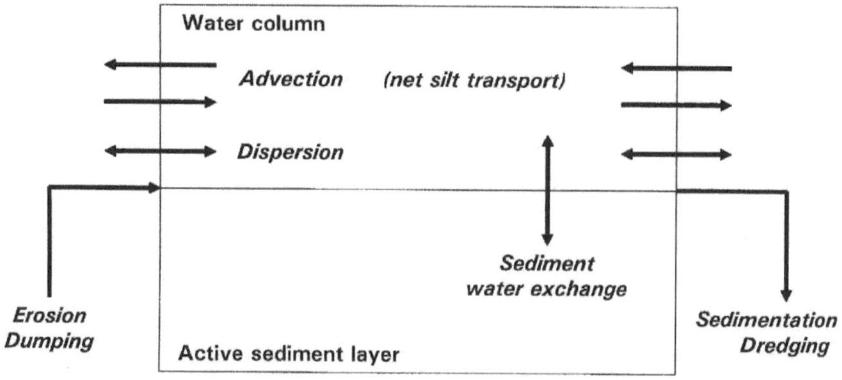

Figure 2. Schematic overview of the transport processes for particulate matter (suspended solids and adsorbed pollutants) which were incorporated in the model.

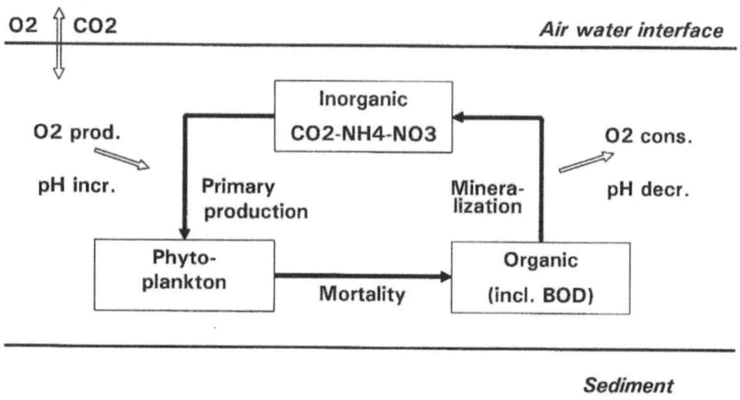

Figure 3. Schematic overview of the general water quality model.

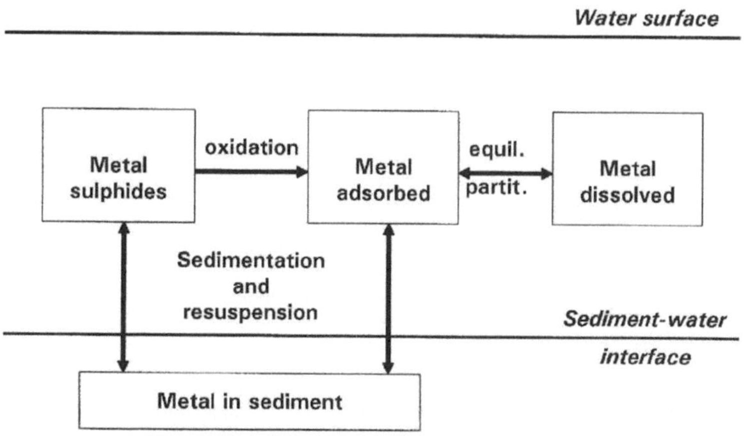

Figure 4. Schematic overview of the model for the behaviour of trace metals.

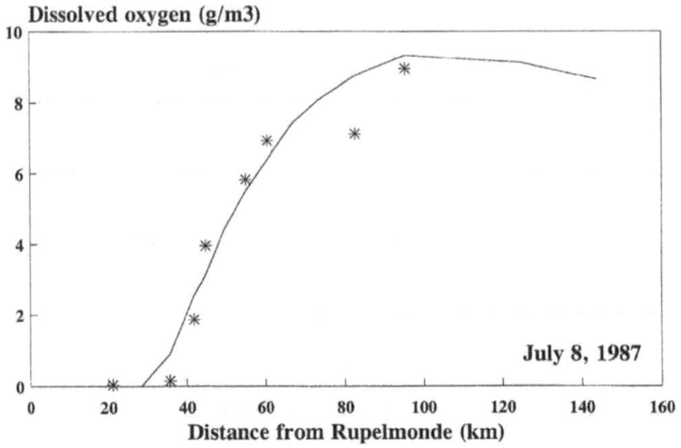

(a)

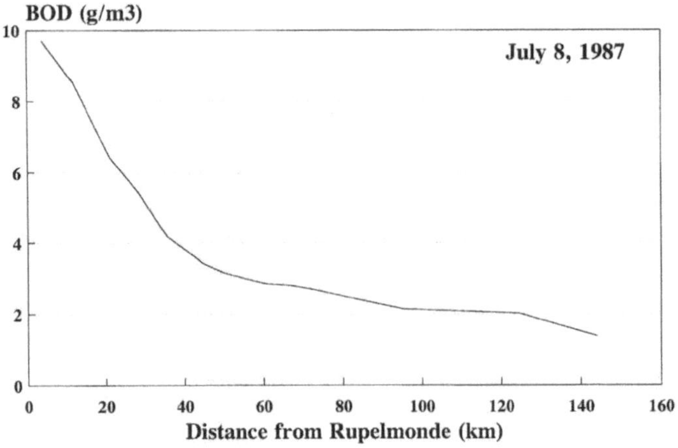

(b)

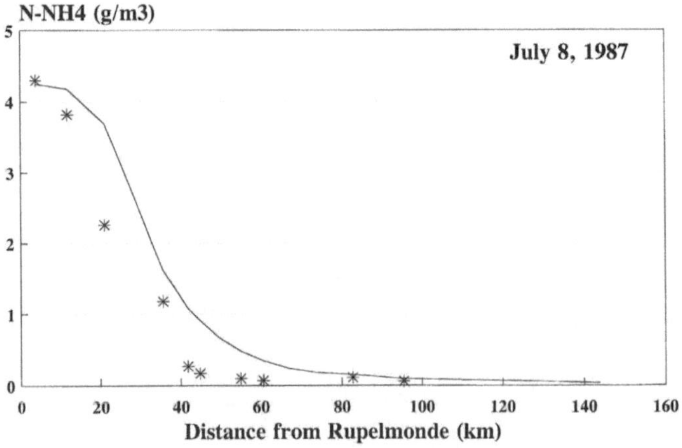

(c)

Figure 5. Calculated concentrations of oxygen (a), BOD (b) and ammonium (c), compared to measured concentrations (*) on July 8 1987, as a function of the distance from Rupelmonde.

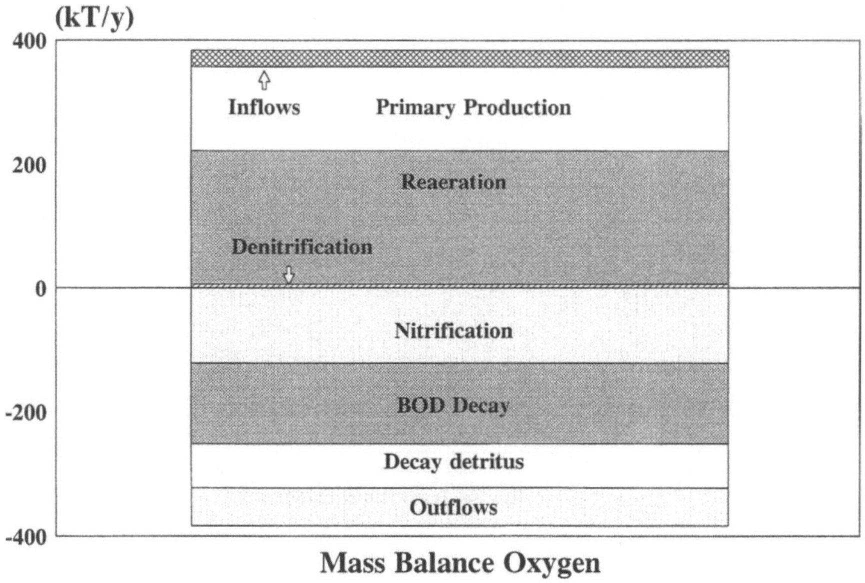

Figure 6. Oxygen balance for the total model area during the final model calibration run: positive terms are above x-axis, negative terms are below the x-axis.

also show that the oxidation is a slow process, dependant on the temperature and oxygen concentration. Therefore oxidation can not be modelled as an equilibrium process and a kinetic formulation for the oxidation of metal sulphides had to be developed. In the water quality model, the trace metals enter the estuary as sulphides and thus are not incorporated in the sorption equilibrium system. This happens only after the oxidation of the sulphides. Figure 4 shows a schematic overview of the model for the behaviour of trace metals.

Calibration of the water quality model

Measurements used for calibration

For calibration of the model, the consistency of the loads to the system, the transport and other processes within the system, and the measured concentrations within the system must be assessed. For calibration of the general water quality model and the model of the behaviour of metals and phosphate, two years are relevant. In 1987, the results of 8 specific monitoring campaigns were available from the SAWES project, providing detailed longitudinal profiles of a wide variety of water quality parameters over a range of seasons. As part of the SAWES project, water quality measurements were also gathered from the standard water quality monitoring program for a more recent year, namely 1991. This year contrasts to 1987 with regard to discharges and large scale dredging in the harbours of Antwerp. Both years are modelled using the same process formulations and process coefficients.

Steps in the calibration process

The model calibration took place in three steps:

- calibration of the dispersion coefficient by fitting to the average chloride profile in the estuary;
- calibration of the parameters in the oxygen and nitrogen related processes, including the algae model; and
- calibration of the parameters in the heavy metal related processes.

The first two steps are important primarily to determine the general water chemistry which affects metal partitioning. They were executed by iteratively running the model and adjusting the coefficients by hand until sufficient similarity between the model results and field data was achieved. A postprocessing facility supported this procedure, enabling the user to observe the magnitude and the spatial and temporal variability of all modelled processes. In this process, the known ranges of the process coefficients as given in the literature were taken into account.

The third step of the calibration procedure was dedicated to the trace metal behaviour. The metal

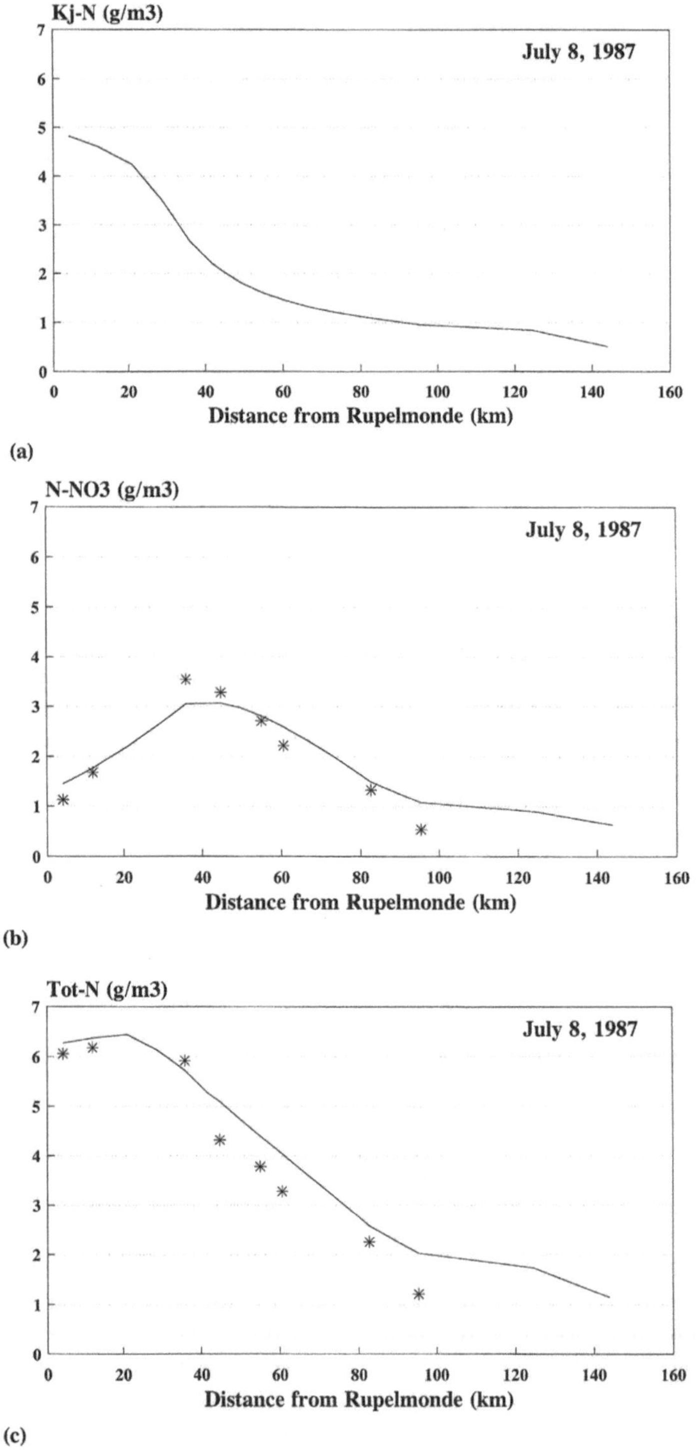

Figure 7. Calculated concentrations of Kjeldahl nitrogen (a), nitrogen in nitrates (b) and total nitrogen (c), compared to measured concentrations (*) on July 8 1987, as a function of the distance from Rupelmonde.

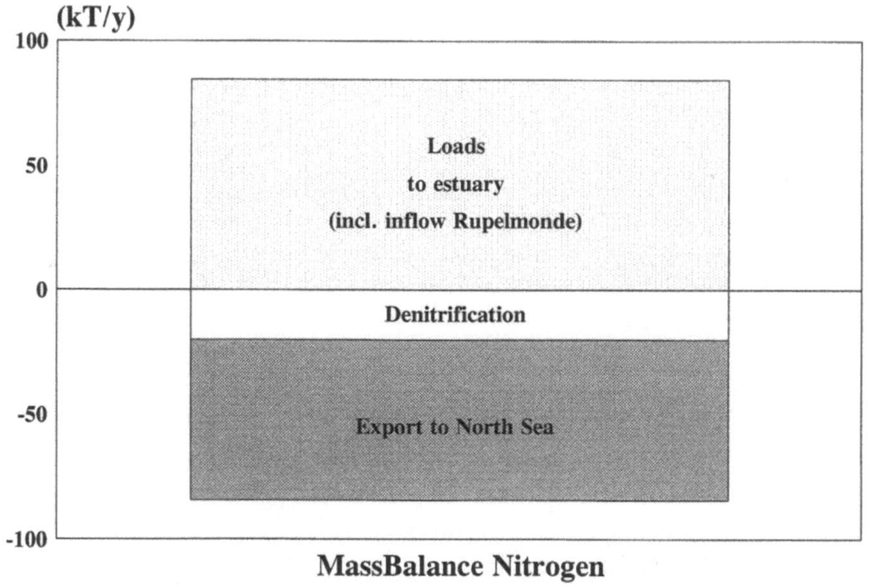

Figure 8. Nitrogen balance for the total model area during the final model calibration run: positive terms are above the x-axis, negative terms are below the x-axis

sulphides of Cd, Cu and Zn have a very poor solubility and their occurrence can be recognised by an extremely high partition coefficient. Lower partition coefficients indicate a 'normal' reversible partition between adsorbed and dissolved metal species.

All samples from the monitoring campaigns in 1987 were divided in samples with and samples without sulphides, based on the magnitude of the partition coefficient. The samples without sulphides were used to calibrate the Gibbs parameters of the adsorbed metal species in the chemical equilibrium model representing the adsorption characteristics of the particles present. For this purpose an automatic calibration procedure was used (Ouboter & De Rooij, 1990). The adsorption of trace metals to fine particles differs between different water systems, due to the different characteristics of the particles. In the Western Scheldt estuary the adsorption is relatively strong, possibly due to the strong adsorption to reactive iron hydroxide FeOOH, which forms at the oxic-anoxic interface in the estuary.

The samples with sulphides were used to calibrate the kinetics of sulphide oxidation. Based on the calibrated equilibrium model, the adsorbed fraction excluding sulphides was calculated and subtracted from the measured particulate fraction in the sample. The difference was assumed to consist of sulphides and thus the sulphide concentration in the samples was estimated. A kinetic formulation for the oxidation of sulphides, dependent on the oxygen concentration and the water temperature was derived from the measurements. Both a high temperature and a high oxygen concentration stimulate the oxidation process.

Discussion of the calibration results

The results of the model calibration give a clear understanding of the functioning of the water quality processes in the Scheldt estuary and coastal zone system. It must be noted that many more measurements of loadings and water quality concentrations were available for the estuary than for the coastal region. The description of the system is thus focused primarily on the estuary and the mouth up to the region of Cadzand.

The loadings with BOD and ammonium cause low oxygen concentrations in the upstream regions of the estuary, especially during the summer (Figure 5). The mass balance of oxygen (Figure 6) shows that consumption of oxygen occurs due to the decay of organic material (mostly anthropogenic of origin) and nitrification. The concentration profiles in the estuary show where the different oxygen consuming processes are dominant. A detailed analysis of the model results shows that in the upper estuary oxygen consumption is mainly due to nitrification, whereas in the lower estuary oxygen is consumed mainly by the decay of BOD. Therefore, nitrification is the main cause of anoxia in the upstream part of the estuary.

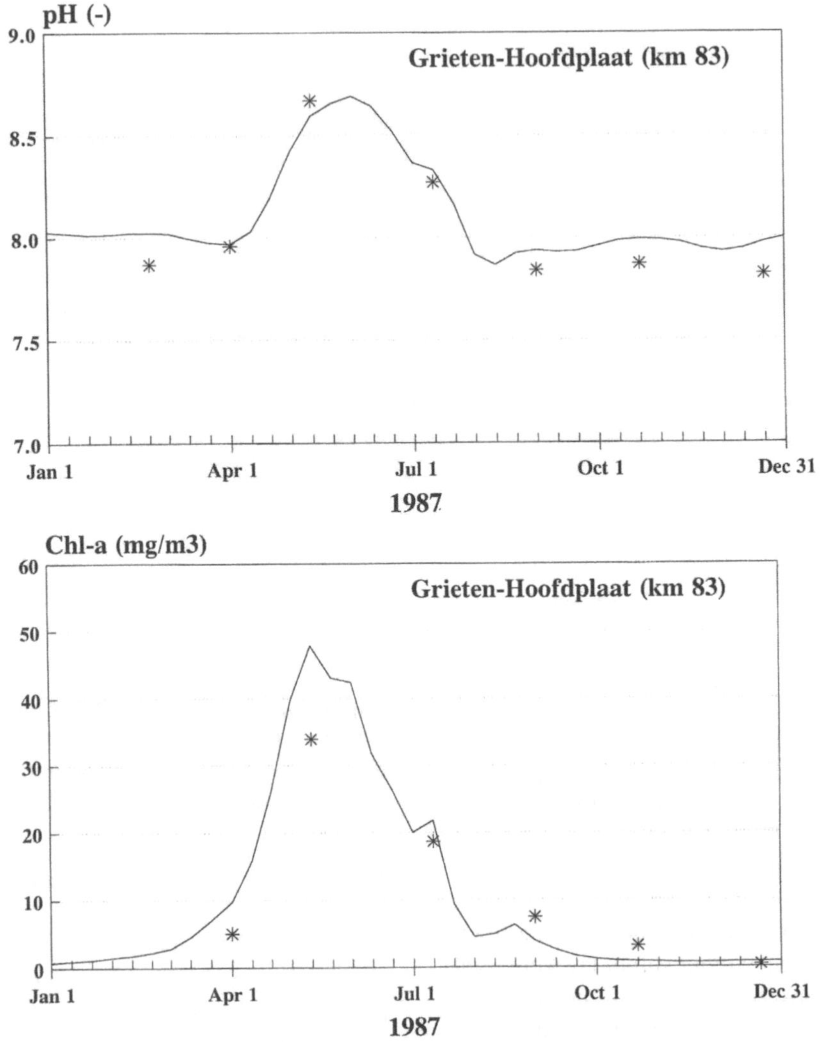

Figure 9. Calculated pH (a) and chlorophyll concentration (b), compared to measured values (*) at the location Grieten-Hoodplaat (83 km downstream of Rupelmonde) for 1987.

The total nitrogen concentration is controlled by waste loads in the river Scheldt and in the upper Scheldt estuary (Figure 7). In the seaward direction the concentration drops, mainly due to dilution with sea water. The overall mass balance for nitrogen shows that about 25% of the total nitrogen load is removed by denitrification (Figure 8). With a method of parameter analysis, it can be determined within which flux ranges the measurements can be reproduced with the model. From this analysis, it is shown that between 15–30% of the nitrogen is denitrified in the estuary.

Looking at the different nitrogen forms, it can be seen that the loads to the system are primarily in the form of Kjeldahl nitrogen, which is the sum of organic nitrogen and ammonium. However, almost all of the organic nitrogen and the ammonium is mineralised and nitrified, especially in the upstream part of the estuary. To illustrate this, Figure 7 shows the Kjeldahl nitrogen profile. The nitrification reaction produces nitrates, and together with the dilution from the sea, this results in a typical maximum in the nitrate concentration.

The transport and concentrations of the trace metals are dominated by the transport of silt in the longitudinal direction in the water column, and in the vertical direction between the active bottom layer and the water column. The dissolved concentration, which is important for bioaccumulation, depends on the sul-

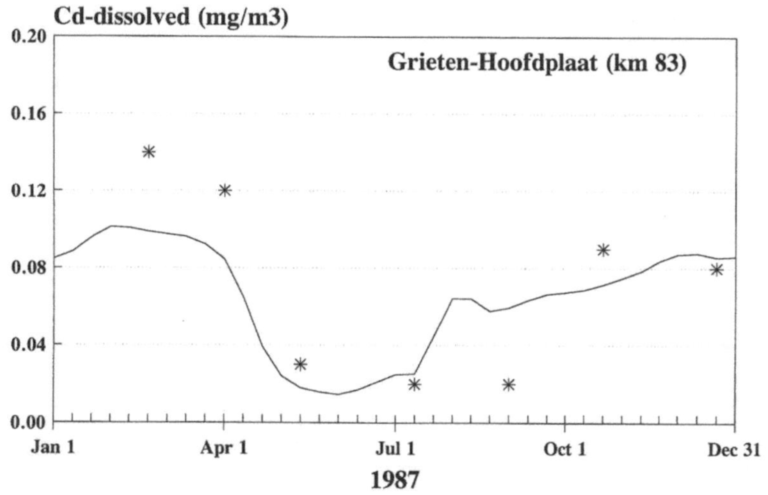

Figure 10. Calculated and measured concentrations of dissolved cadmium at the location Grieten-Hoofdplaat (83 km downstream of Rupelmonde) for 1987.

phide oxidation and after oxidation on the pH, alkalinity, temperature and salinity. The pH is determined by the acid forming processes (nitrification and decay of organic material), and the exchange of CO_2 with the atmosphere. An another important influence on the pH is the production of organic material during an algae bloom. This occurs primarily between mid April and mid July in the mouth of the estuary. Different amounts of production occur in different years, usually due to the turbidity conditions. In 1987 there was a significant amount of production (Figure 9), which was coupled to a noticeable increase in the pH. By high pH, heavy metals are more strongly adsorbed to particulate material. As a result, the concentration of dissolved metals (e.g. cadmium) decreases (Figure 10).

The concentration profile of dissolved cadmium along the estuary is determined by the oxidation of sulphide in the area of the Belgian-Dutch border and from the dilution with sea water. The two processes determine a profile for dissolved metals with a maximum in the central part of the estuary (Figure 11). This maximum is further enhanced by the desorption of cadmium due to the influence of chloride in the sea water.

Following the calibration and verification of the model, a sensitivity analysis has been executed. This analysis focused on about 10 important input parameters. The sensitivity analysis was the last step in a modeling exercise aimed at a better understanding of the water system under consideration. It is also led to a general understanding of the reliability of the model as a decision support tool. This understanding was communicated to decision-makers using the model (DELFT HYDRAULICS et al., 1995).

Model application for policy analysis

Based on the ability of the model to accurately reproduce the most important transport and chemical processes occurring in the estuary and coastal region, it is possible to use the model to assess different management policies and scenarios. One such evaluation can be made of the effect of the large scale dredging in the harbours of Antwerp.

According to international agreements between the Netherlands and Belgium, a large amount of sediment has been removed from the estuaries since 1990 (approximately 300 000 tons per year). Rijkswaterstaat in the Netherlands has made a calculation of the sediment balance in the estuary, in which this amount is taken into account (Vereeke, 1994). The model has been used to calculate the effects of the dredging on the discharge of cadmium from the Scheldt estuary to the North Sea. Two calculations have been made, based on the sediment balances with and without dredging activities in the harbours (as calculated by Vereeke, and Van Maldegem, respectively). Both calculations use the meteorological conditions and loads for 1987. For substances which are adsorbed to particulates, the reaction time of the water quality system is on the order of 10–20 years. This is due to the large amount of material in the bottom sediment, which changes slowly due to kinetics. Therefore, both calculations

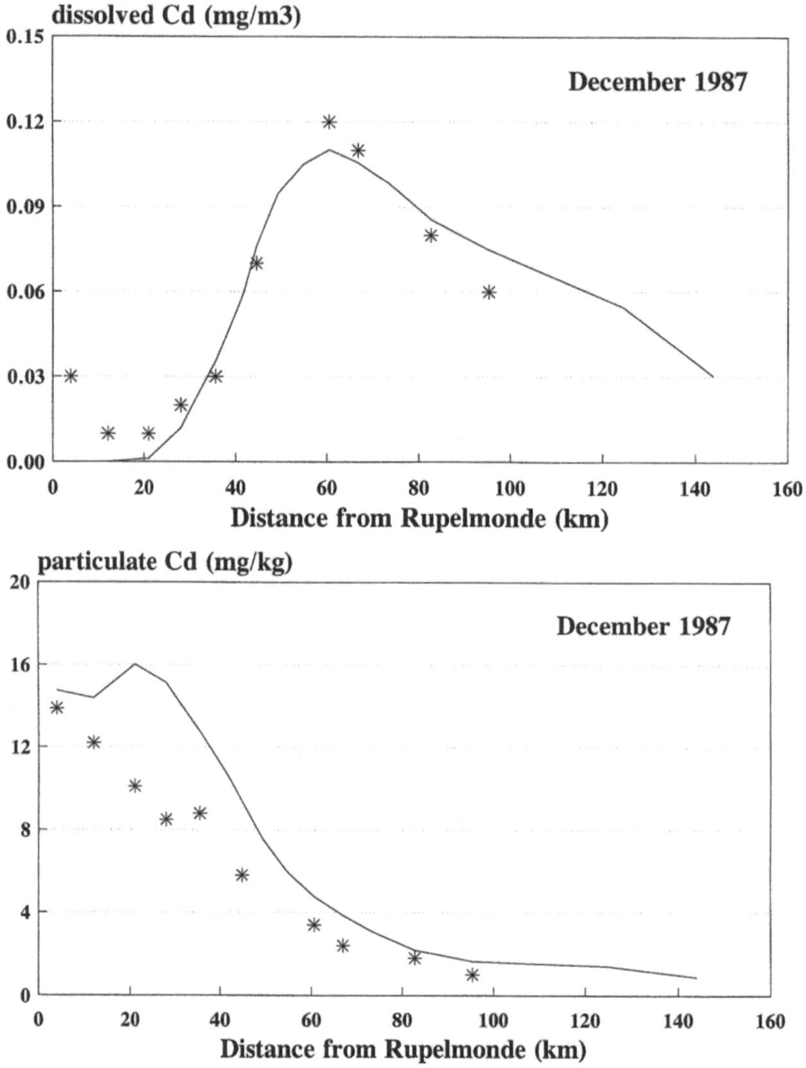

Figure 11. Calculated concentrations of dissolved (a) and particulate (b) cadmium, compared to measured concentrations (*) in December 1987, as a function of the distance from Rupelmonde.

have been made for a period of 10 years. The results for the last year are presented.

Near Vlissingen, the dredging activities are expected to resulted in a 25% decrease in the total cadmium concentration (Figure 12). This decrease implies that approximately 33% of the cadmium enters the North Sea, where as prior to dredging this fraction is computed as 50% (Figure 13). These values are based on the equilibrium between the water column and bottom.

Evaluation and recommendations

The analysis of Western Scheldt estuary by means of a mathematical model was possible due to the availability of a large set of waste load and water quality data. The model system that has been developed seems to match its main objectives. Its set up as an integrated tool, based on available knowledge about the system, increased the understanding of the water system with respect to the relation between waste disposal and water quality. With the calibration, verification, sensitivity analysis and application of the water quality model to the Scheldt estuary it has been shown

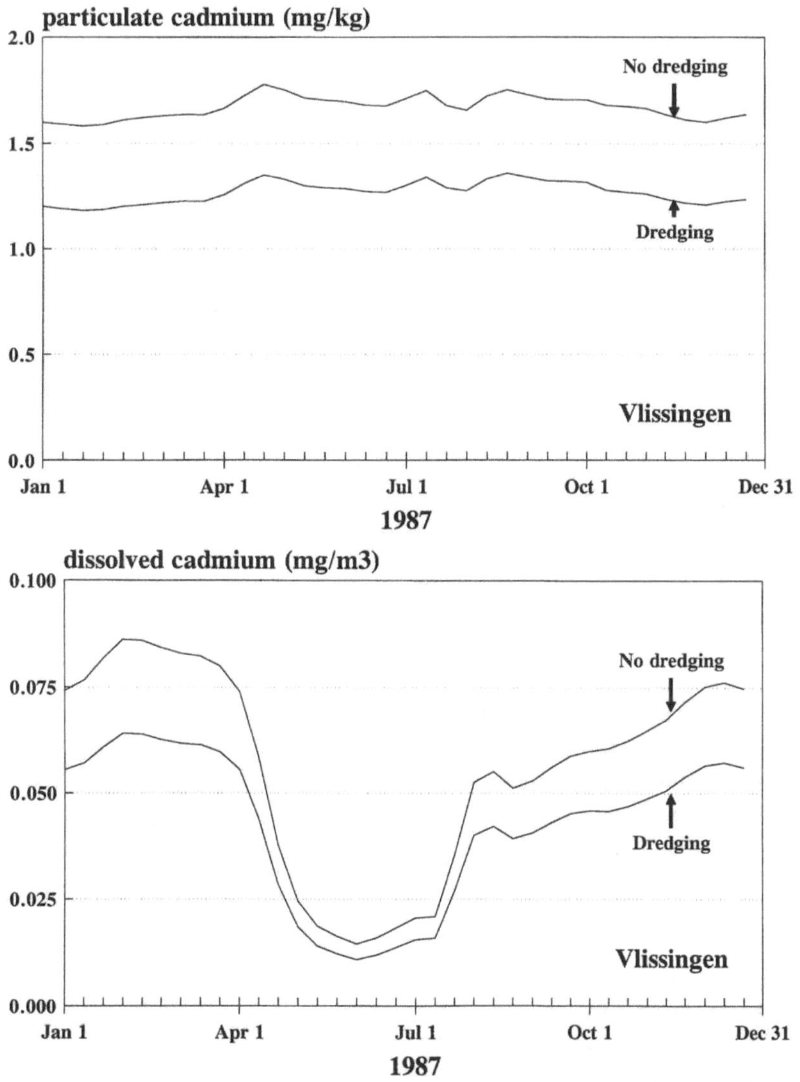

Figure 12. Effect of dredging (starting in 1990) on the concentrations of particulate cadmium (a) and dissolved cadmiuim (b) at the location Vlissingen (pollution loads from 1987).

that a model for the combined eastuary and coastal region can be an important aid for the management of such a system. The use of the model indicated clearly which processes are the most important in this particular estuary. The model also proved to be an important tool to decide about the necessities and the priorities for further investigations. For the estuary, a large set of measurements were available, for both the loads to the system as well as for the water quality in the system. For the coastal region, fewer measurements were available. Especially for the Belgian region of the coast, data availability was the limiting factor in the model development. In this region, the model is based only on large scale transport features, process dynamics in the estuary, on a limited nuumber of nutrient, oxygen, and on some measurements of bottom sediment concentrations of metals.

Further investigations may be aimed at two aspects in particular. Firstly, it is necessary to have more insight on the adsorption characteristics of suspended matter. It must be verified if it is actually iron precipitates that have a high affinity for metals. It must also be checked what is the influence of the loss of the redox boundary caused by a reduction in the loads. This aspect is important in order to make a good estimate of the effect of large scale load reductions on the metal

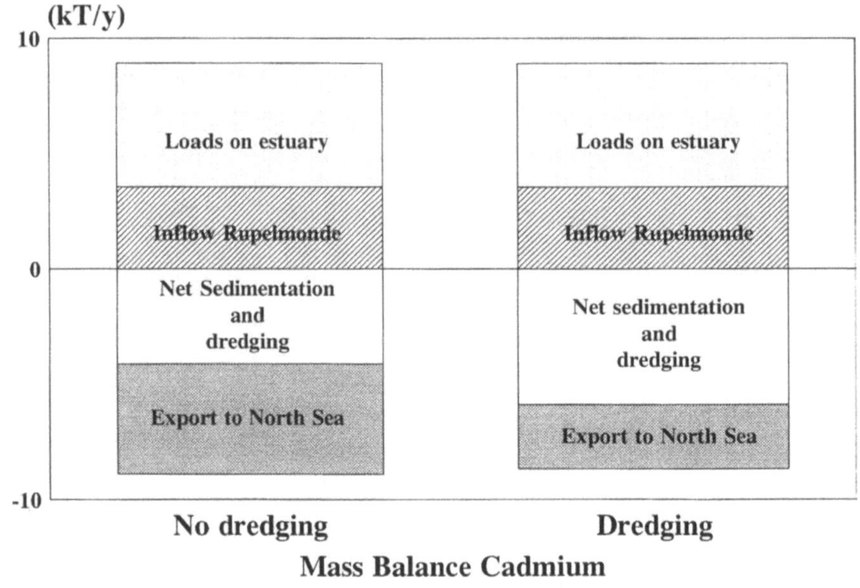

Figure 13. Comparison of cadmium, balances for the total model area, with and without dredging (dredging started in 1990, pollution loads from 1987).

concentrations. In the second place further work may be dedicated to establish more accurate data about the net transport of suspended solids through the estuary, for example by multidimensional modelling.

Acknowledgements

The water quality model for the Scheldt estuary was developed by DELFT HYDRAULICS and the National Institute fort Coastal and Marine Management (RIKZ) in the Netherlands. The extension with the coastal area near the estuary mouth was made during a project in which also participated the Ministry of Health and Environment of Belgium and the Brussels Free University. The authors wish to express their gratitude to all who have contributed to the results presented herein

References

Crank, J., 1975. The Mathematics of Diffusion. Clarendon Press, Oxford, 414 pp.

DiToro, D. M., D. J. O'Connor & J. A. Mueller, 1987. Course on water quality modelling. Manhattan College, New York.

DELFT HYDRAULICS, 1995a. DELWAQ version 4.0 Users Manual, Delft, The Netherlands.

DELFT HYDRAULICS, 1995b. DELWAQ version 4.0 Technical Reference Manual. Delft, The Netherlands

DELFT HYDRAULICS, Rijkswaterstaat & Vrije Universiteit Brussel, 1995. Beleidsondersteunend systeem voor het waterkwaliteitsbeheer van het estuarium van de Schelde en de kustzone (Hoofdrapport, T1140). Delft, The Netherlands (in Dutch).

Ouboter, M. R. L. & N. M. De Rooij, 1990. Use of an autocalibration to evaluate the consistency of a data-set in relation to geochemical model formulations. In ModelCare90; calibration and reliability in Groundwater Modelling. Proceedings of the conference held in the Hague, September 1990, IAHS publication no. 195.

Smits, J. C. G., J. Bril & N. M. De Rooij, 1985. Modelling Heavy Metals in Groundwater near Harbour Sludge Disposal Sites. Envir. Tech. Lett. 6.

Stumm, W. & J. Morgan, 1981. Aquatic Chemistry. Wiley & Sons, New York.

Van Gils, J. A. G., M. R. L. Ouboter & N. M. De Rooij, 1993. Modelling of water and sediment quality in the Scheldt Estuary. Neth. J. Aquat. Ecol. 27: 247–256.

Van Maldegem, D. C., H. P. J. Mulder & A. Langerak, 1993. A cohesive sediment balance for the Scheldt estuary. Neth. J. Aquat. Ecol. 27: 257–265.

Vereeke, S. J. P., 1994. Geactualiseerde slibbalans Schelde-estuarium. Nota AX 94.065, Rijkswaterstaat Directie Zeeland (in Dutch).

Zwolsman, J. J. G., G. T. M. Van Eck & C. H. Van Der Weijden, 1990. Seasonal variability of dissolved trace metal behaviour in the Scheldt estuary, SW Netherlands. Geochem. Cosmochem. Acta.

Zwolsman, J. J. G. & G. T. M. Van Eck, 1992. Dissolved and particulate trace metal geochemistry in the Scheldt estuary, S. W. Netherlands (water column and sediments). Neth. J. Aquat. Ecol. 27: 287–300.

Hydrobiologia **366:** 143–155, 1998.
W. F. J. Baeyens (ed.), Trace Metals in the Westerschelde Estuary.

A time-dependent flow model for heavy metals in the Scheldt estuary

Florimond De Smedt, Veselinka Vuksanovic, Serge Van Meerbeeck & Didier Reyns
Laboratory of Hydrology, Free University Brussels, Pleinlaan 2, 1050 Brussels, Belgium

Abstract

The main processes that determine the behaviour of heavy metals in the Scheldt estuary are tidal hydrodynamics, sediment transport, and sorption of heavy metals on suspended matter. The water quality model WASP is applied to simulate the spatial distribution of five heavy metals in the estuary, under average hydrodynamic and suspended sediment transport regimes. First, the hydrodynamical part of the model is constructed and the results are verified by comparison with measured water levels and flow velocities. Secondly, a salt transport model is set up in order to evaluate the hydrodynamical dispersive mixing characteristics. Thirdly, a suspended sediment transport model is constructed and finally a transport model for heavy metals.

The simulated distributions of the sorbed amounts of heavy metals, suspended sediment and salinity in the estuary agree well with observations. The calculated profiles of dissolved and sorbed concentrations of heavy metals in the water column indicate an accumulation of heavy metals in the zone of the turbidity maximum, while closer to the sea the concentrations diminish due to mixing of the polluted fluvial sediments with unpolluted marine sediments and because of sediment deposition in the estuary. It can be concluded that only a small part of the heavy metals reaches the sea.

Introduction

The Scheldt estuary is known to be highly polluted by industrial and domestic waste waters, containing suspended matter enriched by heavy metals and other toxic substances. This anthropogenic input is either deposited in the estuary or transported to the sea. Because heavy metals are strongly adsorbed by estuarine sediments, the sediments act a reservoir for these metals, such that ecotoxicological risks in the estuary are considerable.

It appears that the behaviour of heavy metals in the Scheldt estuary is governed by dynamic transport of water and sediment, while sorption is the most important physico-chemical reaction. In order to analyse and quantify the distribution of heavy metals in the estuary, a modelling study was conducted using the water quality modelling package WASP (Water quality Analysis Simulation Program), developed by US-EPA (Ambrose et al., 1993). Earlier versions of the present model were used for the prediction of sediment and PCB transport in the Scheldt estuary (Vuksanovic, 1993; Van Meerbeeck, 1994; Vuksanovic et al., 1995).

Hydrodynamic transport

The drainage basin of the Scheldt river and its tributaries has an area of 21 580 km^2. The estuary is about 160 km long and the width increases towards the sea from 20 m to about 5 km. The average water depth is about 10 m. The freshwater inflow to the estuary varies between 20 and 600 m^3 s^{-1}, with a mean value of 110 m^3 s^{-1}, such that the residence time of the fresh water in the estuary varies between 2 to 3 months. The tides are semi-diurnal, with a period of approximately 12 hrs 26 min and an amplitude between 2 and 3 m. The tidal flows at Vlissingen can rise to more than 100 000 m^3 s^{-1}, causing movements of huge water masses, e.g. more than 1 10^9 m^3 on the average per tide (Claessens, 1988).

From the hydrodynamic conditions and salinity distribution, three zones can be clearly distinguished:

(a) the Scheldt estuary extending from the river mouth at Vlissingen to the Dutch-Belgian border, 55 km long; this part is exposed to strong tidal actions such that there is no vertical salinity stratification;

(b) the lower Sea-Scheldt, located between the Dutch-Belgian border and the Rupel tributary, 40 km long;

this zone has a narrow channel and forms the transition region from brackish to fresh water;

(c) the upper, Sea-Scheldt, extending from the Rupel to Ghent, 65 km long; in this part, the salinity is less than 1 g l^{-1}, and freshwater flow conditions dominate over the tide.

Model theory

The hydrodynamic module of WASP is the DYNHYD5 model, which is based on the Saint-Venant equations for unsteady flow in open channels. Expressing the principle of conservation of mass applied to an elemental reach of a prismatic channel with rectangular cross-section, the equation of continuity has following form:

$$\frac{\partial H}{\partial t} + D\frac{\partial U}{\partial x} = 0, \qquad (1)$$

where H is the water surface elevation (head) [L], D is the water depth [L], U is the average longitudinal velocity [L T^{-1}], t is the time [T], and x is the longitudinal distance [L].

The equation of motion can be derived from the principle of conservation of momentum. Taking only into considerations the actions of the gravity and the friction force exerted by the bed, the equation of motion is given by:

$$\frac{\partial U}{\partial t} + U\frac{\partial U}{\partial x} = -g\frac{\partial H}{\partial x} - g\frac{n^2}{R^{3/4}}U|U|, \qquad (2)$$

where g is the acceleration of gravity [L T^{-2}], n is the Manning roughness coefficient [T $L^{-1/3}$], R is the hydraulic radius [L], and $|U|$ is the magnitude of the velocity.

Parameters

In order to solve the flow equations, the water body is discretized in a computational network. The Scheldt estuary is represented by 79 segments as depicted in Figure 1. The first and most landward segment, situated at Ghent, corresponds to a fluvial boundary where tidal effects become insignificant. The last and most seaward segment is located at the river mouth at Vlissingen, where the tides are imposed. The hydrodynamic properties of the network segments are adopted from Laforce et al. (1977). All segments have rectangular cross-sections, an average length of about 2 km, and a Manning roughness coefficient of about 0.028 s $m^{-1/3}$ (small differences are allowed depending upon physical characteristics).

Supply of the fresh water in the estuary is provided by the Scheldt river at Ghent, and four tributaries. The freshwater inflows are imposed in the corresponding segments, and simulated as being constant in time. No other lateral inflows are considered. Computations are performed for three cases: a total inflow, Q [L^3 T^{-1}], equal to the mean annual value (110 $m^3 s^{-1}$), one typically low value (50 m^3 s^{-1}), and one typically high value (250 $m^3 s^{-1}$).

Fluctuations of the water level at the sea boundary are simulated using tidal records from Vlissingen, averaged over the decade 1971–1980. In the simulation, the tidal input wave is fitted by a series of sinusoidal curves. The resulting maximum flow produced by the average tide at Vlissingen amounts to 75 000 m^3 s^{-1}. The calculations are carried out with time steps of 60 s, and cyclic patterns were obtained after a simulation time of 25 hrs.

Results

Results for the mean tidal situation are presented in Figures 2 to 4, showing the instantaneous water levels, flows, and velocities at the marine border (segment 79), and at two other segments (42 and 36) for which measurements are available. The agreement between the calculated and measured water levels is quite good, while calculated velocities and flows are somewhat lower than the observed values. Because, only limited measurements of velocities and flows are available, more information is needed in order to verify the present simulation results.

The residual or mean flow velocities, $\bar{U}$, can be calculated by averaging the simulated velocities over a tidal cycle. In Figure 5, these mean velocity profiles are presented for low, mean, and high freshwater discharge. The mean flow velocities can be directly attributed to the total freshwater inflow, that on the average prevails over the tide. When the profiles are compared, it follows that these residual velocities are higher in the upper part of the estuary and are function of the magnitude of the freshwater inflow only. However, in the middle and lower parts, their significance decreases, and actual hydrodynamic conditions are mainly controlled by the tide.

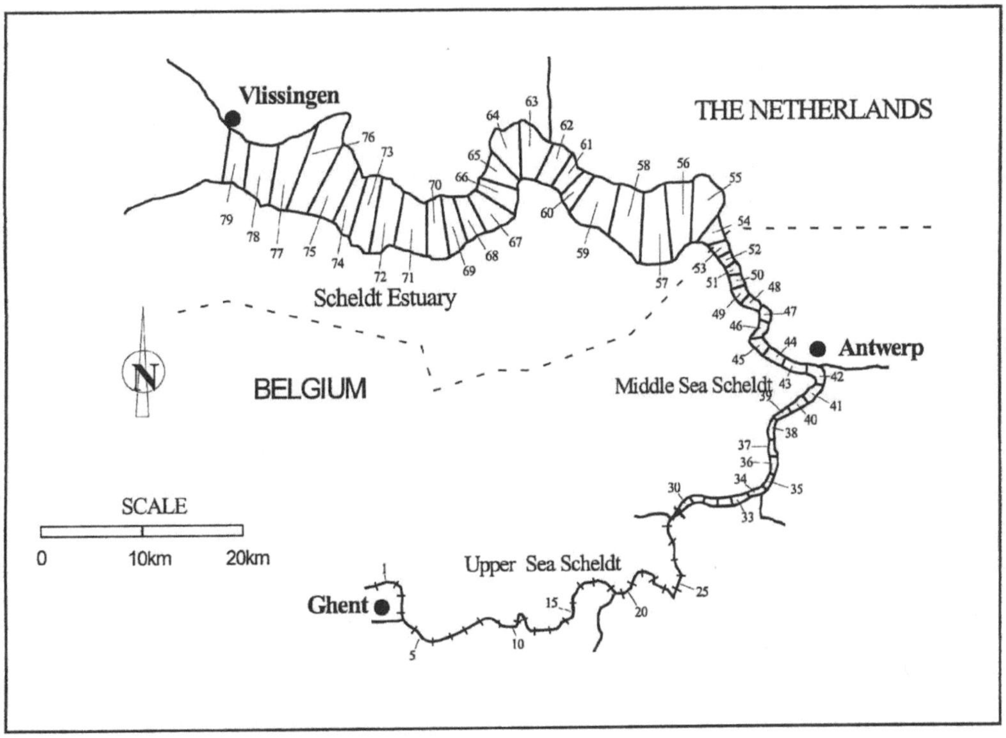

Figure 1. Plan view of the Scheldt estuary and river, with the computation netwrok.

Dispersive transport

Hydrodynamic dispersion is one of the important processes that govern the transport of dissolved or suspended constituents in the water. The mechanisms controlling the dispersive mixing of dissolved and suspended matter in estuaries are numerous and complicated (Chatwin & Allen, 1985). The accurate determination of dispersion coefficients is an essential requirement for the simulation of solute transport. Dispersion in a natural water body as the Scheldt estuary is controlled by differential advection (shear) and turbulent mixing (exchange), while effects of molecular diffusion are normally negligible compared to turbulent dispersion.

Model theory

Longitudinal dispersive transport of a conservative substance is modelled in WASP by a one-dimensional advection-dispersion equation:

$$\frac{\partial C}{\partial t} = -\frac{\partial}{\partial x}(UC) + \frac{\partial}{\partial x}\left(E_x\frac{\partial C}{\partial x}\right), \qquad (3)$$

where C is the cross-sectional average concentration of the constituent [M L^{-3}], and E_x is the effective longitudinal dispersion coefficient [L^2T^{-1}].

Use of the advection-dispersion equation requires properly assigned values for the dispersion coefficient. Various methods for prediction of dispersion coefficients in streams and estuaries have been developed, as for instance reviewed by Bowie et al. (1985). The dispersion coefficient can be evaluated by the well-known formula of Fisher (Fisher et al., 1979):

$$E_x = dng^{1/2}R^{5/6}|U| \qquad (4)$$

where d is the dispersivity [-], which depends upon geometric characteristics, and generally varies between 6 and 600 for natural streams. However, in the WASP model, dispersion coefficients are considered constant in each cell, such that Equation 4 cannot be used explicitly.

Determination of dispersion coefficients

Equation 4 is employed to calculate how dispersion coefficients vary within a tidal cycle in the Scheldt estuary. Since we are mainly interested in mass trans-

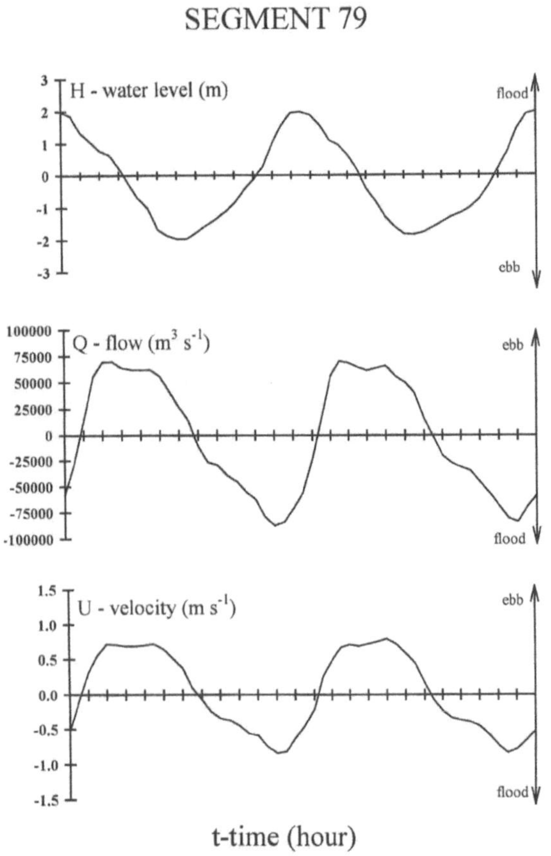

Figure 2. Simulated water levels, flows and velocities in segment 79 for two tidal periods in the Scheldt estuary.

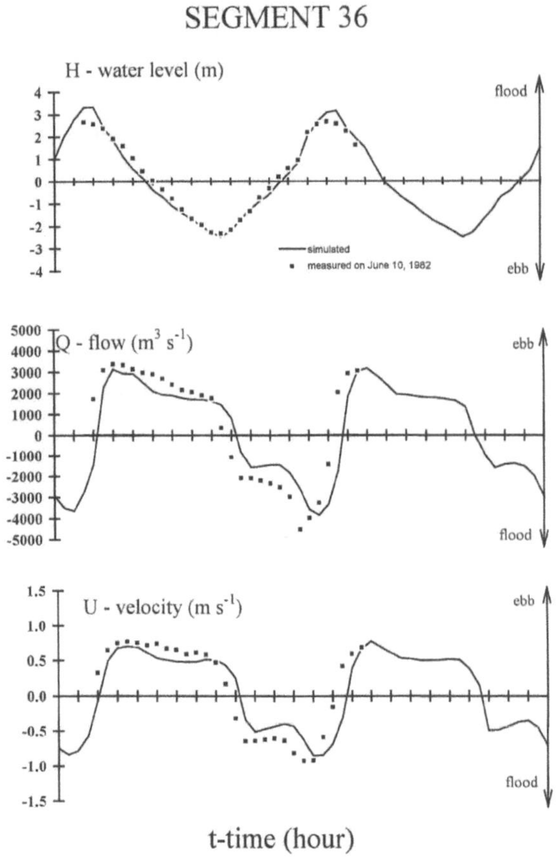

Figure 3. Simulated and observed water levels, flows and velocities in segment 42 for two tidal periods in the Scheldt estuary.

port through the Scheldt estuary and part of the Sea-Scheldt, the computation is done for the region from south of Antwerp to Vlissingen only; this part contains 43 segments numbered from 36 to 79. Values of the necessary hydraulic variables: velocity, U, and hydraulic radius, R, are taken from the tidal simulation. In the present computation, a high value of 600 is considered for the dispersivity, because it is assumed that mixing in the estuary is very intensive due to tidal effects. The resulting effective dispersion coefficients, averaged over a tidal cycle, $\bar{E}_x$, are shown in Figure 6.

The calculated values of the dispersion coefficients range between 150 to 300 $m^2\ s^{-1}$, which is in agreement with typical values for estuaries of 100 to 300 m^2/s, observed by Fisher et al. (1979). Although the coefficients fluctuate from segment to segment as a result of local geometry and friction variations, a slight trend of landward decrease can be noticed. If the same computation is extended to the upper (fluvial) part of the estuary, a decrease in dispersion up to 50 $m^2\ s^{-1}$ results. These average dispersion coefficients obtained for each cell, are used in all subsequent transport simulations with the WASP model.

Results

In order to verify these dispersion coefficients, simulations of salinity S_o [M L^{-1}] are performed. The computed salinities averaged over a tidal cycle, , are compared with measurements (van Eck et al., 1991) in Figure 7 and generally show a good agreement, especially for mean and low freshwater inflows. The simulation with high freshwater inflow are somewhat less accurate, but boundary effects at the sea inlet could be responsible for this. Hence, for further simulations of suspended sediment and heavy metal transport, the obtained dispersion coefficients are accepted as such.

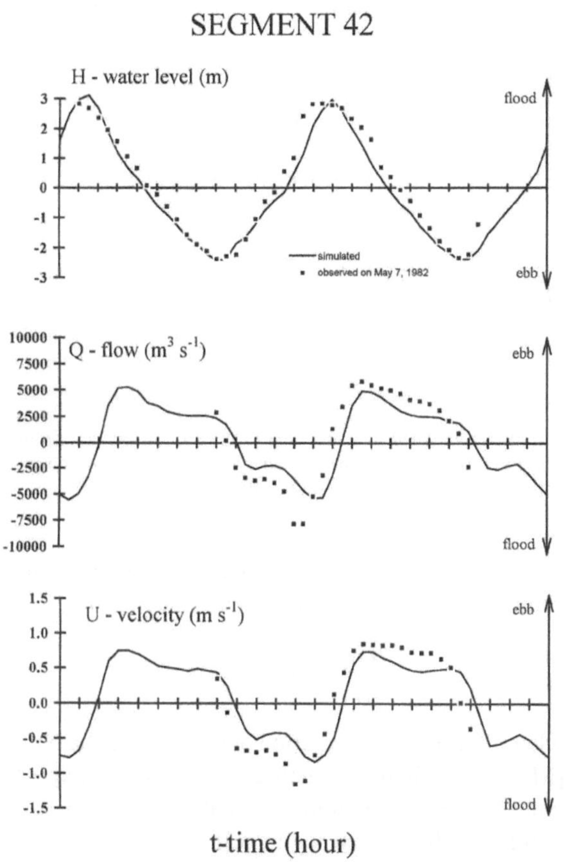

Figure 4. Simulated and observed water levels, flows and velocities in segment 6 for two tidal periods in the Scheldt estuary.

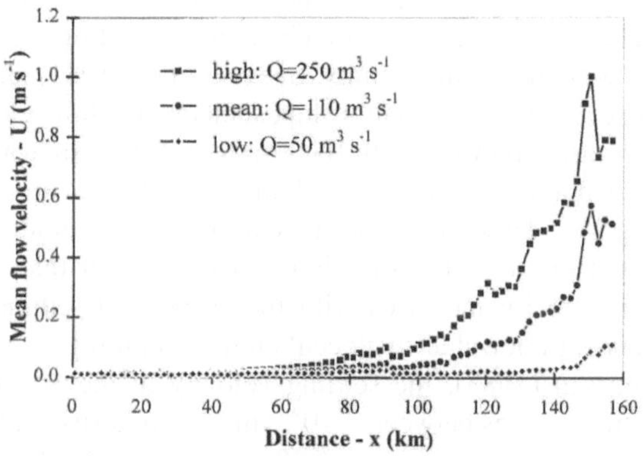

Figure 5. Simulated mean water flow velocity profiles in the Scheldt estuary and part of the lower Sea Scheldt, for low, mean, and high freshwater influx.

Transport of suspended sediment

Suspended sediments are conventionally classified as particles with a diameter smaller than 63 μm. Problems

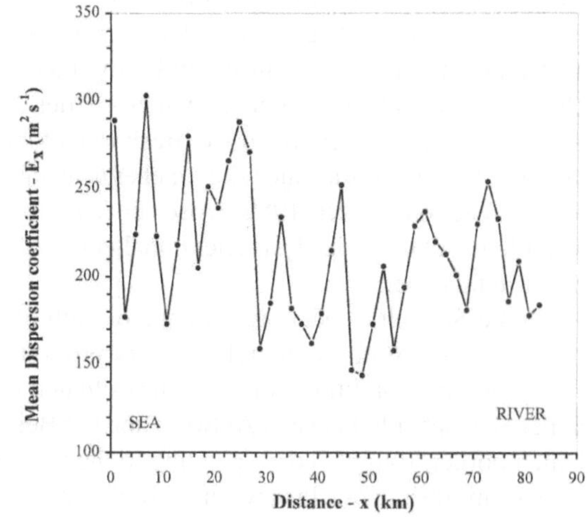

Figure 6. The mean tidal longitudinal dispersion coefficient for the Scheldt estuary calculated by Fisher's equation.

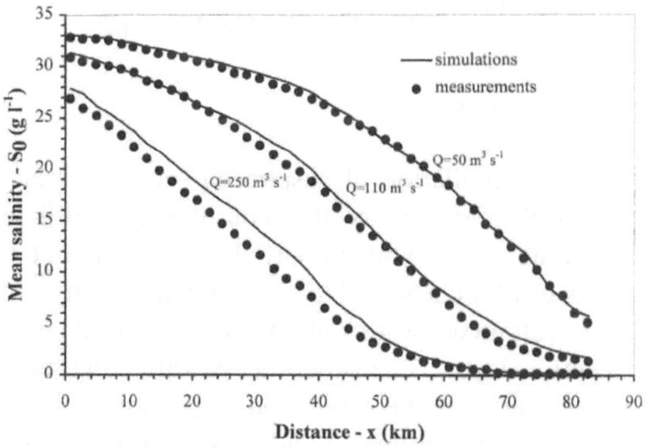

Figure 7. Simulated and observed mean tidal salinity profiles, for low, mean and high fresh water discharges in the Scheldt estuary, and part of the lower Sea Scheldt (data from van Eck et al., 1991).

caused by suspended sediments arise from their ability to adsorb significant quantities of various pollutants. Therefore, prediction of transport, erosion, and deposition of estuarine sediments in itself is very important in order to understand the estuarine water quality characteristics.

An indicator of the magnitude of sediment mobility within estuaries is the so-called turbidity maximum. This zone is characterised by an increased suspended matter concentration which exceeds that of the river, or that of the estuary further seaward. The turbidity maximum is generally located at the head of the salt intrusion. The turbidity maximum responds in a dynamic way to the varying river inflows and the state of the tide by changing its position and density. In estuaries where tidal action is strong and influx of suspended sediment relatively large, the turbidity maximum is a permanent feature. Within the turbidity maximum, physico-chemical and compositional properties of the water change rapidly from those of fresh water to those of sea water, and it is a major site for chemical and biological reactions (Dyer, 1989). Also, flocculation and coagulation of clay-sized particles (smaller than 2 μm) occur in this zone.

In the Scheldt estuary such a zone of high turbidity is clearly present, although not very pronounced. For mean flow conditions, this region is situated in the upper Sea-Scheldt, between Antwerp and the Belgian-Dutch border, roughly corresponding to the transition zone from fresh to brackish water. The suspended matter is mainly composed of colloidal particles that flocculate easily such that pronounced deposition occurs in the part where salinity ranges from 1 to 5 g l^{-1}. According to Wollast (1988), two thirds of the fluvial sediments are being deposited in this zone.

Model theory

For the simulation of suspended sediment transport in the Scheldt estuary, it is considered as sufficiently accurate to deal with all particles smaller than 63 μm as one solid class, and to conceptualise each segment as a well mixed water column bounded from below by a bottom layer. The major processes affecting sediment distribution are advection and dispersion in the water column, and settling to and erosion from the bottom layer. In such case, suspended matter transport can be predicted by the following mass transport equation:

$$\frac{\partial S}{\partial t} = -\frac{\partial}{\partial x}(US) + \frac{\partial}{\partial x}\left(E_x \frac{\partial S}{\partial x}\right) - W_d + W_e, \quad (5)$$

where S is the concentration of suspended sediment [M L^{-3}], W_d is the rate of sediment deposition [M $L^{-3}T^{-1}$], and W_e is the rate of sediment erosion [M $L^{-3}T^{-1}$].

The rate of sediment deposition can be described by:

$$W_d = \frac{wS}{D} \quad (6)$$

where w is the settling velocity [L T^{-1}].

The erosion rate of sediment can be estimated by following equation:

$$W_e = \frac{M}{D} \quad (7)$$

where M is the erosion flux [M L^{-2} T^{-1}].

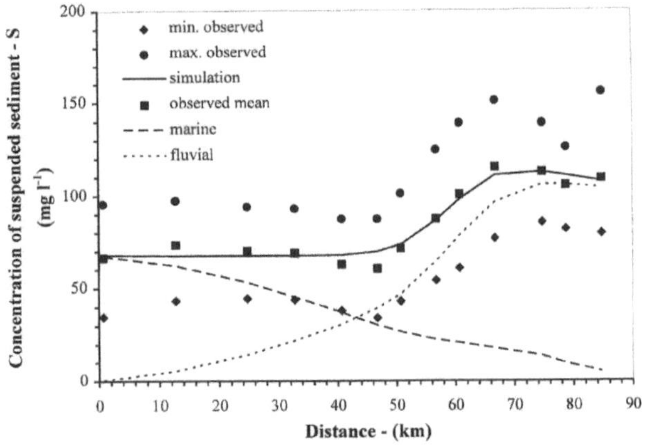

Figure 8. Simulated and observed profiles of suspended sediment concentrations in the Scheldt estuary, and part of the lower Sea Scheldt (data from van Eck et al., 1991).

In the bottom layer, the concentration of sediment changes is given by a mass balance equation:

$$\frac{\partial S_b}{\partial t} = wS\frac{A}{V} - M\frac{A}{V} \quad (8)$$

where S_b is the concentration of sediment at the bottom [M L^{-3}], A is the exchange area between the water column and the bottom segment [L^2], and V is the volume of the bottom segment [L^3].

Parameters

Earlier studies of the sedimentation processes in the Scheldt estuary, consider an increase in settling velocity with increasing salinity. Baeyens et al. (1981), propose an empirical relationship, such that the fall velocity varies between 8.9 10^{-5} m s^{-1} for fresh water to a maximum of 2.7 10^{-4} m s^{-1} for a salinity of 5 g l^{-1}. However, more recent studies are doubtful about the influence of salinity, and consider that the fall velocity increases with the suspended sediment concentration due to flocculation. According to van Leussen (1988), the settling velocity in the Scheldt estuary varies between 4 10^{-5} m s^{-1} for a suspended sediment concentration of 20 mg l^{-1} to 12 10^{-5} m s^{-1} for a concentration of 100 mg l^{-1}. Because, in the WASP model, the settling velocity has to be introduced as a constant parameter for each computational cell, it was decided to use an overall constant value of 9 10^{-5} m s^{-1}.

The value of the erosion rate constant, M, is more difficult to determine. At the moment, there is no adequate instrumentation for direct measurements (Dyer,

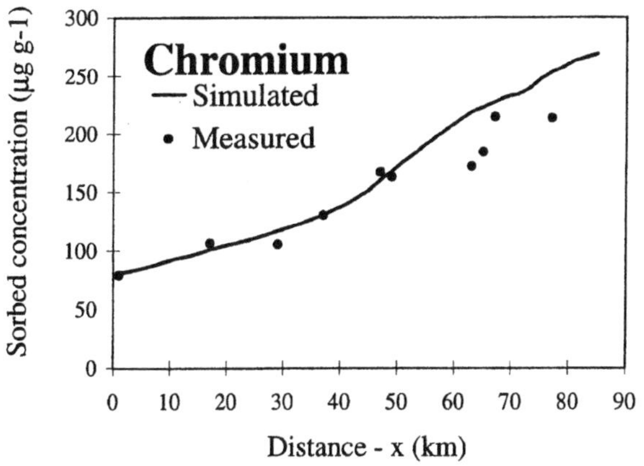

Figure 9. Comparison between simulated and measured concentrations of sorbed Cr in the particulate phase versus distance in the Scheldt estuary, and part of the lower Sea Scheldt (data from Van Alsenoy et al., 1989).

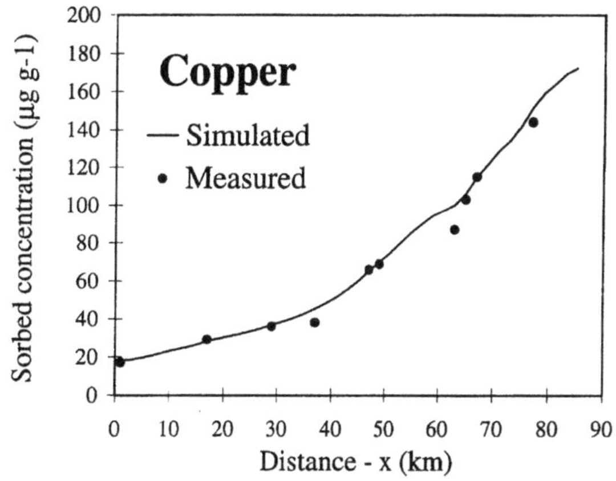

Figure 10. Comparison between simulated and measured concentrations of sorbed Cu in the particulate phase versus distance in the Scheldt estuary, and part of the lower Sea Scheldt (data from Van Alsenoy et al., 1989).

1989), and on the other hand, it is also difficult to reproduce the process under laboratory conditions. Also, data about erosion in the Scheldt estuary are not available in the literature. Therefore, in this work, the erosion rate constant is calibrated, such that observed suspended sediment concentration values are reasonably reproduced, as discussed further on.

Estimations of the influx of suspended sediment from the river are rather inaccurate, with large variations from 320 10^6 kg year^{-1} (Van Zoest & van Eck, 1989) to 750 10^6 kg year^{-1} (Wollast, 1982). For the simulation, an average concentration of fluvial suspended sediment in the fresh water inflows of 106 mg l^{-1} was accepted. When this concentration is multiplied with the average fresh river discharge of 110 m^{3}s^{-1}, the fluvial sediment load becomes 1 10^6 kg d^{-1}, or 340 10^6 kg year^{-1}.

According to Eisma and Kalf (1987), the Belgian-Dutch coastal waters have a high suspended sediment concentration, with a value close to the river mouth of 50 mg l^{-1}. More accurate measurements are presented by van Eck et al. (1991), showing that the average suspended sediment concentration at Vlissingen is around 68 mg l^{-1}. When this concentration is fixed, the sediment load transported by the tides becomes 109 10^6 kg d^{-1}.

The simulation of marine and fluvial sediment transport was carried out only in the Scheldt estuary and lower Sea-Scheldt, under mean tidal conditions with time steps of 10 min. Starting from an initial distribution of suspended matter equal to zero, sediments were gradually introduced from the river and sea boundaries, such that a cyclic pattern of suspended sediment profiles was attained after 60 tides. Results were then integrated over a tidal period.

Results

The model was calibrated by adjusting the erosion rate in different parts of the estuary, but because model parsimony was considered essential, the erosion rate distribution was kept as simple as possible. Good results were obtained by assuming an erosion rate of 0.006 g m^{-2} s^{-1} in the estuary and 0.011 g m^{-2} s^{-1} in the lower Sea-Scheldt, with a transition zone of about 15 km, situated upstream of the Belgian-Dutch border (Reyns, 1995). As there are no direct measurements of erosion rates available, no physical verification is possible of these findings.

In Figure 8 the obtained distribution of suspended sediment in the water column is compared to the mean, minimum and maximum observed values; the data are taken from van Eck et al. (1991) and represent maximum and minimum observed and calculated mean suspended sediment concentrations during the period 1970–1990. The simulated profile fits the observed values and reproduces the zone of the turbidity maximum reasonably.

If a diurnal sedimentary balance is established, it becomes evident that huge quantities are involved: 140 10^6 kg day^{-1} is transported at the mouth of the estuary by the tides, and about 160 10^6 kg is eroded in the estuary every day, while during the same peri-

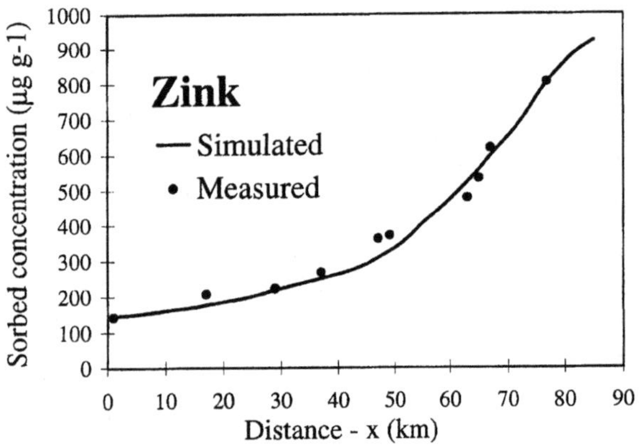

Figure 11. Comparison between simulated and measured concentrations of sorbed Zn in the particulate phase versus distance in the Scheldt estuary, and part of the lower Sea Scheldt (data from Van Alsenoy et al., 1989).

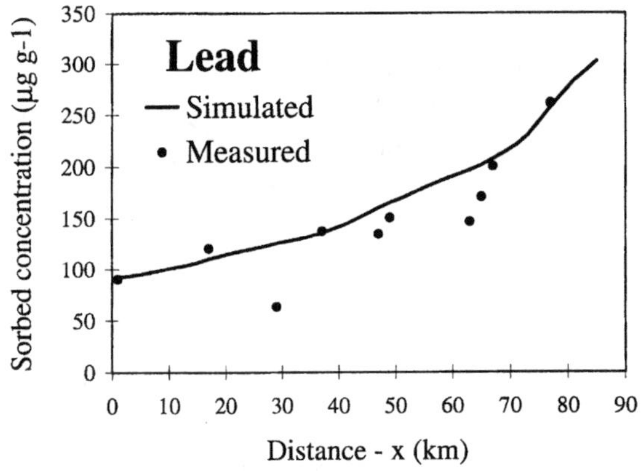

Figure 12. Comparison between simulated and measured concentrations of sorbed Pb in the particulate phase versus distance in the Scheldt estuary, and part of the lower Sea Scheldt (data from Alsenoy et al., 1989).

od only 1 10^6kg of fluvial sediment is carried into the estuary. However, the net result of tidal transport is very small as this movement is cyclic. The same applies for the eroded sediment, as most or all of it is deposited back in the estuary at slack tide. Since, the model WASP can only trace sediment roughly, it is not possible to make a more detailed analysis of the exact behaviour of the marine, fluvial or eroded sediments. It can only be concluded that large amounts of sediment are involved in cyclic erosion-deposition processes, and cyclic exchanges with the sea, while net amounts of transported or deposited sediment are so minor that these become insignificant in the global sediment balance. What can be said with certainty however, is that most of the fluvial sediments are not reaching the sea, but are deposited in the estuary.

Transport of heavy metals

Heavy metals in the Scheldt estuary result from different sources, but especially from domestic and industrial wastes. The Scheldt estuary has the highest Zn and Cr contents compared to other rivers draining to the North sea, while also Cd, Cu and Pb are very high (van Eck et al., 1991). In order to understand the behaviour of heavy metals in the estuarine environment, studies have been conducted, but the information remains fragmentary, because no systematic analysis or measurements have been made. However earlier studies as Duinker et al. (1982), Baeyens et al. (1982), Valenta et al. (1986), Van Alsenoy et al. (1990), and van Eck et al. (1991), show that the behaviour of heavy metals is strongly influenced by adsorption on suspended matter. Heavy metals can become immobile when sediments are settling to the bottom, or can be mobilised again during erosion. Hence, the fate of heavy metals is very much determined by the sediment behaviour and transport.

Possibly, other processes are also involved. Some observations have indicated that very low dissolved heavy metal concentrations occur in summer periods in the lower Sea-Scheldt. This has been explained by anoxic conditions in the lower Sea-Scheldt, especially in summer, enabling the formation of heavy metal sulphides, which precipitate (van Eck et al., 1991; Monteny et al. 1993). The existence of precipitated sulphides of Cd, Cu and Zn has been demonstrated by Zwolsman & van Eck (1990). Further downstream in the seaward direction, the sulphides are mobilised again, when oxygen concentrations increase. But, according to van Eck et al. (1991) the influence on particulate heavy metals remains small.

A water quality model, including heavy metal transport in dissolved and particulate form, was developed by Van Gils et al. (1993), taking into account the effects of precipitating sulphides in the upper Scheldt estuary. However, the spatial resolution of this so-called SAWES (System Analysis WEstern Scheldt) model is limited, also no tidal actions are taken into account, and the sediment transport is not explicitly included in the model; instead a fixed sediment balance is used as a forcing function. In the present approach with the WASP model, the mobility of heavy metals in particulate form is considered as the predominant transport mechanism. Precipitation and dissolution of heavy metal sulphides is not considered, because these are non equilibrium processes (Van Gils et al., 1993), that can not be simulated with WASP model. Hence, the present model ignores anoxic effects, but as the water quality of the estuary is gradually improving, and anoxic conditions are becoming rare, this forms no objection.

Model theory

The total amount of a heavy metal in the water column is given by the amount in dissolved form and the amount adsorbed on the sediment:

$$C = C_d + C_p = C_d + SC_s, \tag{9}$$

where C is the total concentration of the heavy metal in the water [M L^{-3}], C_d is the concentration of dissolved heavy metal [M L^{-3}], C_p is the concentration of particulate heavy metal adsorbed on suspended matter [M L^{-3}], and C_s is the concentration of sorbed heavy metal per mass of suspended sediment [M M^{-1}].

The total concentration of a heavy metal in the water column is obtained from the following mass balance:

$$\frac{\partial C}{\partial t} = -\frac{\partial}{\partial x}(UC) + \frac{\partial}{\partial x}\left(E_x\frac{\partial C}{\partial x}\right) - C_s W_d + \frac{C_b}{S_b} W_e \tag{10}$$

where C_b is the concentration of the sorbed heavy metal in the river bed sediments [M L^{-3}].

The sorption of a heavy metal on suspended sediment is modelled by a linear Freundlich isotherm:

$$C_s = KC_d, \tag{11}$$

where K is a distribution or partitioning coefficient [L^{3}M^{-1}]. In the Freundlich model, the sorption is described as an instantaneous and reversible reaction, where adsorption and desorption follow the same linear isotherm. It is also assumed that when metals are mixed, they sorb independently following their own respective isotherms.

The sediment sorbed concentration of a heavy metal in the bottom segment, C_b, changes according to its mass balance equation:

$$\frac{\partial C_b}{\partial t} = wC_p\frac{A}{V} - M\frac{C_b}{S_b}\frac{a}{V}. \tag{12}$$

Finally the adsorbed concentration of heavy metal on the suspended matter, C_s, can easily be derived from Equations 9 and 11:

$$C_s = \frac{KC}{1 + KS}. \tag{13}$$

Parameters

The transport of heavy metals is simulated for the same region – the Scheldt estuary and part of the lower Sea-Scheldt – similar as for the salt and suspended sediment transport models. Also, the results of these latter models are used for supporting the heavy metal transport model.

Data concerning the presence and distribution of heavy metals in the Scheldt estuary are limited. In this work, data given by Van Alsenoy et al. (1989) are used; these data are also discussed by Van Alsenoy et al. (1990). The observations result from a sampling campaign undertaken in July 1988. Amounts of heavy metals, e. g. Cr, Cu, Zn, Pb and Ni, adsorbed on suspended matter were measured at 20 stations in the North Sea and the Scheldt estuary. For this study, only the measurements in the estuary are considered, which involve 10 locations between Vlissingen and Antwerp. Unfortunately, no measurements of total or dissolved heavy metal concentrations were performed.

As the important processes that determine the transport of sorbed heavy metals in the present model, are the transport and mixing of fluvial and marine sediments, appropriate boundary conditions have to be determined for these parameters, especially their heavy metal contents. At the mouth of the estuary, the sorbed concentrations of heavy metals were fixed and put equal to the measured values at Vlissingen by Van Alsenoy et al. (1989). For the boundary at the upstream section south of Antwerp, appropriate values were obtained by extrapolating the measurements of the most upstream sampling locations of Van Alsenoy et al. (1989). Other inputs of heavy metals, in particular emission along the estuary, were not taken into account.

The only remaining parameters that need to be determined are the adsorption distribution coefficients. Heavy metals in the aquatic environment can form soluble complexes with organic and inorganic ligands, or sorb onto organic and inorganic suspended matter. Partitioning coefficients depend upon the characteristics of the sorbents, including mineralogy, chemical structure, composition and electrical properties, presence of coatings, etc. Hence, site specific values should be used when possible. However, data about distribution

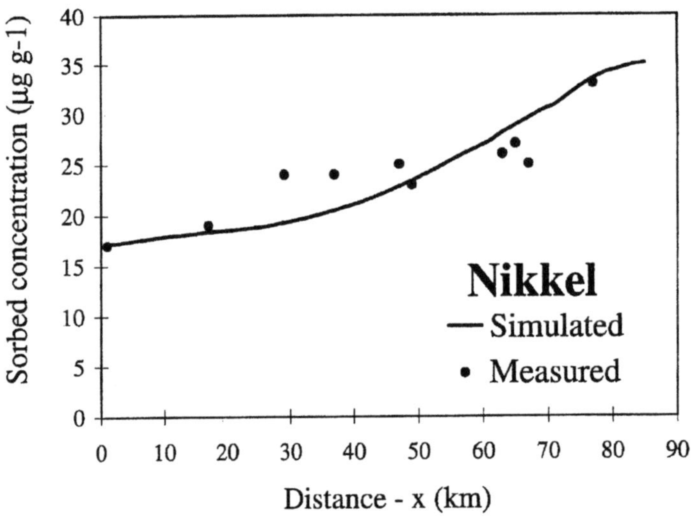

Figure 13. Comparison between simulated and measured concentrations of sorbed Ni in the particulate phase versus distance in the Scheldt estuary, and part of the lower Sea Scheldt (data from Van Alsenoy et al., 1989).

coefficients in the Scheldt estuary are scarce. Monteny et al. (1993) give an average value of 3.1 10^4 l kg^{-1} for Cu and Zn in the downstream Scheldt estuary, and state that the actual *K*-values may vary depending upon the composition of both the solid and the liquid phase; especially the salinity can have marked effects. Similar conclusions were obtained by Van Alsenoy et al. (1989), but due to high experimental variability no precise *K*-values could be given. Another result worth mentioning is that the time required to obtain equilibrium conditions between dissolved and sorbed heavy metal concentrations varied between a couple of hours for river water samples to a few days for sea water samples, such that an instantaneous Langmuir isotherm seems warranted.

Because no precise values for the distribution coefficients could be obtained from literature, it was decided to use general distribution coefficients cited by Ambrose et al. (1991); these lumped *K*-values are given for different heavy metals in function of the suspended sediment concentration. The values used in the present study are represented in Table 1. From this table, appropriate values were interpolated for each computational cell, depending upon the average suspended sediment concentration as obtained with the sediment transport model.

Table 1. Values of distribution coefficients for heavy metal adsorption on suspended sediments, used in the model (data taken from Ambrose et al., 1991)

Suspended sediment concentration S (mg l^{-1})	Distribution coefficient - *K* (l kg^{-1})				
	Cr	Cu	Zn	Pb	Ni
10	4.10^5	2.10^5	2.10^5	2.10^5	1.10^5
100	5.10^4	3.10^5	5.10^4	1.10^5	4.10^4

Results

Starting from a zero initial distribution of heavy metal and with water flow conditions and suspended sediment concentrations as discussed before, the simulation showed that equilibrium conditions for the heavy metals are readily established after a period of about 60 tides. The resulting profiles of sorbed heavy metal concentrations were averaged over a tidal period. The calculated concentrations are plotted versus distance from the sea, and compared with the measurements in Figures 9 to 13, for respectively Cr, Cu, Zn, Pb and Ni.

In general, there appears to be a fair agreement between the simulations and measurements. All sorbed heavy metal concentrations show a pronounced variation in function of the distance to the mouth of the estuary. There is a clear increase of the concentrations from the mouth of the Scheldt estuary to the high turbitidy zone in the Sea-Scheldt around Antwerp. This demon-

Table 2. Annual immission and emission loads of sorbed heavy metals in the Scheldt estuary, and part of the lower Sea Scheldt, estimated with the model.

Heavy metal	Load (kg yr^{-1}) Input	Load (kg yr^{-1}) Output	Output/input (%)
Cr	99,300	18,600	18,7
Cu	62,500	4,010	6,4
Zn	349,000	33,700	9,6
Pb	120,000	21,200	17,7
Ni	12,900	4,010	31,1

strates that the distribution of the sorbed heavy metals primarily depends upon the transport and mixing of the suspended sediments, and the sorption characteristics of the heavy metals.

Similar as in the case of suspended sediment transport, an accurate analysis and quantification of the bottom related process, i.e. net rates of deposition and erosion for the heavy metals, is not possible. Nevertheless, the results indicate that the behaviour of heavy metals in the Scheldt estuary is reproduced in a realistic way, although more data are needed in order to verify the predictability of the simulations. Associating the sorbed quantities of the heavy metals with an average fresh water inflow of 110 $m^3 s^{-1}$ and an average suspended solids concentration of 106 mg l^{-1}, the total annual fluvial influx of sorbed heavy metals, entering the Sea Scheldt south of Antwerp, can be estimated as shown in Table 2. The range of the loads is estimated between 12 900 kg yr^{-1} for Ni, and 349 000 kg yr^{-1} for Zn.

The load of sorbed heavy metals reaching the sea at the mouth of the estuary, can be estimated in a similar way, taking into account a suspended sediment concentration of 68 mg l^{-1} at Vlissingen. These results are also given in Table 2. With these values, it becomes possible to calculate the ratio between emission and immision loads of sorbed heavy metals in the estuary. It appears that only part of the heavy metals in sorbed form reach the sea, as indicated in the last row of Table 2; the values range between 6.4% for Cu and 31.1% for Ni.

With the present model, also dissolved and particulate heavy metal concentrations in the water column can be estimated, but these results have to be interpreted with care, because there are no measurements available for verification, simulations are based on estimated distribution coefficients, and effects of sulphides precipitation have not been taken into account. As an example, estimated total, dissolved and particulate concentrations for Ni are shown in Figure 14. Because the transport of Ni closely follows pathways of suspended sediment, all profiles show significant accumulation in the zone of maximum turbidity. Further seaward quantities of Ni decrease as result of the mixing of fluvial sediment with marine sediment. Only part of the riverborne Ni is transported to the sea, while the remainder is accumulated in the estuary, due to settling of suspended sediment.

It can be concluded that the simulations clearly show that the distribution of sorbed heavy metals can be predicted accurately by tidal and fluvial hydrodynamics, dispersive mixing, transport of suspended material, and adsorption processes, with the WASP model. Hence, when more data become available for better verification, the present model can be used as a tool for water quality management in the Scheldt estuary.

Conclusions

The main processes that govern the transport and behaviour of heavy metals in the Scheldt estuary were studied with the WASP model. Generally observations agree with simulation results for hydrodynamic, salinity and suspended sediment transport. A thorough understanding of the estuarine physics in terms of hydrodynamic, dispersive and sediment transport is a necessity when modelling transport of heavy metals. It appears that sorption of the heavy metals on suspended matter is the predominant process that regulates the heavy metal distribution between sediment and water, and the concentration distributions in the estuary.

The results from the simulations performed using WASP suggest that the model is capable of simulating profiles of sorbed heavy metals satisfactorily. The results indicate a strong accumulation of the heavy metals in the zone of high turbidity at the head of the salt water intrusion front, and less transport to the sea. However, more measurements are needed in order to verify the accuracy and predictability of the present modelling results.

References

Ambrose, R. B., T. A. Wool & J. L. Martin, 1993. The water quality analyses simulation program, Wasp5 – Part A: Model documentation. US-EPA, Athens, Georgia, USA, 209 pp.

Ambrose, R. B., T. A. Wool, J. L. Martin, J. P. Connolly & R. W. Schanz, 1991. WASP5.x, a hydrodynamic and water quali-

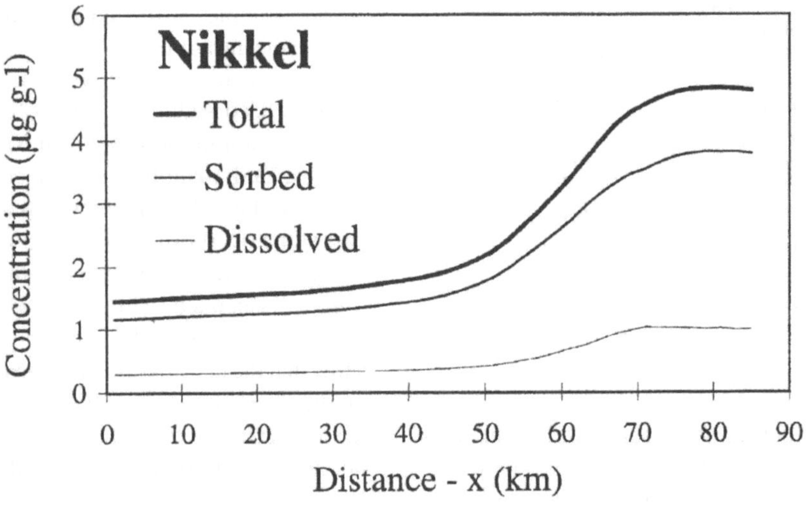

Figure 14. Simulated profiles of total, dissolved and particulate Ni concentrations in the water column in the Scheldt estuary, and part of the lower Sea Scheldt.

ty model – Model theory, user's manual and programmer's guide. US-EPA, Athens, Georgia, USA, 163 pp.

Baeyens, W., Y. Adam, J. P. Mommaerts & G. Pichot, 1981. Numerical simulation of salinity, turbidity and sediment accumulation in the Scheldt estuary. In J. Nihoul (ed.), Ecohydrodynamics, Elsevier, Amsterdam: 319–332.

Baeyens, W., S. Wartel, F. Dehairs, G. Decadt, M. Bogaert, G. Gillain & H. Dedeurwaerder, 1982. The river Scheldt as a transport route for heavy metals in the sea. In I. Elskens & A. Distèche (eds), Distribution, Transport and Fate of Bi, Cu, Cd, Hg, Pb, Sn and Zn in the Belgian Coastal Marine Environment, Actions de Recherche Concertées, Progr. Polit. Scient. Brussels, Belgium, Vol 2, 4: 87–108.

Bowie, G. L., W. B. Mills, D. B. Porcella, C. L. Campbell, J. R. Pagenkopf, G. L. Rupp, K. M. Johnson, P. W. H. Chan, S. A. Gherini & C. E. Chamberlin, 1985. Rates, constants and kinetics formulations in surface water quality modelling. US-EPA, Athens, Georgia, USA, EPA/600/3-85/040, 455 pp.

Chatwin, P. C. & C. M. Allen, 1985. Mathematical models of dispersion in rivers and estuaries. Annu. Rev. Fluid Mech. 17: 119–149.

Claessens, J., 1988. Het hydraulisch regime van de Schelde. Water 43: 163–169.

Duinker, J. C., R. F. Nolting & D. Michel, 1982. Effects of salinity, pH and redox conditions on the behaviour of Cd, Zn, Ni and Mn in the Scheldt Estuary. Thalassia Jugoslavica 18: 191–202.

Dyer, K. R., 1989. The sediment processes in estuaries: future research requirements. J. Geophys. Res., 94-C10: 14,327–14,339.

Eisma, D. & J. Kalf, 1987. Dispersal, concentration and deposition of suspended matter in the North Sea. J. Geol. Soc. London 144: 161–178.

Fisher, H. B., E. J. List, R. C. Y. Koh, J. Imberger & N. H. Brooks, 1979. In: Mixing in inland and coastal waters. Academic Press, New York, 483 pp.

Laforce, E., F. Wens & P. Roovers, 1977. Mathematisch model van het tijgebied der Schelde. Waterbouwkundig Laboratorium Borgerhout, Belgium, 42 pp.

Monteny, F., M. Elskens & W. Baeyens, 1993. The behaviour of copper and zinc in the Scheldt estuary. Neth. J. aquat. Ecol. 27: 279–286.

Reyns, D., 1995. Simulatie van het transport van zware metalen in het Schelde-estuarium met het waterkwaliteitsmodel WASP5. Thesis Fac. Applied Sciences, Vrije Universiteit Brussel, 120 pp.

Valenta, P., E. K. Duursma, A. G. A. Merks, H. Rützel & H. W. Nürnberg, 1986. Distribution of Cd, Pb and Cu between the dissolved and particulate phase in the Eastern and Western Scheldt estuary. Sci. Total Envir. 3: 41–76.

Van Alsenoy, V., F. Araujo, P. Bernard & R. Van Grieken, 1989. Toestand betreffende de verontreiniging van mariene en estuarine sedimenten door metalen van de zwarte en grijze lijst. Eindrapport Contract BH/87/25, Beheerseenheid van het Mathematisch Model van de Noordzee en het Schelde-Estuarium, Ministerie van Volksgezondheid, Belgium, 155 pp.

Van Alsenoy, V., P. Bernard & P. Van Grieken, 1990. Zware metalen in de Noordzee- en Schelde-sedimenten. Wat. 51: 113–121.

van Eck, G. T. M., N. De Pauw, M. Van den Langenbergh & G. Verreet, 1991. Emissies, gehalten, gedrag en effecten van (micro)verontreinigingen in het stroomgebied van de Schelde en Schelde-estuarium. Wat. 60: 164–181.

Van Gils, J. A. G., M. R. L. Ouboter & N. M. De Rooij, 1993. Modelling of water and sediment quality in the Scheldt estuary. Neth. J. aquat. Ecol. 27: 257–265.

van Leussen, W., 1988. Aggregation of particles, settling velocity of mud flocs – A review. In J. Donkers & W. van Leussen (eds), Physical Processes in Estuaries, Springer Verlag, Berlin: 347–403.

Van Meerbeeck, S., 1994. Simulatie van het sedimenttransport in de Schelde met het waterkwaliteitsmodel WASP5. Thesis Fac. Applied Sciences, Vrije Universiteit Brussel, pp 108.

Van Zoest, R. & G. T. M. van Eck, 1989. Gehalten, gedrag en modellering van organische microverontreinigingen in water en bodem van het Schelde estuary – Content, behaviour and modelling of organic micropollutants in water and bottom of the Scheldt estuary. Rijkswateren, Gravenhage, and Rijksuniversiteit Utrecht, internal report in Dutch and English, 211 pp.

Vuksanovic, V., 1993. Simulation of the transport of polychlorinated biphenyls (PCB) in the Scheldt Estuary with the water quality model WASP4. Thesis Interuniversity Postgraduate Programme in Hydrology, Vrije Universiteit Brussel, 111 pp.

Vuksanovic, V., F. De Smedt & S. Van Meerbeeck, 1995. Transport of polychlorinated biphenils (PCB) in the Scheldt Estuary simulated with the water quality model WASP. J. Hydrol. 174: 1–18.

Wollast, R., 1982. Behaviour of Organic Carbon, Nitrogen, and Phosphorus in the Scheldt Estuary. Thalassia Jugoslavica 18: 11–34.

Wollast, R., 1988. The Scheldt estuary. In W. Salomons, B. L. Bayne & E. K. Duursma (eds), Pollution of the North Sea: an Assessment. Springer-Verlag, Berlin: 183–193.

Zwolsman, J. J. G. & G. T. M. van Eck, 1990. The behaviour of dissolved Cd, Cu and Zn in the Scheldt estuary. In W. Michaelis (ed.), Estuarine Water Quality Management, Monitoring, Modelling and Research. Springer-Verlag, Berlin: 413–420.

Hydrobiologia **366:** 157–167, 1998.
W. F. J. Baeyens (ed.), Trace Metals in the Westerschelde Estuary.

Evolution of trace metal concentrations in the Scheldt estuary (1978–1995). A comparison with estuarine and ocean levels

Willy Baeyens
Department of Analytical Chemistry, Free University of Brussels, Pleinlaan 2, 1050 Brussels, Belgium

Key words: Scheldt estuary, trace metals, dissolved, particulate, history, comparison

Abstract

Dissolved and particulate metal concentrations assessed in the Scheldt estuary in the period 1978–1995, yield a clear indication that the levels of all metals decreased significantly. A comparison with other estuaries classifies the Scheldt as being moderately polluted for all metals in the dissolved phase, but as being fairly high polluted in the particulate phase, especially for Cd. However, even the dissolved metal concentration fractions in the Scheldt are still between 10 and 50 times enriched compared to total metal concentrations in unpolluted seas and oceans.

Introduction

In 1970, the belgian government launched the National R & D Programme 'North Sea' which lasted five years and included the assessment of the pollution level in the Scheldt estuary (Figure 1).

The results obtained during this period were published by the Department of Science Policy in a series of 11 volumes, volume 10 being the one related to the Scheldt estuary (Nihoul & Wollast, 1976).

Most of the effort was focused on the understanding of the hydrodynamics, the sediment transport and the degradation of the high load of organic waste entering the estuary via the various tributaries of the Scheldt (see Baeyens et al., 1998d). The assessment of the trace metal concentrations suffered from a number of constraints:

– 1. The lack of sensitivity of the analytical methods and measuring devices. Atomic Absorption Spectrometry was the most current technique used in the early seventies, but it was unsuitable for the determination of natural concentrations of Cd, Cu, Pb and Zn in brackish or sea water.

Graphite Furnace Atomic Absorption Spectrometry (GFAAS) provided enough sensitivity, but background correction, a prerequisite for the analysis of a saline sample, was not as performant as today (e.g. the Zeeman background correction technique) and the use of matrix modifiers allowing thermal separation between the analyte and the salt (e.g. Hoenig et al., 1991) was not yet sufficiently developed.

Therefore, during the period 1970–1975, the analysis of trace metals in the estuary was mainly limited to sediments and suspended matter. An exception was the determination of iron and manganese in the dissolved phase. These two metals are involved in the redox processes governing the degradation of the high loads of organic matter in the Scheldt. The consumption of the oxydants in the upstream estuary and the reaeration in the downstream estuary lead to strong redox condition changes in the Scheldt (from anoxic or suboxic upstream to oxic downstream). Such conditions allow the tracing of redox involved and hence redox sensitive elements such as iron and manganese. In the anoxic/suboxic area, dissolved iron and manganese levels exceed 250 $\mu g\ l^{-1}$ (Duinker et al., 1979; Baeyens et al., 1998b) and can thus be determined with specific, appropriate colorimetric methods.

In addition, methods were developed and improved in order to allow the determination of trace metals in the Scheldt estuary. Amongst these methods: Anodic Stripping Voltammetry (ASV) (e.g. Gillain et al., 1979), a mercury double beam spectrophotometer (Baeyens et al., 1979; Baeyens et al., 1982), the double gold amalgamation system for Hg (Fitzgerald & Gill,

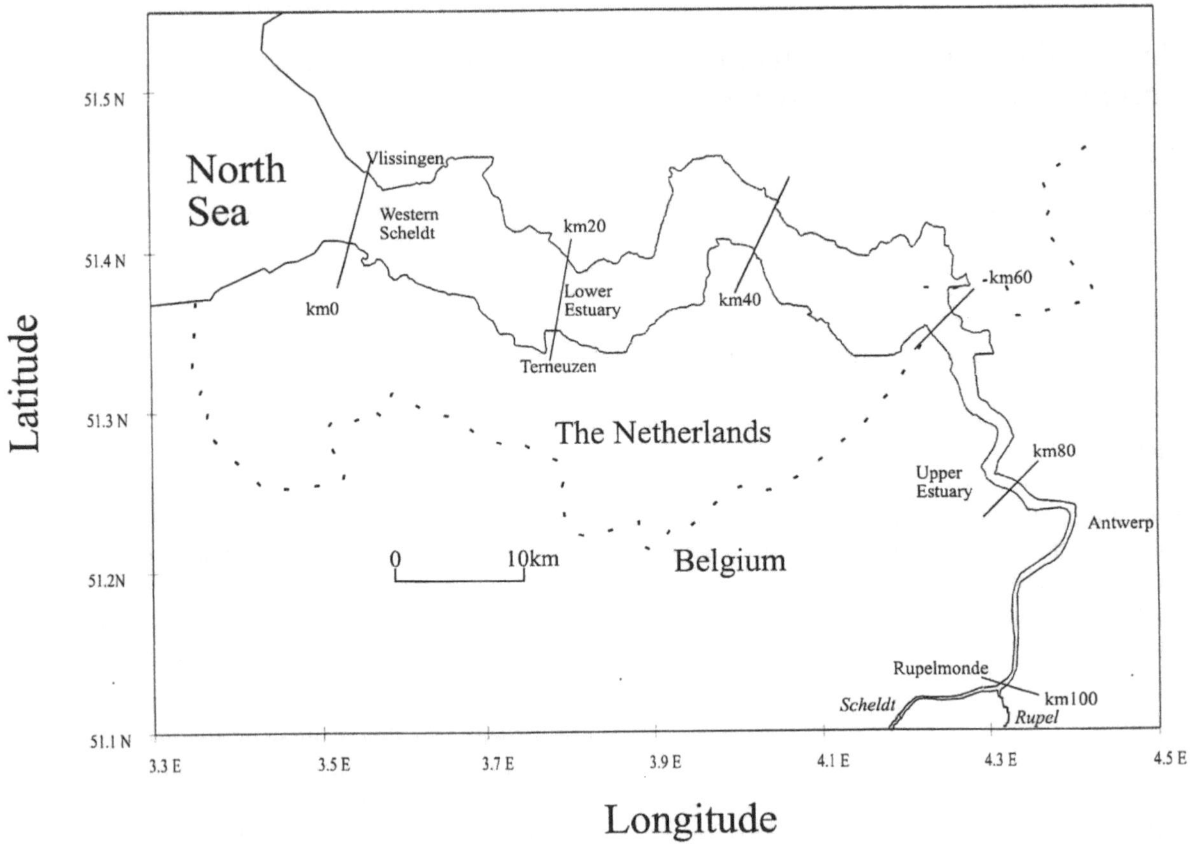

Figure 1. Map of the Scheldt estuary.

1979; Baeyens & Leermakers, 1989), and the testing of matrix modifiers for GFAAS, as mentioned above.

– 2. Many reagents used for the liberation, the transformation, the isolation, . . . of the trace metal prior to the measuring step showed very high blank values, rendering the selected procedure inadequate for the analysis of natural saline samples. Examples are oxydants such as $KMnO_4$, H_2O_2, . . . used for the transformation of the total Hg load into Hg(II) or for the elimination of interfering organic ligands during the electrochemical analyses of trace metals. Also reagents such as acids, bases or other stabilising additives could give rise to high blanks and/or artefacts.

As a consequence of the lack on clean bench space during the early seventies, the storage of samples on board of the research vessel was a necessary, unavoidable process for belgian oceanographers. However, storage is often not carried out in an appropriate way and a limited time, resulting in artefacts such as contamination losses and species transformation. This problem was fully recognised and efforts were made to minimise, even to completely avoid changes in the sample composition (Gillain et al., 1982).

– 3. Contamination during sampling is probably the most serious cause of error. Sampling of estuarine and coastal water in the seventies was carried out from the RV Mechelen, a ship having a wooden hull and a shallow draught, minimising pollution of the surrounding watermasses by the ship. However, during the first years of the study, sampling was performed with a standard Niskin bottle fixed to a stainless steel hydrowire. Even thorough cleaning of the bottles was no guarantee for avoiding contamination. In the late seventies, Gillain & Brihaye (1985) constructed a suitable sampling system for trace metals (see also Baeyens et al., 1998a), which operated completely independent from the other sampling devices on board of the ship. No metallic parts came, at any time, into contact with the water mass. In addition, the whole device was entirely screened from the air, even after filtration was finished

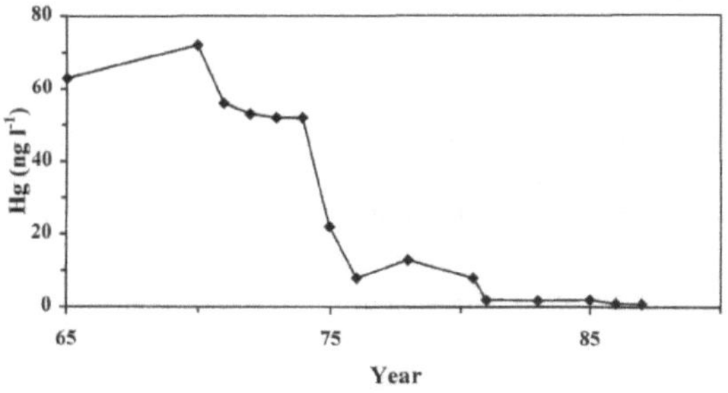

Figure 2. Total mercury concentrations in unpolluted water (adapted from N. Bloom, seminar course).

so that, for example, anoxic samples were not exposed to oxydation. Thorough cleaning of the tubings, the filtration device, the filter, the storage vessel as well as clean bench space in a separate laboratory, provided the necessary conditions for obtaining reliable, non-contaminated samples.

To illustrate how the above mentioned three constraints lead to unreliable results, it is worthwile to make a detailed analysis of the drastic decrease of the dissolved Hg concentrations observed in natural waters since the sixties (Figure 2).

When the natural concentration of a metal decreases by two orders of magnitude (from 60 ng l^{-1} to 0.5 ng l^{-1}), there is more at stake than the sole fact of imposing very severe emission limits or even banning the industrial use of that metal. Indeed for mercury, analytical constraints caused the high levels. The most significant decreases in Figure 2 are due to:

(a) the introduction of the double gold amalgamation technique which lowered the detection limit significantly, so that natural concentrations yield response signals which are several times that of the detection limit;

(b) the progress made in material sciences. Devices constructed in inert, very clean materials such as PTFE or FEP became available and replaced those made in glass, PVC, Polyethylene, etc. . . .

Scheldt surveys

The first reliable results of Cd and Zn in the Scheldt obtained with GFAAS relate to 1978 and were reported by Duinker et al. (1982). They rely on one survey (October-November) covering the whole salinity range in the estuary. In 1980, Cd, Cu and Zn were measured with ASV by Valenta et al. (1986), but the measurements were limited to the salinity range of 10–20 psu. From 1981 to 1983 several surveys were carried out in the North Sea (Baeyens et al., 1987a, 1987b) and in the Scheldt estuary (Baeyens et al., 1998a) with the specific sampling device conceived by Gillain & Brihaye (1985). Analyses of Cd, Cu, Pb and Zn were carried out by ASV, allowing a speciation between labile and non-labile species, and of Hg with the double gold amalgamation technique. In 1986, dissolved Cu and Zn were assessed with ASV in the salinity range of 10–20 psu (Van den Berg et al., 1986). In 1987, Zwolsman & Van Eck (1993) studied the seasonal variations of Cd, Cu and Zn, using extraction-preconcentration prior to GFAAS detection. In 1987–1988, particulate concentrations of Cu, Pb and Zn were reported by V. Alsenoy et al. (1993). In the nineties, Paucot & Wollast (1997) carried out several surveys covering all seasons. They determined concentrations of numerous trace metals including Cd, Cu and Zn with GFAAS without pre-concentration, but using matrix modifiers or special gas mixtures.

The most recent data were produced by Baeyens et al. (1998c), after the installation of a clean laboratory container on board of the R.V. Belgica, and using an extraction-preconcentration technique prior to GFAAS and ICP-MS.

All authors reported separate results for the dissolved phase (in $\mu g\ l^{-1}$ or in nM) and for the particulate phase ($\mu g\ g^{-1}$). If necessary, total metal concentrations can be calculated from these two variables and the suspended matter content. However, it is not very meaningful to compare total metal concentrations in estuaries because these concentrations are too much influenced by the turbidity of the system, which itself is often a function of the sampling conditions (the depth and the selected moment of sampling).

Seasonal variability

The comparison of metal concentration levels between different years in the same estuary or between different estuaries should be made very carefully, taking into account the seasonal variability (Figures 3 and 4) which can be very important in some estuaries like, for example, the Scheldt. Table 1 shows the large concentration differences in the salinity intervals 0–2 psu, 10–20 psu and 30 psu between the early March and late May 1995 surveys. These differences (Figure 3) are very pronounced for Cd, significant for Cu and Zn and almost negligible for Pb (except in the salinity range < 5 psu).

Table 1. Seasonal variability of the metal concentrations in the Scheldt estuary. Comparison of the March and May 1995 cruises

Salinity range (psu)	Cd (nM)	Cu (nM)	Pb (nM)	Zn (nM)
	March-May	March-May	March-May	March-May
0–2	0.28–0.044	14.6–7.5	1.19–0.59	132–49
10–20	0.83–0.261	24.5–18.5	0.40–0.35	88.4–34.6
27.5	0.50–0.080	12–9.2	0.19–0.27	41.6–7.9

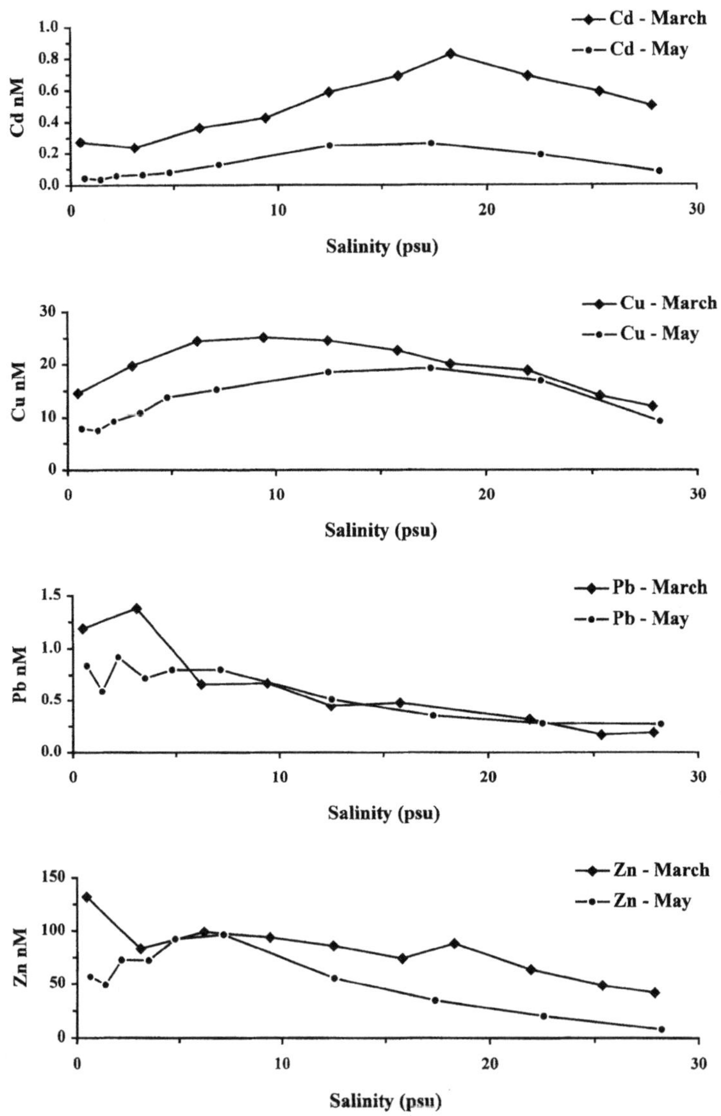

Figure 3. Dissolved trace metal profiles in March and May 1995.

The most important factors involved (see also Baeyens et al., 1998c) are:

(1) the much lower oxygen levels in the upper estuary in May and

(2) the increased plankton activity in May.

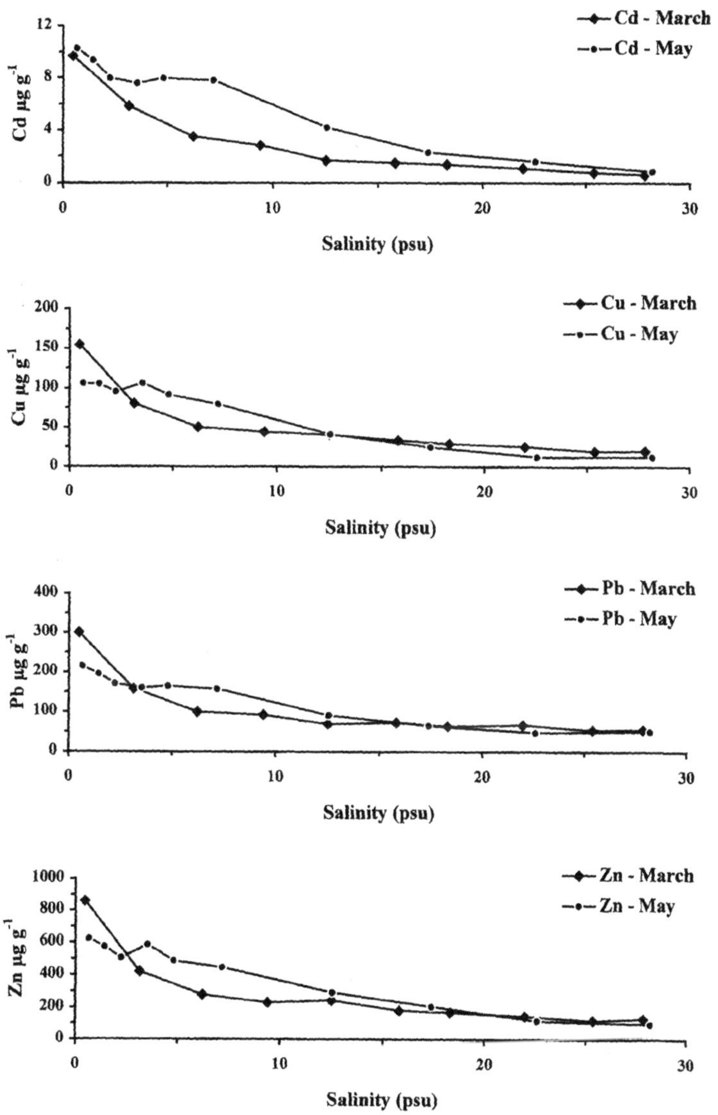

Figure 4. Particulate metal profiles in March and May 1995.

The best approach to smooth out the seasonal variability is to use seasonal averages or to compare only results obtained in the same seasonal period.

In the next section, dealing with the temporal evolution, the K_D concept (the distribution between the dissolved and particulate phases) also helped us to take the effect of seasonal variability better into account.

Evolution between 1978–1995

Only in the periods 1981–1983, 1987, 1991–1994 and 1995, seasonal averages in the different salinity zones of the estuary could be calculated. The additional investigations in the years 1978, 1980 and 1986 are, however, valuable to complete the data set and to confirm certain trends. It appears that the K_D's (Table 2) are comprised in a fairly acceptable range, especially if we take into account the different sampling and analysing methods that have been used, as well as the natural variability. The K_D of Zn, especially at the mouth of the estuary (30 psu), shows the largest fluctuations.

Examining the individual metals, different considerations can be made.

Cadmium: The Cd industry in Belgium is amongst the most important on a worldscale and it is thus not sur-

Table 2. Evolution of the dissolved and particulate metal concentrations in the Scheldt estuary

	Cd			Cu			Pb			Zn			
Salinity:	0–2	10–20	30	0–2	10–20	30	0–2	10–20	30	0–2	10–20	30	
Year	(psu)	(psu)	(psu)	(psu)	(psu)	(psu)	(psu)	(psu)	(psu)	(psu)	(psu)	(psu)	Reference
	Dissolved metals (μ g l^{-1})												
1978	0.056	0.28	0.12	–	–	–	–	–	–	3.3	16.3	6.5	(1)
1980	–	0.30	–	–	1.02	–	–	0.11	–	–	–	–	(2)
1981–1983	0.086	0.19	0.070	1.02	1.9	1.1	0.23	0.24	0.18	11.7	8.8	3.3	(3)
1986	–	–	–	–	3.7	–	–	–	–	–	7.9	–	(4)
1987	0.020	0.077	0.050	0.95	1.34	0.95	–	–	–	9.7	7.77	2.15	(5)
1991–1994	0.019	0.089	0.065	1.57	2.14	1.03	–	–	–	11.9	14.9	8.24	(6)
1995	0.020	0.090	0.046	0.68	1.37	0.70	0.17	0.087	0.050	6.2	4.5	1.65	(7)
	Particulate metals (μ g g^{-1})												
1978	45	15	5	–	–	–	–	–	–	1250	700	350	(1)
1980	–	7.9	–	–	78	–	–	125	–	–	–	–	(2)
1981–1983	27.9	8.2	1.8	278	95	31	288	153	83	800	272	126	(3)
1987	11.9	3.2	0.97	187	50	28.8	–	–	–	928	311	179	(5)
1987–1988	–	–	–	166	65	24	264	130	100	882	360	179	(8)
1991–1994	8.25	4.2	1.26	159	76	32.4	–	–	–	725	447	182	(6)
1995	8.8	3.7	0.80	112	47	16.9	214	105	45	610	275	102	(7)
	K_D (l g^{-1})												
1978	804	53.6	41.7	–	–	–	–	–	–	379?	42.9	53.8	
1980	–	26.3	–	–	76.5	–	–	1140	–	–	–	–	
1981–1983	324	43.2	25.7	273	50	28.2	1250	638	461	68.4	30.9	38.2	
1987	595	41.6	19.4	197	37.3	30.3	–	–	–	95.7	40.0	83.3	
1991–1994	434	47.2	19.4	101	35.5	31.5	–	–	–	60.9	30.0	22.1	
1995	440	41.1	17.4	165	34.3	24.1	1260	1210	900	98.4	61.1	61.8	

References: (1) Duinker et al. (1982), the values mentioned in this table are inferred from the graphs in their publication; (2) Valenta et al. (1986); (3) Baeyens et al. (1998a); (4) van den Berg et al. (1986); Zwolsman & van Eck (1993), only the surveys of February, July, August and December 1987 were selected; (6) Paucot & Wollast (1997), for Cd and Cu only the surveys of March 1993–94 and May and June 1993 were selected; (7) Baeyens et al. (1998c); (8) Van Alsenoy et al. (1993).

prising that the Cd concentrations in the environment, and in particular in the Scheldt, are very high.

However, PARCOM regulations (PARCOM is the Commission of the Convention for the Protection of the Marine Environment of the North-East Atlantic Ocean, the North Sea included. Enforce national authorities to take the necessary actions to reduce all the emissions of Cd into the environment. Hence, the Cd levels in the Scheldt (Table 2) show a drastic reduction between 1981 and now: about 50% in the particulate and 50 to 66% in the dissolved phase.

Copper: The Cu concentrations (dissolved and particulate) in the Scheldt (Table 2) show a decrease of approximately 33% for the dissolved phase and approximately 45% for the particulate phase over the period 1980 to 1995. This decrease is less spectacular than for Cd, but

(1) the Scheldt is not particularly polluted by Cu and
(2) the international regulations concerning Cu are less stringent than for Cd.

Lead: Of the four metals discussed in this section, lead is the least measured in the estuary, probably because of the low levels in solution, the high risk of contamination and its peculiar behaviour. Suspended matter enrichments of Mn and Fe (oxy-hydroxides) are formed in the aerated part of the Scheldt estuary (e.g. Duinker et al., 1979), as shown by the dissolved iron and manganese profiles (Baeyens et al., 1998b) illustrating the precipitation of iron and manganese once oxygen reappears in the estuary. They should represent efficient scavengers of trace metals and specifically of Pb as suggested by Spencer et al. (1980) and Dehairs et al. (1989). These Mn and Fe precipitates are, especially when they are freshly formed, in a colloidal state. Hence, a significant fraction of the dissolved lead content (sample that passed through the 0.45 μm filter) may be associated with colloidal particles. Colloidal Fe, Al, Pb and Zn (but not Cu) account for most of the filter-passing form of these metals in Galveston Bay, one of Texas's major estuaries (Benoit et al., 1994). It is possible that the respective analytical methods used to determine Pb in the Scheldt estuary include this colloidal fraction differently:

- 1980 Surveys (Valenta et al., 1986): DPASV at pH = 2 and UV destruction.
- 1981–1983 Surveys (our results): DPASV at pH = 1, and at pH = 1 and UV destruction.
- 1995 Surveys (our results): GFAAS or ICP-MS after extraction-preconcentration.

Since the K_D's for the period 1981–1983 compare fairly well with those obtained by other investigators, when they are based on dissolved Pb concentrations obtained at pH = 1 without UV destruction, it seemed us logic to consider these dissolved Pb data for comparison purposes.

In the upstream area the levels are decreased by about 25% in the dissolved and the particulate phases from 1980 to now. At the mouth this decrease is much more pronounced (50 to 66%). A possible explanation for this strong decrease can be related to a modification in the atmospheric input of Pb into the sea. A large fraction of the Pb load present in the marine environment, even in coastal areas, originates from atmospheric wet and dry deposition (e.g. Baeyens et al., 1990). However, a gradual change from leaded to unleaded gasoline fuel started in the mid-eighties and led to a substantial decrease of the atmospheric deposition fluxes for Pb. Hence, the sea-end-member concentration of Pb has significantly decreased meanwhile.

Zinc: Zn concentrations are the highest of the four metals in both phases. In the dissolved phase, the Zn concentrations decreased by about 50%, while in the particulate phase this reduction is somewhat smaller (about 20–40%).

Comparison of the Scheldt estuary with other estuaries and marine systems

Table 3 shows metal concentration data in the 10–20 psu salinity range for a series of randomly selected but representative estuaries in the U.S.A., Russia, UK, Germany and France. Most of these data rely on a rather limited number of surveys (sometimes even a single one) and thus do not represent seasonally averaged values. They nevertheless allow to classify the Scheldt estuary for a given metal in the category of strong or weak polluted estuaries.

In the period 1981–1983 the Scheldt estuary belonged to the group of heavily polluted estuaries for both the dissolved and the particulate phases. Meanwhile the situation improved substantially, at least for the dissolved phase. The concentrations measured in the particulate phase in 1995 are still high for Cd and Pb compared to the other estuaries. The least polluted estuaries seem to be the Lena (Russia) and the Galverston (Texas, USA).

When comparing actual dissolved metal concentrations in the Scheldt with background levels in seas and

Table 3. A comparison of the metal concentrations in the Scheldt with estuarine and ocean levels

Estuary	Cd	Cu	Pb	Zn	Reference
Dissolved concentrations ($\mu g\ l^{-1}$) at 10–20 psu					
Delaware (U.S.A.)	0.07	1.9	–	–	Sharp et al. (1982)
Elbe (Germany)	0.25	–	–	15	Ahlf (1983)
Tamar (U.K.)	0.09	–	–	10	Morris et al. (1986)
Seine (France)	0.13	0.65	1.3	8.0	Ouddane et al. (1992)
Lena (Russia)	0..011	0.65	0.010	0.41	Martin et al. (1993)
Galveston (U.S.A.)	–	1.2	0.097	1.5	Benoit et al. (1994)
Humber (U.K.)	0.25	7	–	15	Comber et al. (1995)
Mersey (U.K.)	0.025	1.3	1.2	13	Comber et al. (1995)
Scheldt 1981–1983	0.19	1.9	0.24	8.8	Baeyens et al. (1998a)
Scheldt 1995	0.090	1.37	0.087	4.5	Baeyens et al. (1998c)
Particulate concentrations ($\mu g\ g^{-1}$) at 10–20 psu					
Elbe (Germany)	0.25	–	–	15	Ahlf (1983)
Tamar (U.K.)	0.09	–	–	10	Morris et al. (1986)
Rhine (the Netherlands)	11	147	352	714	Nolting et al. (1989)
Seine (France)	0.67	64	27	184	Ouddane et al. (1992)
Lena (Russia)	–	40	23	130	Martin et al. (1993)
Galveston (U.S.A.)	–	23	14	78	Benoit et al. (1994)
Humber (U.K.)	0.65	34	62	220	Comber et al. (1995)
Mersey (U.K.)	1.65	72	150	380	Comber et al. (1995)
Scheldt 1981–1983	8.2	95	153	272	Baeyens et al. (1998a)
Scheldt 1995	3.7	47	105	275	Baeyens et al. (1998c)
Background levels in seas and oceans ($\mu\ g\ l^{-1}$)					
Ocean/Sea	Cd	Cu	Pb	Zn	Reference
Mediterranean Sea	0.005–0.02	0.04–0.10	0.015–0.08	0.05–0.18	(Gillain & Brihaye, 1985)
Atlantic Ocean			0.033 ± 0.015		(Brüggman et al. (1985)
Atlantic Ocean				0.10–0.16	(Danielsson et al. 1985)
Atlantic Ocean	0.0038	0.093		0.055	(Yeats, 1988)
Arctic Ocean	0.033	0.127			(Yeats & Westerlund, 1991)
North Sea	0.004–0.009	0.070 ± 0.013			(Hydes & Kremling, 1993)

oceans, it appears that Cd and Cu concentrations are 20 times enriched, Pb concentrations 4 times (but 9 times compared to the Lena estuary, which shows the lowest dissolved Pb concentrations) and Zn concentrations 50 times.

Comparison of Hg concentrations in the Scheldt estuary with other estuaries and marine systems

At the end of the eighties, reliable methods for the speciation of Hg compounds became available (Baeyens, 1992). Since then a serious effort was made to assess the concentrations of the various Hg species in estuaries, lakes, coastal seas and open ocean waters. This time period is, however, too short to detect any temporal trend.

Table 4. A comparison of the mercury concentrations in the Scheldt with estuarine and ocean levels

Location	Hg_T ($ng\ l^{-1}$)	Hg_{TD} ($ng\ l^{-1}$)	Hg_R ($ng\ l^{-1}$)	Hg_P ($ng\ l^{-1}$)	Hg_P ($\mu g\ g^{-1}$)	$MMHg_T$ ($pg\ l^{-1}$)	$MMHg_D$ ($pg\ l^{-1}$)	$MMHg_P$ $ng\ g^{-1}$	$MMHg_P$ ($pg\ l^{-1}$)	Hg° ($pg\ l^{-1}$)	DMHg ($pg\ l^{-1}$)	Reference
Lakes												
Little Rock (Winsconsin)	0.4–11					40–100						Bloom & Watras, 1989
Baikal (Siberia)	0.14–0.77					2–38						Meuleman et al., 1995
Rivers & estuaries												
Scheldt estuary		0.7–2.8	0.2–2.0		0.38–1.60		13–600	2–10		20–130		Baeyens et al., 1998b
St. Lawrence (Canada)		0.08–0.9			0.04–0.40							Quemerais et al., 1996
Pettaquamscutt (U.S.A.)		0.2–3.0	0.1–1.6	0.2–3.0			10–400		10–600			Mason et al., 1993
Elbe (Germany)		0.76–3.28	0.16–0.8		0.30–1.41			2.2–9.2		54–122		Coquery & Cossa, 1995
Lena (Russia)		0.90–1.08			0.03–0.21							Coquery et al., 1995
Ob (Russia)		0.48–0.64			0.04–0.06							Coquery et al., 1995
Yenisei (Russia)	2.14–6.15					110–2230						Parkman et al., 1995
Coastal waters												
Alboran Sea											< dl–58	Cossa et al., 1994
North Sea	0.32–4.28	0.18–0.96	0.08–0.36	0.04–3.32	0.116–0.484		< dl	< dl–12	< dl–38	< dl–90	< dl	Coquery & Cossa, 1995
– Belgian coast	0.13–16.1	0.13–1.42			0.046–0.641		32–188	< dl–10		20–160	< dl	our results
English Channel	0.80–4.1	0.30–0.84			0.51–1.77							Cossa & Fileman, 1991
– English Channel	0.15–0.87						15–65					our results
– Dover Straight		0.12–1.34		0.06–5.36								Cossa et al., 1994b
Open Ocean												
North Atlantic	0.31–0.85		0.15–0.21							14–178	< 2–40	Mason et al., 1995
			0.14–0.21									Dalziel, 1995
Equatorial Pacific			0.16–0.40							10–72	< 2–60	Mason et al., 1996

The total dissolved Hg levels in the Scheldt estuary compare very well with those in most other estuaries, but the particulate levels ($\mu g\ g^{-1}$) are clearly amongst the highest ones (Table 4). A possible explanation is the high organic load (POM) carried by the Scheldt and the strong affinity of mercury for functional groups, more particularly the SH-group included in that organic matter. Data about the other Hg species measured in the Scheldt (dissolved and particulate MMHg and Hg°) are very scarce in the other estuaries. Only in the Pettaquamscutt (Mason et al., 1993) and in the Elbe (Coquery & Cossa, 1995) some of these compounds were also measured. The agreement with our data is, however, excellent (Table 4).

The total Hg concentrations in unpolluted areas like the English Channel (our results), the North Atlantic (Mason et al., 1995) or Lake Baikal (Meuleman et al., 1995) are about 3 to 4 times lower than the dissolved Hg concentrations in the Scheldt estuary. The highest MMHg levels in unpolluted areas like the English Channel (our results) or Lake Baikal (Meuleman et al., 1995) are about 10 to 15 times lower than the highest levels in the Scheldt estuary. Only the Hg° concentrations in unpolluted areas like the North Atlantic (Mason et al., 1996) are not substantially different from those measured in estuaries or coastal waters.

Conclusions

As a conclusion, it seems realistic to state that, due to the regulations according to PARCOM and the implementation of the North Sea Declaration, the metal concentrations in the Scheldt estuary decreased significantly between the early eighties and now (1995), although to a various extent per individual metal.

A comparison of the actual dissolved metal concentrations in the Scheldt estuary with background levels in seas and oceans indicates that the Scheldt is still polluted: dissolved Cd, Cu, Pb and Zu concentrations are respectively about 20, 20, 4 (9 times compared to the Lena estuary) and 50 times enriched in the Scheldt; the particulate metal levels ($\mu g\ g^{-1}$) in the Scheldt remain relatively high.

The dissolved Hg, MMHg and Hg° concentrations compare very well with those observed in other estuaries. Particulate mercury levels in the Scheldt are, like for the other metals, high. Still these Hg species, except Hg°, are in the Scheldt estuary enriched compared to unpolluted oceans and lakes.

References

Ahlf, W.: 1983. The river Elbe: behaviour of Cd and Zn during estuarine mixing. Envir. Technol. Lett. 4: 405–410.

Baeyens, W., G. Decadt & I. Elskens, 1979. A study of the distribution of mercury in the various compartments of the North Sea and Scheldt estuary ecosystems. Oceanol. Acta 2: 447–457.

Baeyens, W., G. Decadt, F. Dehairs & L. Goeyens, 1982. An automated method for the assessment of mercury adsorption rates on particulate suspended matter. Oceanol. Acta 5: 261–264.

Baeyens, W., G. Gillain, G. Decadt & I. Elskens, 1987. Trace metals in coastal waters of the eastern North Sea. I. Analyses and short term distributions. Oceanol. Acta 10: 169–179.

Baeyens, W., G. Gillain, G. Decadt & I. Elskens, 1987. Trace metals in coastal waters of the eastern North Sea. II. Flows through the coastal area. Oceanol. Acta 10: 301–309.

Baeyens, W., M. Elskens, G. Gillain & L. Goeyens, 1998a. Biogeochemical behaviour of Cd, Cu, Pb and Zn in the Scheldt estuary during the period 1981–1983. Hydrobiologia 366: 15–44.

Baeyens, W., C. Meuleman & B. Muhaya & M. Leermakers, 1998b. Behaviour and speciation of mercury in the Scheldt estuary (water, sediments and benthic organisms). Hydrobiologia 366: 63–79.

Baeyens, W., K. Parmentier, L. Goeyens, G. Ducastel, M. De Gieter & M. Leermakers, 1998c. The biogeochemical behaviour of Cd, Cu, Pb and Zn in the Scheldt estuary: results of the 1995 surveys. Hydrobiologia 366: 45–62.

Baeyens, W., G. T. M. Van Eck, C. Lambert, R. Wollast & L. Goeyens, 1998d. General description of the Scheldt estuary. Hydrobiologia 366: 1–14.

Benoit, G., S. D. Oktay-Marshall, A. Cantu., E. M. Hood, C. H. Coleman, M. O. Corapcioglu & P. H. Santschi, 1994. Partitioning of Cu, Pb, Ag, Zn, Fe, Al, and Mn between filter-retained particles, colloids, and solution in six Texas estuaries. Mar. Chem. 45: 307–336.

Bloom, N. S. & C. J. Watras, 1989. Observations on methylmercury in precipitation. Envir. Sci. Technol. 87/88: 199–207.

Brugmann, L., L. G. Danielsson, B. Magnusson & S. Westerlund, 1985. Lead in the North Sea and the northeast Atlantic Ocean. Mar. Chem. 16: 47–60.

Comber, S. D. W., A. M. Gunn & C. Whalley, 1995. Comparison of the Partitioning of Trace Metals in the Humber and Mersey Estuaries. Mar. Pollut. Bull.

Coquery, M., D. Cossa & J. M. Martin, 1995. The distribution of dissolved and particulate mercury in 3 Siberian estuaries and adjacent Artic coastal waters. Wat. Air Soil Pollut. 80: 653–664.

Coquery, M. & D. Cossa, 1995. Mercury speciation in surface waters of the North Sea. Neth. J. Sea Res. 34: 245–257.

Cossa, D. & C. Fileman, 1991. Mercury concentrations in surface waters of the English Channel. A cooperative study. Mar. Pollut. Bull. 24: 197–200.

Cossa, D., J. M. Martin & J. Sayman, 1994a. Dimethylmercury formation in the Alboran Sea. Mar. Pollut. Bull. 28: 381–384.

Cossa, D., J. Sayman & J. Noel, 1994b. Mercury transport in waters of the Strait of Dover. Mar. Pollut. Bull. 28: 385–388.

Danielsson, L. G., B. Magnusson & S. Westerlund, 1985. Cadmium, copper, iron, nickel and zinc in the northeast Atlantic Ocean. Mar. Chem. 17: 23–41.

Dalziel, J. A., 1995. Reactive mercury in the eastern North Atlantic and southeast Atlantic. Mar. Chem. 49: 307–314.

Dehairs, F., W. Baeyens & D. Van Gansbeke, 1989. Tight coupling between enrichment of iron and manganese in North Sea suspended matter and sedimentary redox processes; evidence for seasonal variability. Estuar. coast. Shelf Sci. 29: 457–471.

Duinker, J. C., R. Wollast & G. Billen, 1979. Behaviour of Manganese in the Rhine and Scheldt Estuaries. II. Geochemical Cycling. Estuar. coast. Mar. Sci. 9: 727–738.

Duinker, J. C., R. F. Nolting & D. Michel, 1982. Effects of salinity, pH and redox conditions on the behaviour of Cd, Zn, Ni and Mn in the Scheldt Estuary. Thalassia Jugoslavica 18: 191–202.

Fitzgerald, W. F., G. A. Gill, 1979. Subnanogram determination of mercury by two-stage gold analgamation and gas phase detection applied to atmospheric analysis. Analyt. Chem. 51: 1714–1720.

Gillain, G., G. Duyckaerts & A. Distèche, 1979. Direct and simultaneous determination of Zn, Cd, Pb, Cu, Sb and Bi dissolved in sea water by direct pulse anodic stripping voltammetry with a hanging drop mercury electrode. Analyt. chim. Acta 106: 23–27.

Gillain, G., G. Decadt, F. Dehairs, H. Dedeurwaerder, L. Goeyens & W. Baeyens, 1982. Distribution, transport and fate of Bi, Cd, Cu, Hg, Pb, Sb, and Zn in the Belgian coastal marine environment. In A. Distèche & I. Elskens (eds), Distribution, Transport and Fate of Heavy Metals in the Belgian Coastal Marine Environment, Projet Mer, Rapport final, 2. Services du Premier Ministre, Programmation de la Politique Scientifique: 13–39.

Gillain, G. & C. Brihaye, 1985. A system for taking a representative sample from a large volume of turbid surface sea water for trace element determinations. Analyt. chim. Acta 167: 387–391.

Hoenig, M., E. Puskaric, P. Choisy & M. Wartel, 1991a. Direct Determination of high and mid volatile Elements (Cd, Pb, Mn) in Seawater by Electrothermal Atomic Absorption Spectrometry: Existing Approaches and Critical Parameters. Analusis 19: 285–291.

Hoenig, M., P. Regnier & L. Chou, 1991b. Determination of the High Aluminium Content in Suspended Matter Samples Collected in Natural Waters by Slurry Sampling Electrothermal Atomic Absorption Spectrometry. J. Anal. At. Spectrom. 6: 273–275.

Hydes, D. J. & K. Kremling, 1993. Patchiness in dissolved metals (Al, Cd, Co, Cu, Mn, Ni) in North Sea surface waters: Seasonal differences and influence of suspended sediment. Cont. Shelf Res. 13: 1083–1101.

Martin, J. M., D. M. Guan, F. Elbaz-Poulichet, A. J. Thomas & V. V. Gordeev, 1993. Preliminary assessment of the distributions of some trace elements (As, Cd, Cu, Fe, Ni, Pb and Zn) in a pristine aquatic environment: the Lena River estuary (Russia). Mar. Chem. 43: 185–199.

Mason, R. P., W. F. Fitzgerald, J. Huxley, A. K. Hanson, P. L. Donoghay & P. L. Sirburth, 1993. Mercury biogeochemical cycling in a stratified estuary. Limnol. Oceanogr. 38: 1227–1241.

Meuleman, C., M. Leermakers & W. Baeyens, 1995. Mercury speciation in Lake Baikal. Wat. Air Soil Pollut. 80: 539–551.

Nihoul, J. C. J. & R. Wollast, 1976. Eds Projet Mer, Rapport final, 10. Services du Premier Ministre, Programmation de la Politique Scientifique: 239.

Nolting, T. F., B. Sunby & J. C. Duinker, 1989. Minor and major elements in suspended matter in the Rhine and Meuse rivers and estuary. Neth. J. Sea Res. 23: 255–261.

Ouddane, B., J.-C. Fischer & M. Wartel, 1992. Evaluation statistique de la repartition des metaux en traces Cd, Pb, Cu, Zn et Mn dans la Seine et son estuaire. Oceanol. Acta 15: 347–354.

Parkman, H., P. Östlund, M. Samuelsson & A. Iverfeldt P., 1995. Methylmercury in a permanently Stratified Fiord. In. C. J. Watras & J. W. Huckabee (eds), Mercury Pollution, Integration and Synthesis. Lewis Publishers, CRC Press, U.S.A.: 107–118.

Paucot, H. & R. Wollast, 1997. Transport and transformation of trace metals in the Scheldt estuary. Mar. Chem. 58: 229–244.

Quemerais, B., D. Cossa, T.-T. Pham & B. Rondeau, 1996. Concentration and transport of Hg in the St-Lawrence river (Canada). 4th Int Conf. on Hg as a Global Pollutant, 4–8 augustus 1996, Hamburg, Germany.

Spencer, D. W., M. P. Bacon & P. G. Brewer, 1980. The distribution of ^{210}Pb and ^{210}Po in the North Sea. Thalassia Jugoslavica, 16: 125–154.

Valenta, P., E. K. Duursma, A. G. Merks, H. Rützel & H. W. Nürnberg, 1986. Distribution of Cd, Pb and Cu between the dissolved and particulate phase in the eastern Scheldt and western Scheldt estuary. Sci. Total Envir. 53: 41–76.

Van Alsenoy, V., P. Bernard & R. Van Grieken, 1993. Elemental concentrations and heavy metal pollution in sediments and suspended matter from the Belgian North Sea and the Scheldt estuary. Sci. Total Envir. 133: 153–181.

van den Berg, C. M. G., A. G. A. Merks & E. K. Duursma, 1987. Organic Complexation and its Control of the Dissolved Concentrations of Copper and Zinc in the Scheldt Estuary. Estuar. coast. Shelf Sci. 24: 785–797.

Yaets, A. P. & S. Westerlund, 1991. Trace metal distributions at an Arctic Ocean ice island, Mar. Chem. 33: 261–277.

Yeats, P. A., 1988. The distribution of trace metals in ocean waters Sci. Total Envir. 72: 131–149.

Zwolsman, J. J. G. & G. T. M. Van Eck, 1993. Dissolved and particulate trace metal geochemistry in the Scheldt estuary, S.W. Netherlands (water column and sediments). Neth. J. Aquat. Ecol. 27: 287–300.

MIX
Papier aus verantwortungsvollen Quellen
Paper from responsible sources
FSC® C105338

If you have any concerns about our products,
you can contact us on
ProductSafety@springernature.com

In case Publisher is established outside the EU,
the EU authorized representative is:
Springer Nature Customer Service Center GmbH
Europaplatz 3, 69115 Heidelberg, Germany

Printed by Libri Plureos GmbH
in Hamburg, Germany